HOLISTIC PERSPECTIVES ON TRAUMA

Implications for Social Workers and
Health Care Professionals

HOLISTIC PERSPECTIVES ON TRAUMA

Implications for Social Workers and Health Care Professionals

Edited by
Lisa Albers Prock, MD, MPH

Apple Academic Press Inc. | Apple Academic Press Inc.
3333 Mistwell Crescent | 9 Spinnaker Way
Oakville, ON L6L 0A2 | Waretown, NJ 08758
Canada | USA

©2015 by Apple Academic Press, Inc.

First issued in paperback 2021

Exclusive worldwide distribution by CRC Press, a member of Taylor & Francis Group

ISBN 13: 978-1-77463-547-6 (pbk)
ISBN 13: 978-1-77188-126-5 (hbk)

Library and Archives Canada Cataloguing in Publication

Holistic perspectives on trauma : implications for social workers and health care professionals / edited by Lisa Albers Prock, MD, MPH.

Includes bibliographical references and index.
ISBN 978-1-77188-126-5 (bound)
1. Psychic trauma in children. 2. Abused children--Mental health. 3. Abused children--Mental health services. 4. Child abuse--Psychological aspects. I. Prock, Lisa Albers, editor

RJ506.P66H64 2015   618.92'8521   C2014-907731-9

Library of Congress Cataloging-in-Publication Data

Holistic perspectives on trauma : implications for social workers and health care professionals / [edited by] Lisa Albers Prock.

pages cm
Includes bibliographical references and index.
ISBN 978-1-77188-126-5 (alk. paper)
1. Psychic trauma in children. 2. Abused children--Mental health. 3. Abused children--Mental health services. 4. Child abuse--Psychological aspects. I. Prock, Lisa Albers.

RJ506.P66H56 2015   618.92'8521--dc23   2014045335

Apple Academic Press also publishes its books in a variety of electronic formats. Some content that appears in print may not be available in electronic format. For information about Apple Academic Press products, visit our website at **www.appleacademicpress.com** and the CRC Press website at **www.crc-press.com**

ABOUT THE EDITOR

LISA ALBERS PROCK, MD, MPH

Dr. Prock is a Developmental Behavioral Pediatrician at Children's Hospital (Boston) where she co-founded and directs the Adoption Program and works in the Developmental Medicine Center. She is a co-director of the Translational Neuroscience Center at Boston Children's Hospital, and she is the director of Developmental Behavioral Pediatric Services at Children's Hospital and Harvard Medical School, where her responsibilities include Clinical Director of the Developmental Behavioral Pediatrics Fellowship Program. She also is active in international health and resident/fellow education. She attended college at the University of Chicago, medical school at Columbia University and received a master's degree in public health from the Harvard School of Public Health. At the end of her training, she worked as a primary care pediatrician at a community health center and as an inpatient hospital physician. After obtaining a public health degree in International Health, she lived and worked in Cambodia where she taught pediatrics and studied the epidemiology of tuberculosis in children. She returned to Boston for further training in general pediatrics and development as a Dyson Fellow at Children's Hospital.

Dr. Prock is currently involved in translational research efforts as the Principal Investigator for four clinical trials working with adolescents and young adults with Fragile X Syndrome. She is also co-director of the clinical arm of the Translational Neuroscience Center at Boston Children's Hospital, a multidisciplinary collaboration to accelerate the translation of basic science findings into clinical meaning for children with developmental disabilities and their families. Dr. Prock has also combined her clinical interests in child development and international health with advocacy for children, particularly in the areas of foster care and adoption. She has been working with adoptees (both domestic and international) involved in medical, residential and educational settings since 1991.

Her research interests include the long-term developmental, behavioral and emotional concerns of adoptees. She has co-authored several original publications, edited several reference volumes, and written numerous articles. She has been a board member for several nonprofit organizations, including the Center for Family Connections, a family therapy organization specializing in issues for foster-care and adoptive families; Adoptive Families Together, a parent support group specific to adoptive families; and the Treehouse Foundation, a foster family empowerment program. She has received numerous local and national awards for her work with children and families, most recently the 2004 United States Congressional Angel in Adoption Award. She is a past chairperson of the American Academy of Pediatrics Section on Adoption and Foster Care and current executive committee member of the American Academy of Pediatrics Council on Foster Care, Kinship and Adoption.

CONTENTS

ACKNOWLEDGMENT AND HOW TO CITE

The editor and publisher thank each of the authors who contributed to this book. The chapters in this book were previously published in various places in various formats. To cite the work contained in this book and to view the individual permissions, please refer to the citation at the beginning of each chapter. Each chapter was read individually and carefully selected by the editor; the result is a book that provides a nuanced look at the outcomes of childhood trauma. The chapters included examine the following topics:

- Chapter 1 offers some guidance to professionals working with victims of child abuse as to the optimal time and nature of intervention in order to prevent or alleviate social anxiety and post-traumatic stress disorders.
- Chapter 2 explores the effects of trauma that is perpetrated from outside the home rather than from within the family. Despite some limitations in the study, it builds a case for early intervention as the most effective way to prevent long-term psychological issues after a major disaster.
- Chapter 3 investigates the genetic factors that interact with environmental trauma to create long-term psychiatric consequences.
- Chapter 4 examines a specific outcome of trauma—reduced coping skills—and investigates the implications for individuals with bipolar disorder. Again, the value of these studies is the implications for how to focus interventions to best help various groups of individuals.
- Chapter 5 and those that follow focus on the neurodevelopmental connections with trauma, providing the physiological framework for some of the information discussed in Part 1.
- Chapter 6 supports the hypothesis that physiological brain attributes interact with low-level social trauma (chronic social exclusion) to create long-term emotional and behavioral problems.
- Chapter 7 investigates the neural mechanisms that undergird psychiatric consequences of trauma. In this case, the authors focus on neglect (rather than abuse). Their findings indicate that neglect as well as abuse changes the brain.
- Chapter 8 indicates that there are sensitive periods during which specific brain regions are particularly vulnerable to trauma.

- Chapter 9 follows the previous article's investigation, in this case focusing on witnessing parental violence, rather than experiencing violence first-hand. Both studies revealed a potential vulnerable period, which in both cases was prior to puberty (brain structures becoming less plastic after puberty). This is important information for social workers considering the definition of "domestic safety."
- Chapter 10 correlates trauma with lifetime incidence of cancer (of all types). The authors do well to indicate that other factors make this not a one-to-one causal relationship, but rather a stress situation that may alter biological functions, making individuals more predisposed to cancer later in life. With that caveat in mind, the authors make a good case for screening for childhood trauma to identify people who may be more at risk for cancer and other chronic diseases.
- Chapter 11 offers further evidence (based on a wider sample and a longer-term study than the previous article) for the connections between cancer and early traumas. These authors make still clearer that this could operate via two main mechanisms: A direct biological effect and an indirect effect via health behaviors (or a combination of both). What is fascinating is the implication that childhood trauma must be treated holistically, since the consequences are so intertwined: adversity causes stress, which may cause physiological changes in the brain and the rest of the body, which can in turn cause psychiatric issues and chronic disease.
- We have known for some time the connection between emotional stress and cardiovascular health, but Chapter 12 finds interesting correlations between cardiovascular health and early childhood adversity, with connections to obesity as well.
- Chapter 13 explores the association between childhood socio-economic circumstances and the future risk of diabetes and obesity—helpful in de-signing more targeted and more efficient prevention strategies, including interventions for high-risk families, coping skills training, empowerment of social networks and healthy neighborhoods.
- Chapter 14 presents evidence showing that painful events during early life cause long-lasting changes in pain processing and suggests that high expo-sure to painful experiences in early life may contribute to increased pain sensitivity. The study includes birth trauma as well as later childhood ex-periences.
- Chapter 15 examines LPC, a school-based intervention, found to be a promising response strategy for children who have experienced trauma. With reduced resources available for school-based mental health services, LPC is an efficient first-level of defense that identifies children in distress, so that they can be connected with appropriate adjunct professionals, such as counselors and social workers.

LIST OF CONTRIBUTORS

Carl M. Anderson
Department of Psychiatry, Harvard Medical School, Boston, Massachusetts, United States of America, Developmental Biopsychiatry Research Program, McLean Hospital, Belmont, Massachusetts, United States of America, and Brain Imaging Center, McLean Hospital, Belmont, Massachusetts, United States of America

Susanne Bakelaar
Department of Psychiatry, Faculty of Medicine and Health Sciences, Tygerberg Campus, Stellenbosch University, Cape Town 7505, South Africa

Melanie Bartley
Department of Epidemiology and Public Health, University College London, London, UK

Melanie Bishop
Department of Psychiatry, Faculty of Medicine and Health Sciences, Tygerberg Campus, Stellenbosch University, Cape Town 7505, South Africa

David Blane
Department of Primary Care and Public Health, Imperial College London, London, UK

Albert E. Boon
De Jutters, Youth Mental Health Care Center, The Hague, The Netherlands, De Fjord Lucertis, Centre for Orthopsychiatry and Forensic Youth Psychiatry, Capelle aan den IJssel, The Netherlands, and Curium-Leiden University Medical Centre, Department of Child and Adolescent Psychiatry, Leiden, The Netherlands

Elisa Brietzke
Universidade Federal de São Paulo, São Paulo, SP, Brazil

Monique J. Brown
Department of Family Medicine and Population Health, Virginia Commonwealth University School of Medicine, Richmond, Virginia, United States of America

Noriko Cable
Department of Epidemiology and Public Health, University College London, London, UK

John Cairney
Departments of Psychiatry and Behavioural Neuroscience, Family Medicine, Kinesiology, CanChild, Centre for Childhood Disability Research, McMaster University, Hamilton ON, Canada and Health Systems Research and Consulting Unit, Centre for Addiction & Mental Health, Toronto, ON, Canada

Mary Cannon
Department of Psychiatry, Royal College of Surgeons in Ireland, Beaumont Hospital, Dublin, Ireland

Joseph E. Cavanaugh
Injury Prevention Research Center, University of Iowa, Iowa City, IA, USA and Department of Biostatistics, University of Iowa, Iowa City, IA, USA

Mary C. Clarke
Department of Psychiatry, Royal College of Surgeons in Ireland, Beaumont Hospital, Dublin, Ireland

Steven A. Cohen
Department of Family Medicine and Population Health, Virginia Commonwealth University School of Medicine, Richmond, Virginia, United States of America

Claire Cole
School of Sociology Social Policy & Social Work, 6 College Park, Queens University Belfast, Belfast, BT7 1LP, Northern Ireland

Dearbhla Connor
Department of Psychiatry, Royal College of Surgeons in Ireland, Beaumont Hospital, Dublin, Ireland

Filomeno Cortese
The Seaman Family Centre, University of Calgary, Foothills Medical Centre, 1403-29th Street NW, Calgary, AB, Canada T2N 2T9

Eveline A. Crone
Leiden University, Leiden Institute for Brain & Cognition, Leiden, The Netherlands and Leiden University, Institute of Psychology, Developmental Psychology, Leiden, The Netherlands

Ledo Daruy-Filho
Pontifícia Universidade Católica do Rio Grande do Sul, Porto Alegre, RS, Brazil

Cristiane da Silva Fabres
Pontifícia Universidade Católica do Rio Grande do Sul, Porto Alegre, RS, Brazil

Dominique Dedieu
INSERM, U1027, Toulouse F-31300, France and Université Toulouse III Paul-Sabatier, UMR1027, Toulouse F-31300, France

Cyrille Delpierre
INSERM, U1027, Toulouse F-31300, France and Université Toulouse III Paul-Sabatier, UMR1027, Toulouse F-31300, France

Yuqiang Ding
Mental Health Institute, The Second Xiangya Hospital, Key Laboratory of Psychiatry and Mental Health of Hunan Province, Central South University, Changsha, Hunan, China and Key Laboratory of Arrhythmias, Ministry of Education, East Hospital, Tongji University School of Medicine, Shanghai, China

Michael Duffy
School of Sociology Social Policy & Social Work, 6 College Park, Queens University Belfast, Belfast, BT7 1LP, Northern Ireland

Bernet M. Elzinga
Leiden University, Leiden Institute for Brain & Cognition, Leiden, The Netherlands and Leiden University, Institute of Psychology, Clinical Psychology, Leiden, the Netherlands

Carol Fitzpatrick
School of Medicine & Medical Science, University College Dublin, Dublin, Ireland

Michael Fitzgerald
Trinity College Dublin, Dublin, Ireland

Padraig Flannery
Department of Psychiatry, Royal College of Surgeons in Ireland, Beaumont Hospital, Dublin, Ireland

Maisha Frederick
Injury Prevention Research Center, University of Iowa, Iowa City, IA, USA

Weijia Gao
Mental Health Institute, The Second Xiangya Hospital, Key Laboratory of Psychiatry and Mental Health of Hunan Province, Central South University, Changsha, Hunan, China

Ismael Gaxiola
The Seaman Family Centre, University of Calgary, Foothills Medical Centre, 1403-29th Street NW, Calgary, AB, Canada T2N 2T9

Bradley Goodyear
Mathison Centre for Mental Health Research & Education, Department of Psychiatry, University of Calgary, TRW Building, 3280 Hospital Drive NW, Calgary, AB, Canada T2N 4Z6, Hotchkiss Brain Institute, University of Calgary, 3330 Hospital Drive NW, Calgary, AB, Canada T2N 4N1, The Seaman Family Centre, University of Calgary, Foothills Medical Centre, 1403-29th Street NW, Calgary, AB, Canada T2N 2T9, and Department of Clinical Neurosciences, University of Calgary, Foothills Medical Centre, 1403-29th Street NW, Calgary, AB, Canada T2N 2T9

Rodrigo Grassi-Oliveira
Pontifícia Universidade Católica do Rio Grande do Sul, Porto Alegre, RS, Brazil

Pascale Grosclaude
INSERM, U1027, Toulouse F-31300, France, Université Toulouse III Paul-Sabatier, UMR1027, Toulouse F-31300, France, and Institut Claudus Regaud, Toulouse F-31300, France

Karisa Harland
Injury Prevention Research Center, University of Iowa, Iowa City, IA, USA

Michelle Harley
Department of Psychiatry, Royal College of Surgeons in Ireland, Beaumont Hospital, Dublin, Ireland and Department of Child and Adolescent Psychiatry, St Vincent's Hospital, Fairview, Dublin, Ireland

Kirsten Hauber
De Jutters, Youth Mental Health Care Center, The Hague, The Netherlands

Zhong He
Department of Radiology, the Second Xiangya Hospital of Central South University, Changsha, China

Christian Herder
Institute for Clinical Diabetology, German Diabetes Center, Leibniz Center for Diabetes Research at Heinrich-Heine-University, Düsseldorf, Germany

Natalia Jaworska
Mathison Centre for Mental Health Research & Education, Department of Psychiatry, University of Calgary, TRW Building, 3280 Hospital Drive NW, Calgary, AB, Canada T2N 4Z6, Hotchkiss Brain Institute, University of Calgary, 3330 Hospital Drive NW, Calgary, AB, Canada T2N 4N1, and Alberta Children's Hospital Research Institute, University of Calgary, 2888 Shaganappi Trail NW, Calgary, AB, Canada T3B 6A8

Tianzi Jiang
National Laboratory of Pattern Recognition, Institute of Automation, Chinese Academy of Sciences, Beijing, China

Ian Kelleher
Department of Psychiatry, Royal College of Surgeons in Ireland, Beaumont Hospital, Dublin, Ireland and National Centre for Suicide Research and Prevention of Mental Ill-Health, Karolinska Institutet, Stockholm, Sweden

Michelle Kelly-Irving
INSERM, U1027, Toulouse F-31300, France and Université Toulouse III Paul-Sabatier, UMR1027, Toulouse F-31300, France

Bruno Kluwe-Schiavon
Pontificia Universidade Católica do Rio Grande do Sul, Porto Alegre, RS, Brazi

Rebecca Lacey
Department of Epidemiology and Public Health, University College London, London, UKl

Thierry Lang
INSERM, U1027, Toulouse F-31300, France, Université Toulouse III Paul-Sabatier, UMR1027, Toulouse F-31300, France, and CHU Toulouse, Hôpital Purpan, Département, Toulouse F-31300, France

Benoit Lepage
Université Toulouse III Paul-Sabatier, UMR1027, Toulouse F-31300, France and CHU Toulouse, Hôpital Purpan, Département, Toulouse F-31300, France

Lingjiang Li
Mental Health Institute, The Second Xiangya Hospital, Key Laboratory of Psychiatry, Mental Health of Hunan Province, Central South University, Changsha, Hunan, China, and Department of Psychiatry, Chinese University of Hong Kong, Hong Kong, China

Zexuan Li
Department of Psychiatry, the Second Xiangya Hospital of Central South University, Changsha, China

Mei Liao
Mental Health Institute, The Second Xiangya Hospital, Key Laboratory of Psychiatry and Mental Health of Hunan Province, Central South University, Changsha, Hunan, China

Lucie A. Low
Alan Edwards Centre for Research on Pain, McGill University, 3640 University Street, Montreal, QC, Canada H3A 2B2

Shaojia Lu
Mental Health Institute, The Second Xiangya Hospital, Key Laboratory of Psychiatry and Mental Health of Hunan Province, Central South University, Changsha, Hunan, China

Fionnuala Lynch
Lucena Clinic, Tallaght, Dublin, Ireland

Frank P. MacMaster
Mathison Centre for Mental Health Research & Education, Department of Psychiatry, University of Calgary, TRW Building, 3280 Hospital Drive NW, Calgary, AB, Canada T2N 4Z6, Hotchkiss Brain Institute, University of Calgary, 3330 Hospital Drive NW, Calgary, AB, Canada T2N 4N1, Alberta Children's Hospital Research Institute, University of Calgary, 2888 Shaganappi Trail NW, Calgary, AB, Canada T3B 6A8, and Department of Pediatrics, University of Calgary, 2888 Shaganappi Trail NW, Calgary, Canada T3B 6A8

Maura McDermott
Western Health and Social Services Trust, Omagh, Northern Ireland

Bregtje Gunther Moor
Leiden University, Leiden Institute for Brain & Cognition, Leiden, The Netherlands and Leiden University, Institute of Psychology, Developmental Psychology, Leiden, The Netherlands

Derek W. Morris
Neuropsychiatric Genetics Research Group, Department of Psychiatry and Institute of Molecular Medicine, Trinity College Dublin, Dublin, Ireland

Deborah D. O'Leary
Department of Community Health Sciences, Brock University, St Catharines ON, Canada

Andy Percy
Institute of Child Care Research, Queens University Belfast, Belfast, Northern Ireland

Ann Polcari
Department of Psychiatry, Harvard Medical School, Boston, Massachusetts, United States of America, Developmental Biopsychiatry Research Program, McLean Hospital, Belmont, Massachusetts, United States of America, and School of Nursing, Northeastern University, Boston, Massachusetts, United States of America

Chelsea Pretty
Department of Community Health Sciences, Brock University, St Catharines ON, Canada

Rajamannar Ramasubbu
Mathison Centre for Mental Health Research & Education, Department of Psychiatry, University of Calgary, TRW Building, 3280 Hospital Drive NW, Calgary, AB, Canada T2N 4Z6, Hotchkiss Brain Institute, University of Calgary, 3330 Hospital Drive NW, Calgary, AB, Canada T2N 4N1, The Seaman Family Centre, University of Calgary, Foothills Medical Centre, 1403-29th Street NW, Calgary, AB, Canada T2N 2T9, and Department of Clinical Neurosciences, University of Calgary, Foothills Medical Centre, 1403-29th Street NW, Calgary, AB, Canada T2N 2T9

Marizen Ramirez
Department of Occupational and Environmental Health, University of Iowa, 105 S. River St. #318, Iowa City, IA 52242, USA and Injury Prevention Research Center, University of Iowa, Iowa City, IA, USA

Hugh Ramsay
Department of Psychiatry, Royal College of Surgeons in Ireland, Beaumont Hospital, Dublin, Ireland

Wolfgang Rathmann
Institute of Biometrics and Epidemiology, German Diabetes Center, Leibniz Center for Diabetes Research at Heinrich-Heine-University, Düsseldorf, Germany

David Rosenstein
Department of Psychiatry, Faculty of Medicine and Health Sciences, Tygerberg Campus, Stellenbosch University, Cape Town 7505, South Africa

Petra Schweinhardt
Alan Edwards Centre for Research on Pain, McGill University, 3640 University Street, Montreal, QC, Canada H3A 2B2

Soraya Seedat
Department of Psychiatry, Faculty of Medicine and Health Sciences, Tygerberg Campus, Stellenbosch University, Cape Town 7505, South Africa

Rhoda Shepherd
Cedar Rapids Community School District, Cedar Rapids, IA, USA

Ming Song
National Laboratory of Pattern Recognition, Institute of Automation, Chinese Academy of Sciences, Beijing, China

Philip Spinhoven
Leiden University, Leiden Institute for Brain & Cognition, Leiden, The Netherlands, Leiden University, Institute of Psychology, Clinical Psychology, Leiden, the Netherlands, and Leiden University Medical Center, Department of Psychiatry, Leiden, The Netherlands

Linyan Su
Department of Psychiatry, the Second Xiangya Hospital of Central South University, Changsha, China

Teresa Tamayo
Institute of Biometrics and Epidemiology, German Diabetes Center, Leibniz Center for Diabetes Research at Heinrich-Heine-University, Düsseldorf, Germany

Martin H. Teicher
Department of Psychiatry, Harvard Medical School, Boston, Massachusetts, United States of America and Developmental Biopsychiatry Research Program, McLean Hospital, Belmont, Massachusetts, United States of America

Leroy R. Thacker
Department of Biostatistics, Virginia Commonwealth University School of Medicine, Richmond, Virginia, United States of America andCenter for Clinical and Translational Research, Virginia Commonwealth University School of Medicine, Richmond, Virginia, United States of America

Akemi Tomoda
Department of Psychiatry, Harvard Medical School, Boston, Massachusetts, United States of America, Developmental Biopsychiatry Research Program, McLean Hospital, Belmont, Massachusetts, United States of America, and Research Center for Child Mental Development, University of Fukui, Fukui, Japan

Anne-Laura van Harmelen
Leiden University, Leiden Institute for Brain & Cognition, Leiden, The Netherlands, Leiden University, Institute of Psychology, Clinical Psychology, Leiden, the Netherlands, and University of Cambridge, Department of Developmental Psychiatry, Cambridge, United Kingdom

Terrance J. Wade
Department of Community Health Sciences, Brock University, St Catharines ON, Canada

Zhaoguo Wei
Mental Health Institute, The Second Xiangya Hospital, Key Laboratory of Psychiatry and Mental Health of Hunan Province, Central South University, Changsha, Hunan, China and Department of Psychiatry, Shenzhen Kangning Hospital, Shenzhen, Guangdong, China

Marleen Wong
School of Social Work, University of Southern California, Los Angeles, CA, USA

Weiwei Wu
Mental Health Institute, The Second Xiangya Hospital, Key Laboratory of Psychiatry and Mental Health of Hunan Province, Central South University, Changsha, Hunan, China

Fan Yang
Department of Psychiatry, the Second Xiangya Hospital of Central South University, Changsha, China

Yan Zhang
Department of Psychiatry, the Second Xiangya Hospital of Central South University, Changsha, China

Zhijun Zhang
The Department of Neuropsychiatry and Institute of Neuropsychiatric Research Affiliated ZhongDa Hospital of Southeast University, Nanking, Jiangsu, China

INTRODUCTION

Children with a history of significant neglect and/or physical, sexual, and/ or emotional abuse are at higher risk for developing long-term emotional, behavioral and mental health concerns, which have implications past childhood and into adulthood. Early trauma impacts individuals' health in ways that reach far past the obvious and immediate damage. It correlates with higher incidences of various mental health disorders, it can alter brain structures, and it can make individuals more susceptible to a variety of diseases, including cancer, cardiovascular disorders, fybromyalgia, and diabetes, among others.

Understanding the long-term effects of childhood trauma is now integral to the daily clinical practice of professionals from many fields, including educators, social workers, mental health professionals, family and substance abuse counselors, police, caregivers, and criminal justice service providers. The boundary lines between the social worker's realm and that of other professionals are blurred or nonexistent.

Given the prevalence of childhood trauma, all professionals working with children should consider:

- Gathering a focused history of potential exposure to neglect, emotional, physical and sexual abuse.
- The potential risk factor of childhood abuse/neglect for later behavioral emotional/behavioral/mental health concerns.

The research collected in this compendium offers vital guideposts to professionals across a wide spectrum of disciplines. It provides a foundation for ongoing research into this area of study, so vital for the wellbeing of our children and their future. Furthermore, with greater understanding of the connections between trauma and health, we can build more effective cross-disciplinary intervention strategies.

Lisa Albers Prock, MD, MPH

The early contributions of childhood trauma (emotional, physical, sexual, and general) have been hypothesized to play a significant role in the development of anxiety disorders, such as posttraumatic stress disorder (PTSD) and social anxiety disorder (SAD). The aim of Bishop and colleagues in Chapter 1 was to assess childhood trauma differences between PTSD and SAD patients and healthy controls, as measured by the Early Trauma Inventory. The authors examined individuals (N=109) with SAD with moderate/severe early developmental trauma (EDT) (n=32), individuals with SAD with low/no EDT (n=29), individuals with PTSD with EDT (n=17), and healthy controls (n=31). The mean age was 34 years (SD=11). Subjects were screened with the Mini-International Neuropsychiatric Interview (MINI), Liebowitz Social Anxiety Scale (LSAS), Clinician-Administered PTSD Scale (CAPS), and Childhood Trauma Questionnaire (CTQ). Analysis of variance was performed to assess group differences. Correlations were calculated between childhood traumas. Although not statistically significant, individuals with PTSD endorsed more physical and sexual childhood trauma compared with individuals with SAD with moderate/severe EDT who endorsed more emotional trauma. For all groups, physical and emotional abuse occurred between ages 6 and 11, while the occurrence of sexual abuse in individuals with PTSD was at 6–11 years and later (13–18 years) in individuals with SAD with moderate/severe EDT. For emotional abuse in all groups, the perpetrator was mostly a primary female caregiver; for sexual abuse, it was mostly a nonfamilial adult male, while for physical abuse, it was mostly a caregiver (male in PTSD and female in SAD with moderate/severe EDT). The contribution of childhood abuse to the development of PTSD and SAD and the differences between these groups and other anxiety disorders should not be ignored and attention should be given to the frequency and severity of these events. The relationship of the perpetrator(s) and the age of onset of childhood abuse are also important considerations as they provide a useful starting point to assess impact over the life course. This can, in turn, guide clinicians on the optimal timing for the delivery of interventions for the prevention of PTSD and SAD.

In Chapter 2, McDermott and colleagues studied how to assess the extent and nature of psychiatric morbidity among children (aged 8 to 13 years) 15 months after a car bomb explosion in the town of Omagh, Northern Ireland. A survey was conducted of 1945 school children attending 13 schools in the Omagh district. Questionnaires included demographic details, measures of exposure, the Horowitz Impact of Events Scale, the Birleson Self-Rating Depression Scale, and the Spence Children's Anxiety Scale. Children directly exposed to the bomb reported higher levels of probable PTSD (70%), and psychological distress than those not exposed. Direct exposure was more closely associated with an increase in PTSD symptoms than in general psychiatric distress. Significant predictors of increased IES scores included being male, witnessing people injured and reporting a perceived life threat but when co-morbid anxiety and depression are included as potential predictors anxiety remains the only significant predictor of PTSD scores. School-based studies are a potentially valuable means of screening and assessing for PTSD in children after large-scale tragedies. Assessment should consider type of exposure, perceived life threat and other co-morbid anxiety as risk factors for PTSD.

Psychotic experiences occur at a much greater prevalence in the population than psychotic disorders. There has been little research to date, however, on genetic risk for this extended psychosis phenotype. In Chapter 3, Ramsay and colleagues examined whether COMT or BDNF genotypes were associated with psychotic experiences or interacted with childhood trauma in predicting psychotic experiences. Psychiatric interviews and genotyping for COMT-Val158Met and BDNF-Val66Met were carried out on two population-based samples of 237 individuals aged 11-15 years. Logistic regression was used to examine for main effects by genotype and childhood trauma, controlling for important covariates. This was then compared to a model with a term for interaction between genotype and childhood trauma. Where a possible interaction was detected, this was further explored in stratified analyses. While childhood trauma showed a borderline association with psychotic experiences, COMT-Val158Met and BDNF-Val66Met genotypes were not directly associated with psychotic experiences in the population. Testing for gene x environment interaction was borderline significant in the case of COMT-Val158Met with individuals with the COMT-Val158Met Val-Val genotype, who had been exposed to childhood trauma borderline significantly more likely to report psychotic experiences than those with Val-Met or Met-Met genotypes. There was no similar interaction by BDNF-Val66Met genotype. The COMT-Val158Met

Val-Val genotype may be a genetic moderator of risk for psychotic experiences in individuals exposed to childhood traumatic experiences.

A personal history of childhood maltreatment has been associated with unfavorable outcomes in bipolar disorder (BD). The impact of early life stressors on the course of BD may be influenced by individual differences in coping skills. The coping construct relies on neurocognitive mechanisms that are usually influenced by childhood maltreatment. Daruy-Filho and colleagues' objective in Chapter 4 was to verify the association between childhood maltreatment and coping skills in individuals with BD Type 1. Thirty female euthymic outpatients with BD Type 1 were evaluated using the Childhood Trauma Questionnaire and two additional instruments to measure their coping preferences: Ways of Coping Questionnaire (coping strategies) and Brief COPE (coping styles). Reports of physical abuse (B = .64, p < .01) and emotional abuse (B = .44, p = .01) were associated with the use of maladaptive strategies that focused on emotional control. Adaptive strategies and styles of coping, such as focusing on the problem, were chosen less frequently by women who had experienced emotional neglect (B = .53, p < .01) and physical abuse (B = -.48, p < .01) in childhood. The small sample size in the present study prevented subgroup analyses. The sample did not include male BD participants. The results indicate that early traumatic events may have a long-lasting deleterious influence on coping abilities in female BD patients. Future prospective studies may investigate whether the negative impact of childhood maltreatment over the course of BD is mediated by individual differences in coping abilities.

Major depressive disorder (MDD) neural underpinnings may differ based on onset age and childhood trauma. In Chapter 5, Jaworska and colleagues assessed cortical thickness in patients who differed in age of MDD onset and examined trauma history influence. Adults with MDD and controls underwent magnetic resonance imaging. Twenty patients had MDD onset <24 years of age (pediatric onset) and 16 had onset >25 years of age (adult onset). The MDD group was also subdivided into those with and without physical and/or sexual abuse as assessed by the Childhood Trauma Questionnaire (CTQ). Cortical thickness was analyzed with Free-Surfer software. Results. Thicker frontal pole and a tendency for thinner transverse temporal cortices existed in MDD. The former was driven by the pediatric onset group and abuse history (independently), particularly in the right frontal pole. Inverse correlations existed between CTQ scores and frontal pole cortex thickness. A similar inverse relation existed with left inferior and right superior parietal cortex thickness. The superior temporal cortex tended to be thinner in pediatric versus adult onset groups with childhood abuse. Conclusions. This preliminary work suggests neural differences between pediatric and adult MDD onset. Trauma history also contributes to cytoarchitectural modulation. Thickened frontal pole cortices as a compensatory mechanism in MDD warrant evaluation.

Children who have experienced chronic parental rejection and exclusion during childhood, as is the case in childhood emotional maltreatment, may become especially sensitive to social exclusion. Chapter 6, by van Harmelen and colleagues, investigates the neural and emotional responses to social exclusion (with the Cyberball task) in young adults reporting childhood emotional maltreatment. Using functional magnetic resonance imaging, the authors investigated brain responses and self-reported distress to social exclusion in 46 young adult patients and healthy controls (mean age = 19.2±2.16) reporting low to extreme childhood emotional maltreatment. Consistent with prior studies, social exclusion was associated with activity in the ventral medial prefrontal cortex and posterior cingulate cortex. In addition, severity of childhood emotional maltreatment was positively associated with increased dorsal medial prefrontal cortex responsivity to social exclusion. The dorsal medial prefrontal cortex plays a crucial role in self-and other-referential processing, suggesting that the more individuals have been rejected and maltreated in childhood, the more self- and other- processing is elicited by social exclusion in adulthood. Negative self-referential thinking, in itself, enhances cognitive vulnerability for the development of psychiatric disorders. Therefore, the findings of this article may underlie the emotional and behavioural difficulties that have been reported in adults reporting childhood emotional maltreatment.

Preclinical studies have demonstrated the relationship between stress-induced increased cortisol levels and atrophy of specific brain regions, however, this association has been less revealed in clinical samples. The aim of Lu and colleagues in Chapter 7 was to investigate the changes and associations of the hypothalamic-pituitary-adrenal (HPA) axis activity and gray matter volumes in young healthy adults with self-reported childhood trauma exposures. Twenty four healthy adults with childhood trauma and 24 age- and gender-matched individuals without childhood trauma were recruited. Each participant collected salivary samples in the morning at four time points: immediately upon awakening, 30, 45, and 60 min after awakening for the assessment of cortisol awakening response (CAR). The 3D T1-weighted magnetic resonance imaging data were obtained on a Philips 3.0 Tesla scanner. Voxel-based morphometry analyses were conducted to compare the gray matter volume between two groups. Correlations of gray matter volume changes with severity of childhood trauma and CAR data were further analyzed. Adults with self-reported childhood trauma showed an enhanced CAR and decreased gray matter volume in the right middle cingulate gyrus. Moreover, a significant association was observed between salivary cortisol secretions after awaking and the right middle cingulate gyrus volume reduction in subjects with childhood trauma. The present research outcomes suggest that childhood trauma is associated with hyperactivity of the HPA axis and decreased gray matter volume in the right middle cingulate gyrus, which may represent the vulnerability for developing psychosis after childhood trauma experiences. In addition, this study demonstrates that gray matter loss in the cingulate gyrus is related to increased cortisol levels.

Generalized anxiety disorder (GAD) is a common anxiety disorder that usually begins in adolescence. Childhood maltreatment is highly prevalent and increases the possibility for developing a variety of mental disorders including anxiety disorders. An earlier age at onset of GAD is significantly related to maltreatment in childhood. In Chapter 8, Liao and colleagues argue that exploring the underpinnings of the relationship between childhood maltreatment and adolescent onset GAD would be helpful in identifying the potential risk markers of this condition. Twenty-six adolescents with GAD and 25 healthy controls participated in this study. A childhood trauma questionnaire (CTQ) was introduced to assess childhood maltreatment. All subjects underwent high-resolution structural magnetic resonance scans. Voxel-based morphometry (VBM) was used to investigate gray matter alterations. Significantly larger gray matter volumes of the right putamen were observed in GAD patients compared to healthy controls. In addition, a significant diagnosis-by-maltreatment interaction effect for the left thalamic gray matter volume was revealed, as shown by larger volumes of the left thalamic gray matter in GAD patients with childhood maltreatment compared with GAD patients without childhood maltreatment as well as with healthy controls with/without childhood maltreatment. A significant positive association between childhood maltreatment and left thalamic gray matter volume was only seen in GAD patients. These findings revealed an increased volume in the subcortical regions in adolescent GAD, and the alterations in the left thalamus might be involved in the association between childhood maltreatment and the occurrence of GAD.

Exposure to interparental violence is associated with negative outcomes, such as depression, post-traumatic stress disorder and reduced cognitive abilities. However, little is known about the potential effects of witnessing domestic violence during childhood on gray matter volume (GMV) or cortical thickness. High-resolution 3.0 T volumetric scans (Siemens Trio Scanner) were obtained on 52 subjects (18–25 years) including 22 (6 males/16 females) with a history of visually witnessing episodes of domestic violence, and 30 (8 males/22 females) unexposed control subjects, with neither a current nor past DSM-IV Axis I or II disorder. Potential confounding effects of age, gender, level of parental verbal aggression, parental education, financial stress, full scale IQ, and total GMV, or average thickness were modeled using voxel based morphometry and FreeSurfer. Witnessing domestic violence subjects had a 6.1% GMV reduction in the right lingual gyrus (BA18) (P = 0.029, False Discovery Rate corrected peak level). Thickness in this region was also reduced, as was thickness in V2 bilaterally and left occipital pole. Theses regions were maximally sensitive to exposure to witnessing domestic violence between 11–13 years of age. Regional reductions in GMV and thickness were observed in both susceptible and resilient witnessing domestic violence subjects. Results in subjects witnessing domestic violence were similar to previously reported results in subjects with childhood sexual abuse, as the primary region affected was visual cortex. Brain regions that process and convey the adverse sensory input of the abuse may be specifically modified by this experience, particularly in subjects exposed to a single type of maltreatment. Exposure to multiple types of maltreatment is more commonly associated with morphological alterations in corticolimbic regions. These findings fit with preclinical studies showing that visual cortex is a highly plastic structure.

Adverse childhood experiences (ACEs) are linked to multiple adverse health outcomes. Chapter 10, by Brown and colleagues, examined the association between ACEs and cancer diagnosis. Data from the 2010 Behavioral Risk Factor Surveillance System (BRFSS) survey were used. The BRFSS is the largest ongoing telephone health survey, conducted in all US states, the District of Columbia, Puerto Rico, Guam and the U.S. Virgin Islands, and provides data on a variety of health issues among the non-institutionalized adult population. Principal component analysis (PCA) was used to derive components for ACEs. Multivariable logistic regression models were used to provide adjusted odds ratios (OR) and 95% confidence intervals (CI) for the association between ACE components and overall, childhood and adulthood cancer, adjusting for confounders such as age, gender, race/ethnicity, income, educational status, marital status, and insurance status. Approximately 62% of respondents reported being exposed to ACEs and about one in ten respondents reported ever having been diagnosed with cancer. Component 1, which had the sexual abuse variables with the highest weights, was significantly associated with adulthood cancer (adjusted OR: 1.21; 95% CI: 1.03–1.43). The association between ACEs and adulthood cancer may be attributable to disease progression through association of ACEs with risk factors for other chronic diseases. More research should focus on the impact of sexual abuse ACEs and adverse health outcomes.

Chapter 11, by Kelly-Irving and colleagues, aimed to analyse whether Adverse Childhood Experiences (ACE) are associated with an increased risk of cancer. The National Child Development Study (NCDS) is a prospective birth cohort study with data collected over 50 years. The NCDS included all live births during one week in 1958 (n=18558) in Great Britain. Self-reported cancer incidence was based on 444 participants reporting having had cancer at some point and 5694 reporting never having cancer. ACE was measured using reports of: 1) child in care, 2) physical neglect, 3) child's or family's contact with the prison service, 4) parental separation due to divorce, death or other, 5) family experience of mental illness & 6) family experience of substance abuse. The resulting variable had three categories, no ACEs/ one ACE/ 2 +ACEs and was used to test for a relationship with cancer. Information on socioeconomic characteristics, pregnancy and birth were extracted as potential confounders. Information on adult health behaviours, socioeconomic environment, psychological state and age at first pregnancy were added to the models. Multivariate models were run using multiply-imputed data to account for missing data in the cohort. The odds of having a cancer before 50 y among women increased twofold for those who had 2+ ACEs versus those with no ACEs, after adjusting for adult factors and early life confounders (OR: 2.1, 95% CI: 1.42-3.21, p<0.001). These findings suggest that cancer risk may be influenced by exposure to stressful conditions and events early on in life. This is potentially important in furthering our understanding of cancer aetiology, and consequently in redirecting scientific research and developing appropriate prevention policies.

Adverse childhood experiences (ACEs), such as abuse, household dysfunction, and neglect, have been shown to increase adults' risk of developing chronic conditions and risk factors for chronic conditions, including cardiovascular disease (CVD). Much less work has investigated the effect of ACEs on children's physical health status that may lead to adult chronic health conditions. Therefore, Chapter 12, by Pretty and colleagues, examined the relationship between ACEs and early childhood risk factors for adult cardiovascular disease. 1,234 grade six to eight students participated in school-based data collection, which included resting measures of blood pressure (BP), heart rate (HR), body mass index (BMI) and waist circumference (WC). Parents of these children completed an inventory of ACEs taken from the Childhood Trust Events Survey. Linear regression models were used to assess the relationship between experiencing more than 4 ACEs experienced, systolic BP, HR, BMI and WC. In additional analysis, ACEs were assessed ordinally in their relationship with systolic BP, HR, and BMI as well as clinical obesity and hypertension status. After adjustment for family education, income, age, sex, physical activity, and parental history of hypertension, and WC for HR models, four or more ACEs had a significant effect on HR (b=1.8 bpm, 95% CI (0.1-3.6)) BMI (b =1.1 kg/m2, 95% CI (0.5-1.8)), and WC (b=3.6 cm, 95% CI (1.8-5.3)). A dose–response relationship between ACE accumulation and both BMI and WC was also found to be significant. Furthermore, accumulation of 4 or more ACEs was significantly associated with clinical obesity (95th percentile), after controlling for the aforementioned covariates. In a community sample of grade six to eight children, accumulation of 4 or more ACEs significantly increased BMI, WC and resting HR. Therefore, risk factors related to reported associations between ACEs and cardiovascular outcomes among adults are identifiable in childhood suggesting earlier interventions to reduce CVD risk are required.

Psychological factors and socioeconomic status (SES) have a notable impact on health disparities, including type 2 diabetes risk. However, the link between childhood psychosocial factors, such as childhood adversities or parental SES, and metabolic disturbances is less well established. In addition, the lifetime perspective including adult socioeconomic factors remains of further interest.

In Chapter 13, Tamayo and colleagues carried out a systematic review with the main question if there is evidence in population- or community-based studies that childhood adversities (like neglect, traumata and deprivation) have considerable impact on type 2 diabetes incidence and other metabolic disturbances. Also, parental SES was included in the search as risk factor for both, diabetes and adverse childhood experiences. Finally, the authors assumed that obesity might be a mediator for the association of childhood adversities with diabetes incidence. Therefore, they carried out a second review on obesity, applying a similar search strategy. Two systematic reviews were carried out. Longitudinal, population- or community-based studies were included if they contained data on psychosocial factors in childhood and either diabetes incidence or obesity risk. the authors included ten studies comprising a total of 200,381 individuals. Eight out of ten studies indicated that low parental status was associated with type 2 diabetes incidence or the development of metabolic abnormalities. Adjustment for adult SES and obesity tended to attenuate the childhood SES-attributable risk but the association remained. For obesity, eleven studies were included with a total sample size of 70,420 participants. Four out of eleven studies observed an independent association of low childhood SES on the risk for overweight and obesity later in life. Taken together, there is evidence that childhood SES is associated with type 2 diabetes and obesity in later life. The database on the role of psychological factors such as traumata and childhood adversities for the future risk of type 2 diabetes or obesity is too small to draw conclusions. Thus, more population-based longitudinal studies and international standards to assess psychosocial factors are needed to clarify the mechanisms leading to the observed health disparities.

The impact of early life events is increasingly becoming apparent, as studies investigate how early childhood can shape long-term physiology and behaviour. Fibromyalgia (FM), which is characterised by increased pain sensitivity and a number of affective co-morbidities, has an unclear etiology. Chapter 14, by Low and Schweinhardt, discusses risk factors from early life that may increase the occurrence or severity of FM in later life: pain experience during neonatal life causes long-lasting changes in nociceptive circuitry and increases pain sensitivity in the older organism; premature birth and related stressor exposure cause lasting changes in stress responsivity; maternal deprivation affects anxiety-like behaviours that may be partially mediated by epigenetic modulation of the genome— all these adult phenotypes are strikingly similar to symptoms displayed by FM sufferers. In addition, childhood trauma and exposure to substances of abuse may cause lasting changes in developing neurotransmitter and endocrine circuits that are linked to anxiety and stress responses.

Listen Protect Connect (LPC), a school-based program of Psychological First Aid delivered by non-mental health professionals, is intended to support trauma-exposed children. In Chapter 15, the objective of Ramirez and colleagues was to implement LPC in a school setting and assess the effectiveness of LPC on improving psychosocial outcomes associated with trauma. A pilot quasi-experiment was conducted with middle school children self-identified or referred to the school nurse as potentially exposed to stressful life experiences. LPC was provided to students by the school nurse, and questionnaires were administered at baseline, 2-, 4- and 8-weeks to assess life stressors, symptoms of post-traumatic stress disorder and depression, social support, and school connectedness. A total of 71 measurements were collected from 20 children in all. Although a small sample size, multiple measurements allowed for multivariable mixed effects models to analyze changes in the repeated outcomes over time. Students who received the intervention had reduced depressive and posttraumatic stress symptoms from baseline throughout follow-up period. Total social support also increased significantly from baseline through 8-weeks, and school connectedness increased up to 4-weeks post-intervention. This study demonstrates the potential of LPC as a school-based intervention of Psychological First Aid. Future randomized trials of LPC are needed, however.

PART I

EARLY TRAUMA AND MENTAL HEALTH

CHAPTER 1

AN ANALYSIS OF EARLY DEVELOPMENTAL TRAUMA IN SOCIAL ANXIETY DISORDER AND POSTTRAUMATIC STRESS DISORDER

MELANIE BISHOP, DAVID ROSENSTEIN, SUSANNE BAKELAAR, AND SORAYA SEEDAT

1.1 INTRODUCTION AND BACKGROUND

Early developmental trauma (EDT) or childhood trauma may loosely be defined as any traumatic experience that occurs before 18 years of age [1]. EDT has been linked to the development of anxiety disorders in adulthood [2-5]. Among South Africans, anxiety disorders (15.8%) are the most prevalent lifetime disorders according to the South African Stress and Health (SASH) study, with social anxiety disorder (SAD) at 2.8% and posttraumatic stress disorder (PTSD) at 2.3% [6]. It has been estimated that one in ten children, on average, is neglected or psychologically abused annually and that approximately 4% to 16% are physically abused [7]. Stein et al. in 1996 found in their sample that both adult males and females with an anxiety disorder had higher rates of childhood physical abuse than those without an anxiety disorder [8]. In addition, females with anxiety disorders had higher rates of childhood sexual abuse [8]. Also, Prigerson et al. in 1996 found that psychological abuse and parental loss were risk factors for the development of adult psychiatric disorders [5]. In SAD [9] and PTSD [10], childhood traumas include physical abuse [11-13], sexual abuse [12-18], and emotional abuse [19-23].

1.1.1 PHYSICAL AND SEXUAL ABUSE IN PTSD AND SAD

The association between different subtypes of EDT and anxiety disorders in adults is a complex one. For example, in a national study by Cougle et al. in 2010, childhood physical abuse was specifically associated with PTSD in adults [11]. Furthermore, there was a relationship between sexual and physical abuse experienced in childhood and adult PTSD [11]. Sexual abuse in childhood has also been established to be a risk factor for the development of SAD [18] and PTSD [12,13,15,16] in adulthood. In a study of women from an American cohort, childhood sexual abuse occurred in approximately 10% of women diagnosed with PTSD [17]. This was supported by the findings of Cutajar et al. in 2010 who found sexual abuse was most strongly correlated with PTSD [16]. Also, significantly more females than males who experienced childhood sexual abuse were diagnosed with PTSD [16]. In a separate study, adults with SAD reported higher rates of sexual abuse in childhood compared to healthy controls [14].

1.1.2 EMOTIONAL AND OTHER ABUSE IN PTSD AND SAD

Childhood emotional abuse has been reported to correlate more strongly, than either physical or sexual abuse, with a diagnosis of SAD and depression [20]. According to Kuo et al. in 2011 individuals with SAD had higher rates of childhood emotional abuse and emotional neglect compared to healthy controls [21]. In one study conducted by Simon et al. in 2009, 56% and 39% of individuals with SAD, respectively, experienced childhood emotional abuse and neglect [22]. Kuo et al. in 2011 investigated the frequency of childhood traumas (sexual abuse, physical abuse, physical neglect, emotional abuse, and emotional neglect) as measured by the Childhood Trauma Questionnaire (CTQ) [21]. As mentioned previously, they found that childhood emotional abuse and neglect were more frequently reported by individuals with SAD [21]. Other childhood traumas have been reported in PTSD and SAD. Afifi et al. in 2009 found that a combination of childhood abuse and parental divorce significantly increased the likelihood of a diagnosis of lifetime PTSD in adulthood [24]. Binelli et al. in 2012 found only one significant positive correlation between family violence and SAD, although the assessment of childhood trauma was based on five closed questions (yes/no) relating to five childhood adversities, namely, the loss of someone close, emotional abuse, physical abuse, family violence, and sexual abuse [25].

1.1.3 AGE OF ONSET AND RELATIONSHIP TO PERPETRATOR IN PTSD AND SAD

The association between EDT subtypes, age of occurrence, and relationship to the perpetrator has not received much attention in the literature [26]. In individuals with PTSD, the onset of childhood sexual abuse (CSA) seems to be between the ages of 6 and 13 years, with a mean duration of 7 years [27]. In one study the age of onset of CSA was approximately 7 years [26]. Also, most perpetrators (87%) of CSA in adults with PTSD were males and were nuclear family members (37%), followed by non-family members (35%), and extended family members (28%) [27]. In another study, the age of onset of abuse by a family friend/other perpetrators was significantly associated with social anxiety in adulthood [28]. Most were abused by other perpetrators (e.g., acquaintances, boyfriends, and babysitters) (44%), followed by strangers (40%), other relatives or family members (29%), family friends (16%), and least by their fathers (14%). Furthermore, the average age of onset was 9.85 years [28].

1.1.4 RATIONALE AND AIMS

Although EDT and other early childhood adversities have been examined in SAD and PTSD, there have been no comparative studies of the frequency and association of EDT subtypes in these disorders. The clinical value of such a comparison between childhood trauma in PTSD and SAD is important as it provides clinicians with information that can be used in interventions with patients with PTSD and SAD. The aim of this study was to assess EDT differences between PTSD and SAD patients and healthy controls. Secondly, we wanted to establish the frequency of events occurring in childhood, as measured by the Early Trauma Inventory (ETI). Thirdly, we wanted to establish the age of onset of childhood abuse and type of perpetrator (relationship to victim and gender) in this sample with SAD and PTSD. Lastly, we correlated childhood traumas as measured by the ETI and childhood traumas measured by the CTQ. We hypothesized that there would be significant group differences, with more childhood trauma overall, and significantly more sexual trauma, reported by the PTSD group. We further hypothesized that (i) the age of onset of EDT and the perpetrators involved would differ between the groups and (ii) childhood trauma as measured by the CTQ would be significantly correlated on the ETI.

1.2 METHOD

1.2.1 DESIGN

This study used clinical data obtained from a larger imaging genetics project whose primary aim was to evaluate group differences in functional magnetic imaging responses in the amygdala by genotype.

1.2.2 PARTICIPANTS

Participants were recruited from various community clinics, psychiatric institutions, hospitals, and NGOs. Participants were also recruited through advertising (print, electronic, and media). Participants were recruited over a 3-year period (2010–2013). A total of 109 participants (31 healthy controls, 32 with SAD and moderate/severe EDT, 29 with SAD with low/no EDT, and 17 with PTSD resulting from EDT) were selected in the basis of the presence or absence of SAD/PTSD and the presence/absence of early childhood trauma, using the Mini-International Neuropsychiatric Interview (MINI), the Liebowitz Social Anxiety Scale (LSAS), the Clinician-Administered PTSD Scale (CAPS), and the CTQ. Healthy controls were selected on the basis of absence of any psychiatric disorder and early childhood trauma, as assessed by the MINI and CTQ (a score below 40 on the CTQ), respectively.

1.2.3 ASSESSMENTS

The MINI was administered to assess for SAD and PTSD, and exclude co-morbid psychiatric conditions, including substance abuse within the previous 12 months of assessment [29]. The LSAS, with a cut-off score of 60, was used to assess for SAD and distinguish between generalized SAD and specific SAD [30]. This clinician-administered questionnaire assesses current social anxiety disorder through 24 questions on a 4-point Likert scale. The CAPS was used to assess for current and lifetime PTSD [31]. The frequency and intensity of PTSD symptoms (as described by the DSM-IV diagnostic criteria) are measured by CAPS on a separate 5-point Likert scales, each ranging from 0 to 4. These ratings are then summarized into a 9-point severity score for each symptom, ranging from 0–8 [32].

The CTQ and ETI were used to assess EDT and to differentiate the low/no and moderate/severe EDT groups with SAD. A score of less than 40 on the CTQ indicated low/no trauma, a score between 40 and 46 was used as a threshold for exclusion, and a score of more than 46 indicated moderate/severe EDT [1]. The CTQ is a robust screening measure of early childhood trauma. It is a retrospective self-report inventory that assesses the five main areas of early developmental trauma: physical abuse, physical neglect, emotional abuse, emotional neglect, and sexual abuse [1]. The ETI is a detailed clinician-administered structured interview that assesses various aspects of developmental trauma: childhood physical abuse, sexual abuse, emotional abuse, and early major adverse life events such as parental loss or serious illness [33]. The ETI provides a structure to systematically assess specific forms of childhood maltreatment. The advantages of using the ETI over the CTQ are that it is able to assess the age of onset, duration, and severity of each trauma, as well as the relationship of the perpetrator to the victim.

1.2.4 ETHICAL CONSIDERATIONS

The protocol received ethical clearance from the Health Research Ethics Committee at Stellenbosch University, South Africa. The study was conducted according to internationally and locally accepted ethical guidelines, namely, the Declaration of Helsinki and the South African Department of Health's 2004 Guidelines: *Ethics in Health Research Structures and Processes*. Participation was completely voluntary, and written informed consent was obtained from all participants.

1.2.5 STATISTICAL ANALYSIS

A one-way analysis of variance (ANOVA) was conducted to assess group differences of childhood trauma as measured by the ETI. For significant group differences, Fisher's least significant difference (LSD) post hoc testing was administered. Cross-tabulations between groups (control, SAD with low/no EDT, SAD with moderate/severe EDT, and PTSD) within childhood trauma domains as measured by the ETI were done to calculate the frequency of events experienced, age of onset of abuse, and relationship to perpetrators. Furthermore, nonparametric correlations were performed to assess the significance between childhood trauma on the CTQ (emotional abuse, sexual abuse, physical abuse, emotional neglect, physical neglect, and total childhood trauma) and ETI (general abuse, emotional abuse, physical abuse, sexual abuse, and total childhood trauma). All tests were two-tailed for significance, and significance (p value) was set at .05.

1.3 RESULTS

1.3.1 DEMOGRAPHIC CHARACTERISTICS

The majority of participants were female (n=60, 55%), of white ethnicity (n=72, 66.1%), and single (n=59, 54.1%). Their age ranged from 20 to 72 years, and the majority had an annual income of more than R60,000 per year (n=76.8%). The average number of years of education was 15 (SD=3.2), with a range of 5–24 years, and the majority was employed (n=71, 65.1%). All participants were fluent in English. With regards to previously diagnosed psychiatric disorder(s), 64 participants had not previously been diagnosed (57.8%), compared to 45 who were previously diagnosed with a psychiatric disorder (41.3%). One participant did not complete the question (0.9%).

1.3.2 ABUSE CHARACTERISTICS

Significant group differences (PTSD, SAD with moderate/severe EDT, and controls) were found on the ETI in the domains of physical abuse (p=0.00), emotional abuse (p=0.00), sexual abuse (p=0.00), other childhood trauma (p=0.00) and total scores (p=0.00) (Table 1). Fisher's LSD post hoc testing was done to establish between-group differences. There were no significant differences between SAD and PTSD groups. Significant differences in each of the trauma domains (physical abuse, emotional abuse, sexual abuse, other childhood trauma, and total trauma) were found between the PTSD and control groups as well as between SAD and control groups (see Table 2). Although not statistically significant, more general, physical, sexual, and total childhood trauma was experienced by the PTSD group. Only emotional abuse was experienced more in the SAD group.

TABLE 1: Results of the ANOVAs for differences between groups (N=80)

Variable	F value	p value
Childhood physical abuse (ETI physical abuse total)	6.77	0.00**
Childhood emotional abuse (ETI emotional abuse total)	22.11	0.00**
Childhood sexual abuse (ETI sexual abuse)	0.88	0.00**
Total childhood trauma experienced (ETI total score)	1.34	0.00**
Childhood general trauma (ETI general trauma total)	10.6	0.00**

$**p < 0.01.$

TABLE 2: LSD post hoc results of ANOVA between groups (PTSD, SAD with moderate/severe EDT, controls)

	Controla								SAD with EDTb			
	SAD with EDT				PTSD				PTSD			
	Significance	Mean difference	95% CI		Significance	Mean difference	95% CI		Significance	Mean difference	95% CI	
			Lower bound	Upper bound			Lower bound	Upper bound			Lower bound	Upper bound
Physical abuse (ETI physical abuse total score)	0.00**	-2.003	-3.04	-0.97	0.00**	-2.214	-3.46	-0.97	0.73	-0.211	-1.45	1.02
Emotional abuse (ETI emotional abuse total score)	0.00**	-3.492	-4.46	-2.52	0.00**	-2.918	-4.08	-1.76	0.33	0.574	-0.58	1.73
Sexual abuse (ETI sexual abuse total score)	0.00**	-2.074	-3.37	-0.77	0.00**	-2.767	-4.32	-1.21	0.38	-0.693	-2.24	.85
Total childhood abuse (ETI Total score)	0.00**	-11.016	-15.25	-6.78	0.00**	-13.751	-18.82	-8.68	0.28	-2.735	-7.78	2.31
General trauma (ETI general trauma total score)	0.00**	-4.198	-6.60	-1.80	0.00**	-5.852	-8.73	-2.98	0.25	-1.654	-4.52	1.21

**p < 0.01. aFirst the control group is compared to the SAD with moderate/severe EDT and PTSD group, respectively; bthe SAD with moderate/severe EDT group is compared to the PTSD group.

TABLE 3: The frequency of events as measured by the ETI

Item		Controls (n=31)	SAD with no/low EDT (n=29)	SAD with moderate/severe EDT (n=32)	PTSD (n=17)	Total (N=109)
Physical abuse						
P1	Spanked with hand	26 (83.9%)	26 (89.7%)	29 (90.6%)	15 (88.2%)	96 (88.1%)
P2	Slapped in face	5 (16.1%)	5 (17.2%)	10 (31.3%)	5 (29.4%)	25 (22.9%)
P3	Burned with cigarette	9 (29.0%)	20 (69%)	15 (46.9%)	9 (52.9%)	53 (48.6%)
P4	Punched or kicked	7 (22.6%)	6 (20.7%)	14 (43.8%)	6 (35.4%)	33 (30.3%)
P5	Hit or spanked with object	18 (58.1%)	21 (72.4%)	25 (78.1%)	16 (94.1%)	80 (73.4%)
P6	Hit with thrown object	3 (9.7%)	3 (10.3%)	10 (31.3%)	6 (35.3%)	22 (20.2%)
P7	Choked	0 (0%)	2 (6.9%)	7 (21.9%)	3 (17.7%)	12 (11%)
P8	Pushed or shoved	9 (29.0%)	9 (31.0%)	15 (46.9%)	9 (52.9%)	42 (38.5%)
P9	Tied up or locked in closet	1 (3.2%)	1 (3.5%)	3 (9.4%)	2 (11.8%)	7 (6.4%)
Emotional abuse						
E1	Often put down or ridiculed	7 (22.6%)	11 (37.9%)	26 (81.3%)	11 (64.7%)	55 (50.5%)
E2	Often ignored or made to feel you didn't count	4 (12.9%)	6 (20.7%)	24 (75%)	11 (64.7%)	45 (41.3%)
E3	Often told you are no good	5 (16.1%)	5 (17.2%)	16 (50%)	9 (52.9%)	35 (32.1%)
E4	Often shouted or yelled at	9 (29.0%)	15 (51.7%)	22 (68.8%)	12 (70.6%)	58 (53.2%)
E5	Most of the time treated in cold or uncaring way	2 (6.5%)	2 (6.9%)	19 (59.4%)	9 (52.9%)	32 (29.4%)
E6	Parents control areas of your life	8 (25.8%)	13 (44.8%)	19 (59.4%)	9 (52.9%)	49 (45%)
E7	Parents fail to understand your needs	4 (12.9%)	13 (44.8%)	26 (81.3%)	10 (58.8%)	53 (48.6%)

TABLE 3: *Cont.*

Item		Controls (n=31)	SAD with no/low EDT (n=29)	SAD with moderate/severe EDT (n=32)	PTSD (n=17)	Total (N=109)
Sexual abuse						
S1	Exposed to inappropriate comments about sex	3 (9.7%)	4 (13.8%)	13 (40.6%)	8 (47.1%)	28 (25.7%)
S2	Exposed to flashing	2 (6.5%)	2 (6.9%)	12 (37.5%)	6 (35.3%)	22 (20.2%)
S3	Spy on you dressing/bathroom	3 (9.7%)	0 (0%)	8 (25%)	2 (11.8%)	13 (11.9%)
S4	Forced to watch sexual acts	0 (0%)	0 (0%)	5 (15.6%)	1 (5.9%)	6 (5.5%)
S5	Touched in intimate parts in way that was uncomfortable	6 (19.4%)	2 (6.9%)	12 (37.5%)	10 (58.8%)	30 (27.5%)
S6	Someone rubbing genitals against you	2 (6.5%)	0 (0%)	8 (25%)	5 (29.4%)	15 (13.8%)
S7	Forced to touch intimate parts	1 (3.2%)	2 (6.9%)	3 (9.7%)	5 (29.4%)	11 (10.1%)
S8	Someone had genital sex against your will	1 (3.2%)	0 (0%)	3 (9.4%)	4 (23.5%)	8 (7.3%)
S9	Forced to perform oral sex	0 (0%)	1 (3.5%)	1 (3.1%)	2 (11.8%)	4 (3.7%)
S10	Someone performed oral sex on you against your will	0 (0%)	0 (0%)	2 (6.5%)	2 (11.8%)	4 (3.7%)
S11	Someone had anal sex with you against your will	0 (0%)	0 (0%)	2 (6.3%)	2 (11.8%)	4 (3.7%)
S12	Someone tried to have sex but didn't do so	1 (3.2%)	0 (0%)	8 (25.8%)	7 (41.2%)	16 (14.7%)
S13	Forced to pose for sexy photographs	1 (3.2%)	0 (0%)	1 (3.1%)	2 (11.8%)	4 (3.7%)
S14	Forced to perform sex acts for money	-	-	-	-	-
S15	Forced to kiss someone in sexual way	0 (0%)	0 (0%)	9 (28.1%)	2 (11.8%)	11 (10.1%)

TABLE 3: *Cont.*

Item	Controls (n=31)	SAD with no/low EDT (n=29)	SAD with moderate/severe EDT (n=32)	PTSD (n=17)	Total (N=109)
General trauma					
T1 Natural trauma	1 (3.2%)	1 (3.5%)	4 (12.5%)	4 (23.5%)	10 (9.2%)
T2 Serious accidents	6 (19.4%)	6 (20.7%)	7 (21.9%)	7 (41.2%)	26 (23.9%)
T3 Serious personal injury	11 (35.5%)	4 (13.8%)	10 (31.3%)	8 (47.1%)	33 (30.3%)
T4 Serious personal illness	4 (12.9%)	2 (6.9%)	8 (25%)	9 (52.9%)	23 (21.2%)
T5 Death of parent	5 (16.1%)	4 (13.8%)	11 (34.4%)	7 (41.2%)	27 (24.8%)
T6 Serious injury/illness of parent	7 (22.6%)	8 (27.6%)	16 (50%)	10(58.9%)	41 (37.6%)
T7 Separation of parents	3 (9.7%)	5 (17.2%)	13 (40.6%)	6 (35.3%)	27 (24.8%)
T8 Raised in home other than parents	2 (6.5%)	2 (6.9%)	12 (37.5%)	5 (29.4%)	21 (19.3%)
T9 Death of Sibling	2 (6.5%)	4 (13.8%)	7 (21.9%)	3 (17.7%)	16 (14.7%)
T10 Serious illness/injury of sibling	5 (16.1%)	5 (17.2%)	7 (21.9%)	5 (29.4%)	22 (20.2%)
T11 Death of friend	12 (38.7%)	5 (17.2%)	10 (31.3%)	7 (41.2%)	34 (31.2%)
T12 Serious injury of friend	7 (22.6%)	4 (13.8%)	7 (21.9%)	7 (41.2%)	25 (22.9%)
T13 Observe death/serious injury of others	4 (12.9%)	6 (20.7%)	14 (43.8%)	9 (52.9%)	33 (30.3%)
T14 Divorce/separation of parents	2 (6.5%)	5 (17.2%)	8 (25%)	4 (23.5%)	19 (17.4%)
T15 Witnessing violence	8 (25.8%)	9 (31.0%)	22 (68.8%)	14(82.4%)	53 (48.6%)
T16 Family mental illness	8 (25.8%)	10 (34.5%)	17 (53.1%)	7 (41.2%)	42 (38.5%)
T17 Alcoholic parents	3 (9.7%)	4 (13.8%)	11 (34.4%)	7 (41.2%)	25 (22.9%)
T18 Drug abuse in parents	2 (6.5%)	1 (3.5%)	2 (6.3%)	0 (0%)	5 (4.6%)
T19 Victim of major theft	12 (38.7%)	13 (44.8%)	17 (53.1%)	9 (52.9%)	51 (46.8%)

TABLE 3: *Cont.*

Item		Controls (n=31)	SAD with no/low EDT (n=29)	SAD with moderate/ severe EDT (n=32)	PTSD (n=17)	Total (N=109)
T20	Victim of armed robbery	3 (9.7%)	0 (0%)	7 (21.9%)	6 (35.3%)	16 (14.7%)
T21	Victim of assault	1 (3.2%)	4 (13.8%)	13 (40.6%)	5 (29.4%)	23 (21.1%)
T22	Victim of rape	1 (3.2%)	0 (0%)	4 (12.5%)	3 (17.7%)	8 (7.3%)
T23	See someone murdered	1 (3.2%)	0 (0%)	2 (6.3%)	2 (11.8%)	5 (4.6%)
T24	Someone close to you murdered	1 (3.2%)	1 (3.5%)	3 (9.4%)	4 (23.5%)	9 (8.3%)
T25	Someone close to you raped	2 (6.5%)	1 (3.5%)	4 (12.5%)	4 (23.5%)	11 (10.1%)
T26	Work in stressful job	1 (3.2%)	4 (13.8%)	9 (28.1%)	5 (29.4%)	19 (17.4%)
T27	POW/hostage	0 (0%)	0 (0%)	0 (0%)	2 (11.8%)	2 (1.8%)
T28	Combat	0 (0%)	1 (3.5%)	1 (3.1%)	2 (11.8%)	4 (3.7%)
T29	Death of child	0 (0%)	1 (3.5%)	2 (6.3%)	0 (0%)	3 (2.8%)
T30	Miscarriage	0 (0%)	0 (0%)	4 (12.5%)	1 (5.9%)	5 (4.6%)
T31	Death of spouse	-	-	-	-	-

1.3.3 PHYSICAL ABUSE

In the total sample, the most prevalent form of physical abuse reported was being spanked with a hand (n=96, 88.1%), followed by being hit or spanked with an object (n=80, 73.4%). This was followed by being burned with a cigarette (n=53, 48.6%) and being pushed or shoved (n=42, 38.5%). This is the same order of frequency reported by the SAD with moderate/severe EDT, SAD low/no EDT, and control groups, respectively. For the PTSD group, the most prevalent childhood physical abuse was being hit or spanked with an object (n=16, 94.1%), followed by being hit with a hand (n=15, 88.2%), and being burned with a cigarette (n=9, 52.9%) or pushed or shoved (n=9, 52.9%) (see Table 3).

1.3.4 EMOTIONAL ABUSE

The most prevalent childhood emotional abuse experienced in the sample was being often shouted or yelled at (n=58, 53.2%), followed by being often put down or ridiculed (n=55, 50.5%), and parental failure to understand the participant's needs as a child (n=53, 48.6%). The most prevalent trauma in the SAD with moderate/severe EDT group, (n=26, 81.3%)was a lack of understanding of the participant's needs and being often put down or ridiculed by \a parent (n=26). This was followed by often being ignored or made to feel that the person did not count (n=24, 75%), and often being shouted or yelled at (n=22, 68.8%). Within the PTSD group, the most prevalent form of emotional abuse experienced as a child was often being shouted or yelled at (n=12, 70.6%), followed by being put down or ridiculed (n=11, 64.7%), and often ignored or made to feel like the person did not count (n=11, 64.7%). The most prevalent emotional abuse experienced in the SAD with low/no EDT group and control groups was being often shouted or yelled at 29% (n=9) and 51.7% (n=15), respectively (see Table 3).

1.3.5 SEXUAL ABUSE

The most prevalent form of sexual abuse experienced was being touched in intimate parts in a way that was uncomfortable (n=30, 27.5%), followed by being exposed to inappropriate comments about sex (n=28, 25.7%), and being exposed to flashing (n=22, 20.2). The most prevalent childhood sexual abuse experienced by the SAD with moderate/severe EDT group was exposure to inappropriate comments about sex (n=13, 40.6%), followed by exposure to flashing (n=12, 37.5%), and being touched in intimate parts in a way that was uncomfortable (n=12, 37.5%). The most prevalent sexual abuse experienced by the PTSD group was being touched in intimate parts in a way that was uncomfortable (n=10, 58.8%), followed by exposure to inappropriate comments about sex (n=47.1%), and someone trying to have sex with the individual, but not doing so (n=7, 41.2%). In the SAD group with low/no EDT, the most prevalent childhood sexual abuse experienced was being exposed to inappropriate comments about sex (n=4, 13.8%). The most prevalent sexual abuse experienced by the control group was being touched in intimate parts in a way that was uncomfortable (n=6, 19.35%) (see Table 3).

1.3.6 OTHER CHILDHOOD TRAUMATIC EVENTS (DEFINED AS GENERAL TRAUMA ON THE ETI)

In the sample as a whole, the most prevalent 'other' childhood traumatic life event was witnessing violence (n=52, 48.6%), followed by being a victim of major theft (n=51, 46.5%). This was followed by family mental illness (n=42, 38.5%) and serious injury/illness of a parent (n=41, 37.6%). Witnessing violence was the most prevalent traumatic life event in both the SAD with moderate/severe EDT (n=22, 68.7%) and the PTSD (n=14, 82.35%) groups. In the SAD with low/no EDT group, the most prevalent traumatic childhood event was being a victim of major theft (n=13, 44.85), followed by family mental illness (n=10, 34.5%), and then witnessing violence (n=9, 31%). In the control group, the most prevalent traumatic childhood event was being a victim of major theft (n=12, 38.7%) and experiencing a serious personal injury (n=11, 35.5%) (see Table 3).

TABLE 4: Age of onset of childhood traumatic events as measured by the ETI

	Controls				SAD with EDT				PTSD			
	Never	0–5[a]	6–11[b]	13–18[c]	Never	0–5[a]	6–11[b]	13–18[c]	Never	0–5[a]	6–11[b]	13–18[c]
Physical abuse events												
Spanked with hand	5	11	14	2	14	14	2	2	6	8	6	1
Slapped in face	23	2	3	3	9	1	11	2	9	1	3	4
Burned with cigarette	31	0	0	0	29	2	1	0	13	2	2	0
Punched or kicked	24	0	1	6	18	1	9	4	11	1	2	3
Hit or spanked with object	13	4	9	5	7	10	12	3	1	4	11	1
Hit with thrown object	28	0	1	2	22	1	5	4	11	0	5	1
Choked	31	0	0	0	25	1	2	4	14	1	0	2
Pushed or shoved	22	0	4	5	17	1	7	7	8	0	5	4
Tied up or locked in closet	30	1	0	0	29	1	2	0	15	1	1	0
Total scores	207	18	32	22	158	32	63	35	84	18	35	16
Emotional abuse events												
Often put down or ridiculed	24	0	2	5	6	5	14	7	6	2	5	4
Often ignored or made to feel you didn't count	27	0	3	1	8	8	9	7	6	5	6	0
Often told you are no good	26	0	3	2	16	0	9	7	8	2	4	3
Often shouted or yelled at	22	2	5	2	9	8	11	4	5	4	6	2

TABLE 4: *Cont.*

	Controls				SAD with EDT				PTSD			
	Never	0–5ᵃ	6–11ᵇ	13–18ᶜ	Never	0–5ᵃ	6–11ᵇ	13–18ᶜ	Never	0–5ᵃ	6–11ᵇ	13–18ᶜ
Most of the time treated in cold or uncaring way	29	1	0	1	13	2	12	5	8	3	5	1
Parents control areas of your life	22	2	6	1	13	5	10	4	8	2	6	1
Parents fail to understand your needs	27	1	1	2	6	6	12	8	7	2	5	3
Total scores	177	6	20	14	71	34	77	42	48	20	37	14
Sexual abuse events												
Exposed to inappropriate comments about sex	28	0	3	0	19	3	5	5	9	0	4	4
Exposed to flashing	29	1	0	1	20	4	4	4	11	0	3	3
Spy on you dressing/bathroom	28	1	0	2	24	2	4	2	15	0	2	0
Forced to watch sexual acts	31	0	0	0	27	0	3	2	16	0	1	0
Touched in intimate parts in way that was uncomfortable	25	0	3	3	20	2	5	5	7	1	4	5
Someone rubbing genitals against you	29	0	1	1	24	2	3	3	12	0	3	2
Forced to touch intimate parts	30	0	0	1	28	1	1	1	12	0	3	2
Someone had genital sex against your will	30	0	0	1	29	1	2	0	13	1	1	2

TABLE 4: *Cont.*

	Controls				SAD with EDT				PTSD			
	Never	0–5[a]	6–11[b]	13–18[c]	Never	0–5[a]	6–11[b]	13–18[c]	Never	0–5[a]	6–11[b]	13–18[c]
Forced to perform oral sex	31	0	0	0	31	1	0	0	15	0	2	2
Someone performed oral sex on you against your will	31	0	0	0	29	0	1	1	15	0	2	0
Someone had anal sex with you against your will	31	0	0	0	30	1	0	1	15	0	1	1
Someone tried to have sex but didn't do so	30	0	1	0	23	1	3	4	10	1	3	3
Forced to pose for sexy photographs	31	0	0	0	31	0	0	1	15	0	1	1
Forced to perform sex acts for money	31	0	0	0	0	0	0	0	0	0	0	0
Forced to kiss someone in sexual way	31	0	0	0	23	1	2	6	15	0	0	2
Total scores	446	2	7	10	358	18	34	35	180	3	30	27

[a]*Between and including the ages 0–5 years;* [b]*between and including the ages 6–11 years;* [c]*between and including the ages 13–18 years.*

TABLE 5: Relationship to perpetrator as measured by the ETI

	Controls				SAD with EDT				PTSD			
	Caregiver	Other adult	Sibling	Stranger	Caregiver	Other adult	Sibling	Stranger	Caregiver	Other adult	Sibling	Stranger
Physical abuse events												
Spanked with hand	25	1	0	0	28	1	1	0	14	1	0	0
Slapped in face	8	0	0	0	15	6	1	2	6	2	0	0
Burned with cigarette	0	0	0	0	3	0	0	0	1	1	0	2
Punched or kicked	2	0	2	3	4	1	3	6	2	3	1	0
Hit or spanked with object	12	4	1	1	23	0	1	1	16	0	0	0
Hit with thrown object	1	1	1	0	7	0	2	1	3	0	2	1
Choked	0	0	0	0	1	1	2	3	2	1	0	0
Pushed or shoved	1	1	2	5	5	3	2	5	2	1	4	2
Tied up or locked in closet	0	1	0	0	2	1	0	0	2	0	0	0
Total scores	49	8	6	9	88	13	12	18	48	9	7	5
Emotional abuse events												
Often put down or ridiculed	3	1	2	1	16	3	4	4	8	1	1	1
Often ignored or made to feel you didn't count	1	0	3	0	14	6	3	1	9	2	0	0
Often told you are no good	2	1	2	0	11	2	1	2	7	0	2	0

TABLE 5: *Cont.*

	Controls				SAD with EDT				PTSD			
	Caregiver	Other adult	Sibling	Stranger	Caregiver	Other adult	Sibling	Stranger	Caregiver	Other adult	Sibling	Stranger
Often shouted or yelled at	7	1	1	0	20	1	1	1	11	5	4	1
Most of the time treated in cold or uncaring way	0	1	1	0	16	1	2	0	8	1	0	0
Parents control areas of your life	9	0	0	0	18	1	0	0	8	1	0	0
Parents fail to understand your needs	4	0	0	0	25	1	0	0	9	1	0	0
Total scores	26	4	9	1	120	15	11	8	60	11	7	2
Sexual abuse events												
Exposed to inappropriate comments about sex	0	3	0	0	5	7	0	1	0	7	1	0
Exposed to flashing	1	0	0	1	1	7	0	5	0	4	1	1
Spy on you dressing/ bathroom	2	0	0	1	3	3	1	1	0	2	0	0
Forced to watch sexual acts	0	0	0	0	1	3	0	2	0	1	0	0
Touched in intimate parts in way that was uncomfortable	1	4	0	1	1	5	0	6	0	8	1	1
Someone rubbing genitals against you	1	0	0	0	1	4	0	3	0	4	0	1

TABLE 5: *Cont.*

	Controls				SAD with EDT				PTSD			
	Caregiver	Other adult	Sibling	Stranger	Caregiver	Other adult	Sibling	Stranger	Caregiver	Other adult	Sibling	Stranger
Forced to touch intimate parts	0	1	0	0	0	2	0	1	0	5	0	0
Someone had genital sex against your will	0	1	0	0	0	1	0	2	0	4	0	0
Forced to perform oral sex	0	0	0	0	0	1	0	0	0	2	0	0
Someone performed oral sex on you against your will	0	0	0	0	0	2	0	0	0	1	0	1
Someone had anal sex with you against your will	0	0	0	0	0	1	0	1	0	1	0	1
Someone tried to have sex but didn't do so	0	1	0	0	1	4	0	3	0	7	0	0
Forced to pose for sexy photographs	0	0	0	0	0	1	0	0	0	2	0	0
Forced to perform sex acts for money	0	0	0	0	0	0	0	0	0	0	0	0
Forced to kiss someone in sexual way	0	0	0	2	0	5	0	2	0	2	0	0
Total scores	5	10	0	3	15	46	1	27	0	50	3	5

TABLE 6: Gender of perpetrators (primary caregivers and other adults)

	Controls				SAD with EDT				PTSD			
	Primary caregivers		Other adults		Primary caregivers		Other adults		Primary caregivers		Other adults	
	F	M	F	M	F	M	F	M	F	M	F	M
Physical abuse events												
Spanked with hand	16	9	1	0	17	11	0	1	7	7	0	1
Slapped in face	6	2	0	0	9	6	2	4	1	5	0	2
Burned with cigarette	0	0	0	0	0	3	0	0	0	1	0	1
Punched or kicked	0	2	0	0	1	3	0	1	2	2	0	3
Hit or spanked with object	8	4	3	1	13	10	0	0	12	4	0	0
Hit with thrown object	1	0	0	1	4	3	0	0	2	1	0	0
Choked	0	0	0	0	1	0	0	1	0	2	0	1
Pushed or shoved	0	1	0	1	2	3	1	2	2	2	0	1
Tied up or locked in closet	0	0	0	1	1	1	0	1	0	1	0	0
Total scores	31	18	4	4	49	40	3	10	22	25	0	9
Emotional abuse events												
Often put down or ridiculed	1	2	1	0	9	7	1	2	4	4	1	1
Often ignored or made to feel you didn't count	1	0	0	0	7	7	3	3	5	4	1	1
Often told you are no good	2	0	1	0	4	7	0	2	3	4	0	0
Often shouted or yelled at	4	3	1	0	15	5	0	1	7	4	1	0
Most of the time treated in cold or uncaring way	0	0	0	1	11	5	0	1	4	4	0	0

TABLE 6: Cont.

	Controls				SAD with EDT				PTSD			
	Primary caregivers		Other adults		Primary caregivers		Other adults		Primary caregivers		Other adults	
	F	M	F	M	F	M	F	M	F	M	F	M
Parents control areas of your life	8	1	0	0	17	1	0	1	6	2	1	0
Parents fail to understand your needs	3	1	0	0	18	7	0	1	5	4	1	0
Total scores	19	7	3	1	81	39	4	11	34	26	5	2
Sexual abuse events												
Exposed to inappropriate comments about sex	0	0	0	3	1	4	2	5	0	0	2	5
Exposed to flashing	1	0	0	0	0	1	1	6	0	0	1	3
Spy on you dressing/bathroom	2	0	0	0	2	1	0	3	0	0	0	2
Forced to watch sexual acts	0	0	0	0	0	1	0	3	0	0	1	0
Touched in intimate parts in way that was uncomfortable	0	1	0	4	0	1	0	5	0	0	1	7
Someone rubbing genitals against you	0	1	0	0	0	1	0	4	0	0	1	3
Forced to touch intimate parts	0	0	0	1	0	0	0	2	0	0	1	4
Someone had genital sex against your will	0	0	0	1	0	0	0	1	0	0	1	3

TABLE 6: *Cont.*

	Controls				SAD with EDT				PTSD			
	Primary caregivers		Other adults		Primary caregivers		Other adults		Primary caregivers		Other adults	
	F	M	F	M	F	M	F	M	F	M	F	M
Forced to perform oral sex	0	0	0	0	0	0	0	1	0	0	1	1
Someone performed oral sex on you against your will	0	0	0	0	0	0	0	2	0	0	1	0
Someone had anal sex with you against your will	0	0	0	0	0	0	0	1	0	0	0	1
Someone tried to have sex but didn't do so	0	0	1	0	1	1	0	4	0	0	1	6
Forced to pose for sexy photographs	0	0	0	0	0	0	0	1	0	0	1	1
Forced to perform sex acts for money	0	0	0	0	0	0	0	0	0	0	0	0
Forced to kiss someone in sexual way	0	0	0	1	1	1	0	5	0	0	1	1
Total scores	3	2	0	10	4	11	3	43	0	0	13	37

F female, M male.

TABLE 7: Spearman's nonparametric correlations between childhood traumas measured by CTQ and ETI

Variable	Spearman's rho (r value) and p values	Physical abuse (CTQ)	Sexual abuse (CTQ)	Emotional neglect (CTQ)	Physical neglect (CTQ)	Total childhood trauma (CTQ)	General trauma (ETI)	Physical abuse (ETI)	Emotional abuse (ETI)	Sexual abuse (ETI)	Total childhood trauma (ETI total)
Emotional abuse (CTQ)	r value	0.65	0.47	0.66	0.54	0.87	0.48	0.47	0.69	0.46	0.68
	p value	0.00***	0.00**	0.00**	0.00**	0.00**	0.00**	0.00***	0.00**	0.00**	
Physical abuse (CTQ)	r value		0.37	0.52	0.47	0.74	0.35	0.56	0.45	0.33	0.53
	p value		0.00**	0.00**	0.00**	0.00**	0.00**	0.00***	0.00**	0.00**	
Sexual abuse (CTQ)	r value			0.38	0.44	0.60	0.41	0.28	0.26	0.71	0.51
	p value			0.00**	0.00**	0.00**	0.00**	0.01**	0.00**	0.00**	
Emotional neglect (CTQ)	r value				0.58	0.85	0.27	0.37	0.63	0.37	0.51
	p value				0.00**	0.01**	0.00**	0.00***	0.00**	0.00**	
Physical neglect (CTQ)	r value					0.73	0.43	0.29	0.42	0.39	0.49
	p value					0.00**	0.00**	0.00***	0.00**	0.00**	
Total childhood trauma (CTQ)	r value						0.49	0.51	0.70	0.54	0.72
	p value						0.00**	0.00**	0.00**	0.00**	0.00**

TABLE 7: *Cont.*

Variable	Spearman's rho (r value) and p values	Physical abuse (CTQ)	Sexual abuse (CTQ)	Emotional neglect (CTQ)	Physical neglect (CTQ)	Total childhood trauma (CTQ)	General trauma (ETI)	Physical abuse (ETI)	Emotional abuse (ETI)	Sexual abuse (ETI total)	Total childhood trauma (ETI total)
General trauma (ETI)	r value							0.45	0.48	0.49	0.85
	p value							0.00**	0.00**	0.00**	0.00**
Physical abuse (ETI)	r value								0.47	0.27	0.69
	p value								0.00**	0.00**	
Emotional abuse (ETI)	r value									0.36	0.76
	p value									0.00**	
Sexual abuse (ETI)	r value										0.63
	p value										0.00**

**Correlations are significant at the 0.01 level (two-tailed). CTQ, n = 108; ETI, N = 109.

1.3.7 AGE OF ONSET

In all groups the age of onset of physical and emotional abuse was generally between the age of 6 and 11 years. Sexual abuse in the SAD with moderate/severe EDT and control groups tended to occur between 13 and 18 years of age, while in the PTSD group the age of onset was earlier (6–11 years) (Table 4).

1.3.8 PERPETRATORS

For physical and emotional abuse, in all groups, the perpetrator was mostly the primary caregiver. In both the SAD with moderate/severe EDT and control groups this was a female caregiver, while for the PTSD group a male primary caregiver was mostly reported. For emotional abuse, the perpetrator most commonly reported by all groups was a female primary caregiver. With regards to the perpetrator of sexual abuse, the most frequently reported perpetrator by all groups was another male adult (Tables 5 and 6).

1.3.9 CORRELATION BETWEEN ETI AND CTQ

Significant positive correlations were found between all childhood traumas measured on the CTQ (physical abuse, physical neglect, emotional abuse, emotional neglect, and sexual abuse) and the ETI (physical abuse, general abuse, sexual abuse, and emotional abuse) (see Table 7). Nine high correlations (identified as correlation coefficients $r > 0.70$) were found and are reported in descending order from the strongest correlations to a correlation coefficient of 0.70 [34]. These were (i) childhood emotional abuse (CTQ EA) and total trauma (CTQ total score) ($r = 0.87$, $p = 0.00$), (ii) childhood emotional neglect (CTQ EN) and total trauma (CTQ total score) ($r = 0.85$, $p = 0.00$), (iii) other traumas experienced during childhood (ETI general trauma score) and total trauma (ETI total score) ($r = 0.85$, $p = 0.00$), (iv) emotional abuse during childhood (ETI EA score) and total childhood trauma (ETI total score) ($r = 0.76$, $p = 0.00$), (v) childhood physical abuse (CTQ PA) and total childhood trauma (CTQ total score) (r=0.74, p=0.00), (vi) childhood physical neglect (CTQ PN score) and total trauma (CTQ total score) (r=0.73, p=0.00), (vii) childhood total trauma (CTQ total score) and total trauma on the ETI total score (r=0.72, p=0.00), (viii) childhood sexual abuse (CTQ SA) and childhood sexual abuse measured by the ETI (r=0.71, p=0.00), and (ix) childhood emotional abuse on the ETI and total trauma on the CTQ (r=0.70, p=0.00) (see Table 7).

1.4 DISCUSSION AND CONCLUSIONS

The main objective of this study was to investigate differences in the type and amount of childhood trauma in adults with PTSD and SAD. No statistically significant differences were found for these disorders.

This finding suggests that in this sample, childhood traumas (major ALE, physical, sexual, emotional, and total) are not significantly different in individuals with PTSD and SAD with moderate/severe EDT. Previous studies have found that physical abuse [10,12,13], sexual abuse [12-18], emotional abuse [19-23], as well as other childhood adversities [24,25] are linked to the development of both PTSD and SAD. Childhood emotional abuse has been found to correlate more strongly with a diagnosis of social anxiety than either physical or sexual abuse [20]. Previous studies have also documented higher rates of childhood emotional abuse and neglect in adults with SAD compared to healthy controls [21,22]. Emotional abuse and neglect significantly impact on the development of SAD and PTSD [23]. The results of our study regarding emotional abuse indicates that parents' emotional expression toward their children can have long-lasting effects and contribute to PTSD and SAD in later life. The aforementioned studies, however, did not specifically compare EDT exposure, by severity and type, in these disorders.

1.4.1 AGE OF ONSET

With regards to the onset of childhood abuse, Rodriguez et al. in 1996 also found the age of onset of sexual abuse in individuals with PTSD to be around 6 years of age and termination at approximately 13 years, which is similar to our finding in the PTSD group of between 6 and 11 years [27]. This finding is supported by two other studies that found the age of onset of childhood sexual abuse to be approximately 7 years [26] and 9.85 years [28], respectively. There is a paucity of studies on the age of onset of EDT in people with PTSD and SAD, underscoring the need for more investigation in this area to better inform the nature and timing of interventions.

1.4.2 PERPETRATORS

With regards to the perpetrators, our finding of sexual abuse mostly being another nonfamilial male adult is supported by the findings in a previous study of perpetrators of EDT [27]. According to Rodriguez et al. in 1996, perpetrators of CSA in individuals with PTSD were mostly male (87%), nuclear family members, followed by nonfamily members, and lastly by extended family members [27]. Also, Feerick and Snow in 2005 showed in their study that participants were mostly sexually abused by other perpetrators, such as acquaintances, boyfriends, and babysitters, i.e., nonfamilial persons and strangers [28]. Ackerman et al. in 1998 found that if a child was physically abused by a male, the likelihood of a psychiatric diagnosis, including PTSD, was higher than if abused by a female [26]. Overall, physically and sexually abused children were more likely to develop a psychiatric disorder which included PTSD in later life. No further studies focusing on perpetrators of EDT in individuals with SAD or PTSD were found, highlighting a gap that warrants attention in future studies.

1.4.3 ETI AND CTQ

All traumas measured by the CTQ were positively correlated with all traumas measured by the ETI. Kuo et al. in 2011 also found that all scales on the CTQ had a significant positive correlation with all subscales, except for sexual abuse and emotional abuse or neglect [21]. Bremner et al. in 2000 tested the convergent validity of the ETI by comparing its domains (physical, sexual, and general) with the components of the Childhood Trauma Severity Index (CLTE) [33]. Significant correlations were calculated between the total score of the ETI and the total score of the CLTE, the physical abuse domain of the ETI and the physical abuse component of the CLTE, as well as between the sexual abuse domain of the ETI and the sexual abuse component of the CLTE [33].

1.4.4 LIMITATIONS AND RECOMMENDATIONS

A number of limitations warrant mention. Firstly, the results of the LSD post hoc testing should be interpreted with caution given that this type of analysis is associated with type 1 error. Secondly, group sizes were small. Thirdly, endorsement of EDT was reliant on the retrospective accounts of participants. This may not always be very accurate with regards to memory recall of traumatic early life events. Recall bias is, therefore, a concern. Recall may also have been affected by the way in which childhood events were assessed [25]. It has been suggested that rather than an exaggeration in the rates, childhood traumas are more likely to be under-reported [35]. In addition, participants varied in age and older participants in particular may have had more difficulty recalling childhood events [36]. The sample may not be demographically representative of the South African population with SAD and PTSD with early trauma. Future studies should include more Black African participants. Furthermore, studies of larger samples that include more qualitative assessments of childhood trauma are needed. In sum, the contribution of EDT to the development of PTSD and SAD and the differences in terms of childhood trauma between these groups, as well as other anxiety disorders, should not be ignored, and attention should be given to the frequency and severity of these events. The relationship between victims of EDT and perpetrators and the age of onset of childhood abuse is another important facet, as it provides a timeline against which the course of abuse and its impact can be tracked over the life trajectory. This can, in turn, provide some guidance to clinicians on the optimal timing and nature of interventions for the prevention of PTSD and SAD.

REFERENCES

1. Bernstein DP, Fink L: Childhood Trauma Questionnaire: a Retrospective Self-Report. Manual. San Antonio, TX: Psychological Corporation; 1998.
2. Brown GW, Harris TO: Aetiology of anxiety and depressive disorders in an inner-city population. Psychol Med 1993, 23:143-154.
3. Heim C, Nemeroff CB: The role of childhood trauma in the neurobiology of mood and anxiety disorders: preclinical and clinical studies. Biol Psychiatry 2001, 49:1023-1039. doi:S0006-3223(01)01157-X
4. Kendler KS, Hetteman JM, Butera F, Gardner CO, Prescott CA: Life events dimensions of loss, humiliation, entrapment, and danger in the prediction of onsets of major depression and generalized anxiety. Arch Gen Psychiatry 2003, 60:789-796.
5. Prigerson HG, Shear MK, Bierhals AJ, Zonarich DL, Reynolds CF: Childhood adversity, attachment and personality styles as predictors of anxiety among elderly caregivers. Anxiety 1996, 2:234-241.
6. Herman A, Stein D, Seedat S, Heeringa S, Moomal H, Williams D: The South African Stress and Health [SASH] study: 12 month and lifetime prevalence of common mental disorders. S Afr Med J 2009, 99(5):339-344.
7. Gilbert R, Widom CS, Browne K, Fergusson D, Webb E, Janson S: Burden and consequences of child maltreatment in high- income countries. Lancet 2008, 373:68-81. doi:10.1016/S0140-6736(08)61706-7
8. Stein MB, Walker JR, Anderson G, Hazen AL, Ross CA, Eldridge G, Forde DR: Childhood physical and sexual abuse in patients with anxiety disorders and in a community sample. Am J Psychiatry 1996, 153(2):275-277.
9. Lieb R, Wittchen HU, Hofler M, Fuetsch M, Stein MB, Marikangas KR: Parental psychopathology, parenting styles and the risk of social phobia in offspring. Arch Gen Psychiatry 2000, 57:859-866.
10. Koenen KC, Moffitt TE, Poulton R, Martin J, Caspi A: Early childhood factors associated with the development of post-traumatic stress disorder: results from a longitudinal birth cohort. Psychol Med 2007, 37(2):181-192. http://dx.doi.org/10.1017/S0033291706009019
11. Cougle JR, Timpano KR, Sachs-Ericsson N, Keough ME, Riccardi CJ: Examining the unique relationships between anxiety disorders and childhood physical and sexual abuse in the National Comorbidity Survey-Replication. Psychiatry Res 2010, 177:150-155.
12. Gelinas D: The persisting negative effects of incest. Psychiatry 1983, 46:312-322.
13. Goodwin J: Post-traumatic stress in incest victims. In Post-traumatic Stress Disorder in Children. Edited by Eth S, Pynoos RS. Washington, DC: American Psychiatric Association; 1985.
14. Bandelow B, Torrente A, Wedekind D, Broocks A, Hajak G, Rutter E: Early traumatic life events, parental rearing styles, family history of mental disorders, and birth risk factors in patients with social anxiety disorder. Eur Arch Psychiatry Clin Neurosci 2004, 254:397-405.
15. Briere JN, Elliott DM: Immediate and long-term impacts of child sexual abuse. Future Child 1994, 4(2):54-69. http://dx.doi.org/10.2307/1602523

16. Cutajar MC, Mullen PE, Ogloff JRP, Thomas SD, Wells DL, Spataro J: Psychopathology in a large cohort of sexually abused children followed up to 43 years. Child Abuse Negl Int J 2010, 34(11):813-822.

17. Kendler KS, Kessler RC, Walters E: Stressful life events, genetic liability and onset of an episode of major depression. Am J Psychiatr 1995, 152:833-842.

18. Magee WJ: Effects of negative life experiences on phobia onset. Soc Psychiatry Psychiatr Epidemiol 1999, 34:343-351.

19. Bruch MA, Heimberg RG: Differences in perceptions of parental and personal characteristics between generalized and nongeneralised social phobics. J Anxiety Disord 1994, 8:155-168.

20. Gibb B, Chelminiski I, Zimmerman M: Childhood emotional, physical, and sexual abuse, and diagnoses of depressive and anxiety disorders in adult psychiatric outpatients. Depress Anxiety 2007, 24:256-263.

21. Kuo JR, Goldin PR, Werner K, Heimberg RG, Gross JJ: Childhood trauma and current psychological functioning in adults with social anxiety disorder. J Anxiety Disord 2011, 25(4):467-473. doi:10.1016/j.janxdis.2010.11.011

22. Simon NM, Herlands NN, Marks EH, Macini C, Letamendi A, Li Z, Pollack MH, Van Ameringen M, Stein MB: Childhood maltreatment linked to greater symptom severity and poorer quality of life and function in social anxiety disorder. Depress Anxiety 2009, 26:1027-1032.

23. Suliman S, Mkabile SG, Fincham DS, Ahmed R, Stein DJ, Seedat S: Cumulative effect of multiple trauma on symptoms of posttraumatic stress disorder, anxiety, and depression in adolescents. Compr Psychiatry 2009, 50(2):121-127. http://www.sciencedirect.com/science/article/pii/S0010440X08000916

24. Afifi TO, Boman J, Fleisher W, Sareen J: The relationship between child abuse, parental divorce, and lifetime mental disorders and suicidality in a nationally representative adult sample. Child Abuse Negl 2009, 33(3):139-147. http://www.sciencedirect.com/science/article/pii/S0145213409000301

25. Binelli C, Ortiz A, Muñiz A, Gelabert E, Ferraz L, Filho AS, Crippa JA, Nardi AE, Subira S, Martin-Santos R: Social anxiety and negative early life events in university students. Rev Bras Psiquiatr 2012, 34:S69-S74.

26. Ackerman P, Newton J, McPherson W, Jones J, Dykman R: Prevalence of post traumatic stress disorder and other psychiatric diagnoses in three groups of abused children (sexual, physical, and both). Child Abuse Negl 1998, 22(8):759-774.

27. Rodriguez N, Ryan S, Rowan A, Foy D: Posttraumatic stress disorder in a clinical sample of adult survivors of childhood sexual abuse. Child Abuse Negl 1996, 20:943-952.

28. Feerick MM, Snow KL: The relationship between childhood sexual abuse, social anxiety disorder, and symptoms of posttraumatic stress disorder in women. J Fam Violence 2005, 20(6):409-419. doi:10.1007/s10896-005-7802-z

29. Sheehan DV, Lecrubier Y, Sheehan KH, Amorim P, Janavs J, Weiller E, Hergueta T, Baker R, Dunbar GC: The Mini-International Neuropsychiatric Interview (M.I.N.I.): the development and validation of a structured diagnostic psychiatric interview for DSM-IV and ICD-10. J Clin Psychiatry 1998, 59:22-33.

30. Baker S, Heinrichs N, Kim H, Hofmann S: The Liebowitz social anxiety scale as a self-report instrument: a preliminary psychometric analysis. Behav Res Ther 2002, 40(6):701-715.

31. Blake DD, Weathers F, Nagy LM, Kaloupek DG, Klauminzer G, Charney DS, Keane TM: A clinician rating scale for assessing current and life-time PTSD: the CAPS-1. Behav Ther 1990, 13:187-188.
32. Weathers FW, Keane TM, Davidson JT: Clinician-administered PTSD scale: a review of the first ten years of research. Depress Anxiety 2001, 13(3):132-156.
33. Bremner JD, Vermetten E, Mazure CM: Development and preliminary psychometric properties of an instrument for the measurement of childhood trauma: the early trauma inventory. Depress Anxiety 2000, 12(1):1-12.
34. Howell DC: Fundamental Statistics for the Behavioural Sciences. 6th edition. Australia: Thomson Wadsworth; 2008.
35. Hardt J, Rutter M: Validity of adult retrospective reports of adverse childhood experiences: review of the evidence. J Child Psychol 2004, 45(2):260-273.
36. Spila B, Makara M, Kozak G, Urbańska A: Abuse in Childhood and Mental Disorder in Adult Life. Canada: Wiley; 2008.

CHAPTER 2

A SCHOOL-BASED STUDY OF PSYCHOLOGICAL DISTURBANCE IN CHILDREN FOLLOWING THE OMAGH BOMB

MAURA MCDERMOTT, MICHAEL DUFFY, ANDY PERCY, MICHAEL FITZGERALD, AND CLAIRE COLE

2.1 BACKGROUND

Children experience a range of psychological reactions to traumatic events including anxiety, depression and behaviour problems. It is now recognised that the broad categories of PTSD symptoms (re-experiencing, avoidance/numbing and increased arousal) are present in children as well as in adults [1]. In children from the age of 8–10 years post traumatic reactions are similar to those of adults [2] although the DSM diagnostic criteria descriptors are more age appropriate [3]. The reactions in children below 8 years of age and particularly below the age of 5 years to traumatic events are less clear [4]. The purpose of this study was to consider the emotional reactions of children from the age of 8–13 fifteen months after the Omagh bomb.

A School Based Study of Psychological Disturbance in Children Following the Omagh Bomb. © McDermott M, Duffy M, Percy A, Fitzgerald M, and Cole C; licensee BioMed Central Ltd. Child and Adolescent Psychiatry and Mental Health, **7,**36 (2013). *doi:10.1186/1753-2000-7-36. Licensed under Creative Commons Attribution 2.0 Generic License, http://creativecommons.org/licenses/by/2.0/.*

2.1.1 THE OMAGH BOMBING

On 15 August 1998, the largest single atrocity of the Northern Ireland conflict took place in Omagh, a market town with a population of 26,000, when a car bomb exploded in the town centre. Thirty-one people, including two unborn children (twins) were killed, 382 people were injured of which 135 were hospitalised. Twenty-six families were bereaved. Of those killed, 15 were aged 17 years or under. The bomb had a devastating effect on the community. A large number of those killed or injured were children and young people or adults with young families. Many children and young people sustained injuries resulting in the loss of limbs, loss of soft tissue, scarring and disfigurement. Many more were exposed to scenes of intense horror and suffering.

The first aim of this study was to assess the extent of psychiatric morbidity among children (aged 8 to 13 years) in a community following a car bomb explosion in the town centre on a busy Saturday afternoon. Children under eight were not included because of the different presentation of trauma reactions in these younger age groups [4]. Children and adolescents over the age of thirteen were included in another study to be reported at a later stage with more age appropriate measures. Secondly, we consider if type of exposure to a traumatic event increases PTSD symptoms in children to a greater extent than symptoms of general emotional distress. Thirdly, we investigate which individual and trauma characteristics identified within this study predict PTSD, depression and anxiety, and consider how our findings compare with the risk factors for PTSD in children and adolescents reported in Trickey and colleagues' recent meta-analysis [5] and other studies.

In relation to the first aim, most epidemiological studies have been of adults and older young people, such as the U.S. National Comorbidity Survey [6] that reported a 10% lifetime prevalence rate. In the U.K. National Mental Health Survey [7] a PTSD rate of 0.4% was found in children aged between 11-15 but scarcely registered below the age of 10 years. However the U.K. study reported a point prevalence estimate and the screening instrument used was not PTSD specific. Fletcher [8] in a meta-analysis of 34 studies reported that 36% of children who had experienced a range of traumas met criteria for PTSD. However, the rates of PTSD associated with traumatic events vary considerably from 0% to 100% [9]. In one review of natural disasters [10] 5-10% of children and adolescents met full criteria for PTSD and after road traffic accidents rates of 25 -30% have been recorded [11].

It has been established in many studies that increased exposure is associated with increased mental health problems including PTSD. In a review of 25 studies Foy and colleagues [12] found exposure to be one of three factors (severity of trauma exposure, trauma-related parental distress, and temporal proximity to trauma) that consistently mediated PTSD development in children. A relationship between level of exposure and PTSD has been found in studies of natural disasters [13-15] community violence [16,17] and political conflict [18-20]. Higher PTSD rates have been reported in relation to specific characteristics of traumatic events, for example rates of 90% have been recorded following exposure to gruesome scenes [21]. In warfare studies of PTSD in children, incidence rates between 25% to 70% are reported depending on type of exposure and type of warfare [2,22]. A number of studies have reported level of exposure and trauma severity as two main risk factors of PTSD [12,23-25]. Trickey and colleagues [5] have identified trauma severity as the trauma characteristic most strongly associated with risk of PTSD in children and adolescents but suggest that trauma severity may be difficult to differentiate from trauma exposure. This poses the possibility of a range of psychological effects associated with a wider range of exposure categories including sub categories of direct exposure based on characteristics like proximity to the potentially traumatic event or being present at the time as opposed to just after an incident. Other established peri traumatic risk factors for PTSD such as physical injury [5], exposure to dead bodies [26] and perceived life threat [5] are theoretically more likely with more "direct" exposure such as being present at the time of a bombing compared with less direct exposure witnessing the immediate aftermath of a bomb. There is also evidence that other forms of indirect exposure such as exposure by media [27,28] are linked to increased risk of PTSD. One concept that previous research does not appear to have systematically addressed is the psychological impact on children who are in the vicinity of an event such as a bomb but narrowly miss being at the precise location during or immediately after the event. We have defined this as a "Near Miss" category for analysis in this paper.

With respect to the third aim of this paper we consider how pre, peri and post trauma factors predict psychological reactions, particularly PTSD, in children following the Omagh bomb. In a recent comprehensive meta-analysis of risk factors for PTSD in children, Trickey and colleague's [5] reported risk factors for PTSD as follows: a small effect size for race and younger age; a small to medium-sized effect for female gender, low intelligence, low SES, pre and post-trauma life events, pre-trauma psychological problems in the individual and parent, pre-trauma low self-esteem, post-trauma parental psychological problems, bereavement, time post-trauma, trauma severity, and exposure to the event by media; and a large effect for low social support, peri trauma fear, perceived life threat, social withdrawal, comorbid psychological problem, poor family functioning, distraction, PTSD at time 1, and thought suppression.

In terms of pre-trauma factors, there have been contradictory findings from studies in relation to age [23,29-31]. Trickey and colleagues [5] reported that younger age is largely unrelated to whether a young person develops PTSD but moderator analysis discovered that there was a statistically significant stronger relationship when the trauma was unintentional although the population effect size remained non-significant regardless of whether the trauma was intentional or non-intentional. Trickey and colleagues [5] also reported that younger age was a significant risk factor, with a small effect, if the index trauma was a group event rather than an individual one. There have also been conflicting findings regarding the relationship between gender and PTSD with some studies recording PTSD in girls at twice the rate as in boys [7]. Whist several studies have reported gender as a significant risk factor [12,21,24,29,32], Trickey and colleagues [5] reported female gender to be a consistent although statistically small risk factor and a stronger risk factor in older children and adolescents and also when the trauma is unintentional. Whilst girls seem more vulnerable to internalizing stress reactions, boys display more externalizing behaviour disturbance [24,33]. Several studies have identified a number of pre-trauma risk factors including; prior traumas [20] prior psychiatric problems [25,32,34] and family cohesion [35]. Whilst type and severity of exposure are recognised as important predictors of PTSD in adults and children, studies have reported other specific peri-trauma factors including: a strong acute trauma response [23,36,37], witnessing dead people

[26], being physically injured [10] and perceived life threat [24,36,37]. Post trauma factors associated with PTSD in children include: social support [25] and co-morbidity, especially depression and generalised anxiety [38-40].

2.2 METHOD

Full ethical approval for the survey was granted by the Sperrin Lakeland Health & Social Care Trust which was the relevant ethical and institutional body at the time (1999). The Trust secured the agreement and assistance of the Western Education & Library Board, the main regulatory body for schools in the Omagh area and school principals to survey children in the classrooms. A passive consent procedure was used to obtain parental consent, that is to say all parents were informed of the study and asked to reply, via prepaid envelope, if they wished their child to be excluded from the study. Parents who consented to their child's inclusion did not have to reply. The parents of bereaved children, children who were hospitalised or children already receiving therapy were contacted directly by members of the Omagh Trauma and Recovery Team and informed of the study. The Omagh Trauma and Recovery Team received 130 referrals for clients aged under 18 between August 1998 and May 2001 [41].

Data was collected 15 months after the car bomb and involved close collaboration between local education and health authorities. All school children aged between 8 and 13 years who were registered within mainstream primary schools within the Omagh area were eligible for inclusion. Thirteen schools participated in the study, with only one school refusing, providing a response rate in excess of 90 per cent. Data was collected via a self-completion booklet and completed by children in their classrooms within schools. All fieldwork was undertaken and supervised by a professional survey organisation and local child and adolescent mental health professionals were available in each school at the time of completion. Table 1 provides details of the characteristics of the children who participated in the survey (n = 1945). The mean age of respondents was 11, and contains slightly more girls than boys. The majority of children lived with both parents (85.3%) and in family units where both parents were employed (75.1%) (Table 1).

TABLE 1: Sample characteristics

Characteristic	Mean	SD	Proportion
Age (Mean)	11.4	1.44	
IES (Mean)	15.65	9.73	
BDS (Mean)	8.67	5.22	
SCAS (Mean)	27.42	17.26	
Gender			
Male			48.7
Female			51.3
Previous psychological treatment (yes)			2.9
Physically injured (yes)			1.2
Perceived life threat (yes)			1.5
Witnessed serious injury (yes)			11.1
Witness people dying (yes)			7.6
Witnessed people dead (yes)			5.6
Post-event support (yes)			2.3
Family structure			
Living with both parents			85.3
Living with single parent			10.8
Reconstituted family			3.1
In state or foster care			0.7
Parental employment			
Both parents employed			75.1
Mother employed - father unemployed			1.5
Father employed - mother unemployed			17.7
Both parents not employed			5.8

2.2.1 MEASURES

Exposure to the bomb: Eight items covered various aspects of exposure to the bombing (see Table 8 in Appendix 1). On the basis of responses to these items, respondents were classified as belonging to one of five mutually exclusive exposure categories. "Exposed - in town at time" means was in Omagh town when the bomb exploded and witnessed injury or death of others or was directly harmed. "Exposed—in town after" means was in Omagh town shortly after the bomb exploded and witnessed injury or death of others or was directly harmed. "Loss" means did not witness injury or death of others, not injured but experienced loss or injury of someone close (family, relative or friend). "Near miss" means was in Omagh town when the bomb exploded but did not witness injury or death of others, was not directly harmed and did not experience loss. "No exposure" means was not in Omagh town when or after the bomb exploded, was not a witness and did not experience loss. In addition, children reported whether they had received any physical injuries (physically injured) or thought they were going to die (perceived life threat).

The Impact of Event Scale (IES) [42] is a widely used screening test for PTSD in children. In this study, the 8 item CRIES-8 (which lacks any arousal items) was used (α=0.82) as it was found to be as efficient as the CRIES-13 (which includes arousal items) in classifying children with and without PTSD [43]. It provides a continuous score for overall PSTD, and two sub-scales each consisting of four items: (1) intrusive thoughts, memories and images and (2) avoidance of thoughts and reminders. Items were grounded in the Omagh Bombing and referenced to experiences within the previous seven days.

The Birleson Depression Self-Rating Scale for Children (BDS) [44] is an 18-item scale assessing the level of depression in children (α=0.82). Items were scored on a three point scale (0,1,2). Responses include 'most', 'sometimes' and 'never'. A score of 0 indicated a healthy response and a score of 2 indicated an unhealthy or depressed response.

The Spence Children's Anxiety Scale (SCAS) [45] consists of 38 items on specific anxiety symptoms with a further six filler items (α=0.94). Responses include 'never', 'sometimes', 'often' and 'always' and are recorded on a four-point scale (0,1,2,3). The scale provides a global anxiety rating together with scores on six individual subscales covering specific anxiety symptoms, namely separation anxiety, social phobia, obsessive-compulsive disorder, panic/agoraphobia, generalised anxiety, and, fears of physical injury.

Socio-demographics: Each respondent provided details of their age and gender, as well as information on family structure (living with both parents/living with single parent/reconstituted family/in state or foster care) and parental employment (both parents employed/mother employed and father not employed/father employed and mother not employed/both parents not employed) (Table 1). Post event support was measured by asking if help was received because of difficulties experienced following the bomb and a checklist of sources of help was provided to identify the provider(s).

TABLE 2: Probable caseness rates for PTSD (IES), depression (BDS) and anxiety (SCAS)

Type of exposure	IES			BDS			SCAS		
	Low	High	%	Low	High	%	Low	High	%
No exposure	603	330	35.4	862	41	4.5	896	37	4.0
Near miss	10	10	50.0	19	1	5.0	20	0	0.0
Loss	353	404	53.4	683	46	6.3	716	42	5.5
Exposed—in town after	43	87	66.9	111	12	9.8	118	12	9.2
Exposed—in town at time	26	75	74.3	86	11	11.3	87	14	13.9
	$r=113.911, p<0.001$			$=11.664, p<0.05$			$=22.791, p<0.001$		

2.2.2 STATISTICAL ANALYSIS

A series of OLS regression models were estimated to examine the predictors of PTSD, anxiety and depression. A three step hierarchical regression was conducted with the predictor variable included in blocks corresponding to pre-, peri- and post-trauma variables. These models were restricted to those individuals who were in town on the day of the bombing and/or witnessed traumatic events. As the sample was clustered at the school level, school dummy variables were included in the model to account for the lack of independence due to school clustering. This ensures that the regression standard errors are adjusted for the lack of independence at the school level. While these dummy variables were included within the model they were not reported within the presented regression tables. None of the school level dummies were significant within the various models.

2.3 RESULTS

2.3.1 PSYCHIATRIC MORBIDITY

Forty seven per cent of the sample met probable clinical PTSD caseness according to IES scores. Using a BDS score of 18 or above, 6% of children in the study met clinical caseness for probable depression and using a cut off score of 60 or more on the SCAS responses 5.7% of the children met clinical caseness for probable anxiety (Table 2).

2.3.2 TYPE OF EXPOSURE: ASSOCIATIONS WITH PTSD AND OTHER PSYCHIATRIC DISORDERS

Over half the children surveyed had some form of exposure to the bombing (52%) (Table 3). This was mainly in the form of loss of a family member, relative or friend (39%), however, over one in ten children did witness the aftermath of the bomb blast. Around one per cent of children were directly injured in the blast, with two per cent thinking they were actually going to die (Table 1).

TABLE 3: Type of exposure experienced by participants

Type of exposure	%
No exposure	48.1
Near miss	1.0
Loss	39.0
Exposed - in town after	6.7
Exposed - in town at time	5.2

No age or gender variations were noted across the levels of exposure (Table 4). The mean scores on the IES, BDS and the SCAS were 15.65, 8.67 and 27.42 respectively (Table 1). The PTSD, depression and anxiety scores varied significantly across types of exposure, with increased exposure associated with higher scores on the IES, BDS and SCAS (Table 4). There were significant differences between the level of exposure and PTSD symptoms ($F(4,1856)=37.698$, $p<0.01$), depression ($F(4,1867)=8.138$, $p<0.01$) and anxiety ($F(4,1778)=18.179$, $p<0.01$). Figure 1 shows the IES, SCAS and BDS standardised symptom scores for each type of exposure. An increase in level of exposure is associated with increased levels of PTSD. However, those in the near miss group exhibited higher levels of anxiety and depression than the loss group. Direct exposure (those present at the time of the explosion and those present after the explosion) was associated with larger increases for PTSD symptoms than for general psychiatric distress. Paired comparisons of these differences showed that standardised IES scores of the two groups directly exposed differed significantly compared to the loss group ($p<0.01$), no exposure ($p<0.01$) and the near miss group ($p<0.05$). The differences between the two groups directly exposed to the bomb scenes, (those present at the time of explosion and those present after the explosion) were not significant on the IES ($p=0.255$), SCAS ($p=0.663$) and depression measures ($p=0.604$). The anxiety scores (SCAS) of those in the two exposure groups were significantly different to those in the no exposure ($p<0.01$), loss ($p<0.01$) groups but not the near miss ($p=0.334$) group. On the depression measure (BDS) those directly exposed differed significantly compared to the loss group ($p<0.01$) and the no exposure group ($p<0.01$), but not the near miss group ($p=0.494$).

2.3.3 PREDICTORS OF PTSD AND OTHER PSYCHIATRIC DISORDERS

Significant predictors of increased IES scores included being male, witnessing people injured and reporting a perceived life threat (Table 5; model 2). However, when co-morbid anxiety and depression are included as potential predictors (see Table 5; model 3), gender, exposure to injury and life threat no longer remain significant predictors. In model 3, anxiety remains the only significant predictor of PTSD scores.

TABLE 4: Sample characteristics by exposure to the bombing

Characteristic	Type of exposure					p
	No exposure	Near miss	Loss	Exposed—in town after	Exposed—in town at time	
Age (Mean)	11.5	11.3	11.4	11.3	11.2	0.233
IES (Mean)	13.17	16.05	17.07	20.29	21.73	0.00**
BDS (Mean)	8.08	9.35	8.92	10.02	10.38	0.00**
SCAS (Mean)	24.23	30.37	29.19	33.85	34.89	0.00**
Female (%)	50.7	55.0	51.3	53.8	52.5	0.959
Previous psychological treatment (yes)	1.3	0.0	3.6	6.9	8.2	0.00**
Post-event support (yes)	0.7	0.0	1.9	7.0	15.0	0.00**
Perceived life threat (yes)	-	0.0	-	6.0	21.1	0.003**
Physically injured (yes)	-		-	10.9	9.9	
Witnessed serious injury (yes)	-		-	92.1	98.0	
Witness people dying (yes)	-		-	61.1	70.3	
Witnessed people dead (yes)	-		-	39.8	57.4	

*Note: "Exposed - in town at time" means in Omagh town when the bomb exploded and witnessed injury or death of others or was directly harmed. "Exposed - in town after" means in Omagh town shortly after the bomb exploded and witnessed injury or death of others or was directly harmed. "Near miss" means in Omagh town when the bomb exploded but did not witness injury or death of others and was not directly harmed. "Loss" means experienced loss or injury of someone close (family, relative or friend) but no direct harm. "No exposure" means not in Omagh town at the time or shortly after the bomb exploded, not a witness and did not experience loss. "IES": Impact of Events Scale; "BDS": Birleson Depression Scale; "SCAS": Spence Children's Anxiety Scale. P: *significant at the 0.05 level; **significant at the 0.01 level.*

Age and gender were significant predictors of probable anxiety, with younger children and girls reporting significantly higher anxiety scores (Table 6). Perceived life threat, witnessing injuries and receiving post bombing psychological support were also significantly associated with higher levels of overall anxiety.

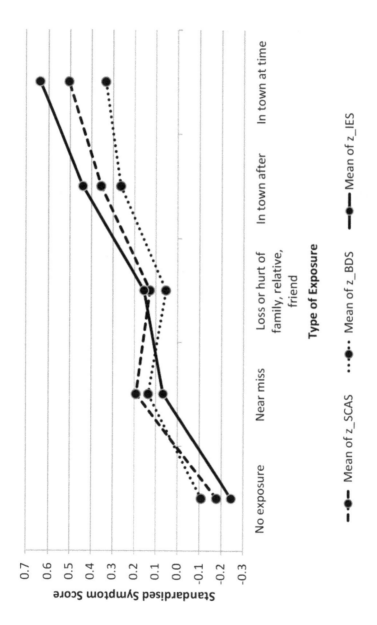

FIGURE 1: SCAS BDS and IES Standard Scores for types of exposure. A graphical representation of the Impact of Events (IES), Spence Children's Anxiety Scale (SCAS) and Birleson Depression Self Rating Scale for Children standardized symptom scores for each type of exposure to the Omagh Bomb.

TABLE 5: Predictors of PTSD symptoms 15 months after explosion among children present in Omagh (N= 212)

Variable	Model 1 β	p	Model 2 β	p	Model 3 β	p
Pre-trauma						
Age	0.229	0.692	0.190	0.734	0.554	0.285
Male	−4.172	0.011*	−3.297	0.036*	−0.495	0.741
Previous psychological treatment	3.473	0.171	3.543	0.148	1.472	0.518
Peri-trauma						
Perceived life threat			5.692	0.015*	2.541	0.250
Physically injured			−1.153	0.679	−0.331	0.899
Witnessed serious injury			4.077	0.015*	2.519	0.103
Witness people dying			0.952	0.589	1.712	0.295
Witnessed people dead			1.764	0.315	0.650	0.688
Post-trauma						
Post-event support					−1.710	0.506
Depression (BDS score)					0.152	0.270
Anxiety (SCAS score)					0.173	0.000**

Notes:
1. PTSD symptoms measured by Impact of Events Scale (IES) score.
2. Those present in Omagh includes those in Omagh town centre when the bomb exploded or shortly after and/or witnessed related traumatic events.
3. Dummy variables for school included in model but excluded from the table.
4. Model 1 Adjusted R2 =0.004; Model 2 Adjusted R2 = 0.107; Model 3 Adjusted R2 = 0.252.
*5. P: *significant at the 0.05 level; **significant at the 0.01 level.*

Being female was also a significant predictor of higher depression score, as was witnessing injury (Table 7). However, even after controlling for witnessing injury and death, the experience of witnessing people you thought were dying was associated with lower depression scores.

Of those directly exposed to the bomb approximately one in ten received post-event psychological/psychiatric interventions. Post-event support significantly predicted probable anxiety (p<0.01; Table 6) but not PTSD (Table 5) or depression (Table 7). Those who received support had significantly higher levels of depression (t(216)=3.007, p<0.01) and anxiety (t(201)=3.656, p<0.01). However, no significant differences in PTSD scores were observed (p=0.057).

TABLE 6: Predictors of anxiety 15 months after explosion among children present in Omagh (N= 222)

Variable	Model 1		Model 2		Model 3	
	β	p	β	p	β	p
Pre-trauma						
Age	−2.244	0.059	−2.396	0.039*	−2.331	0.041*
Male	−15.313	0.000**	−14.067	0.000**	−14.271	0.000**
Previous psychological treatment	10.317	0.052	10.607	0.039*	8.604	0.091
Peri-trauma						
Perceived life threat			15.467	0.001**	13.151	0.006**
Physically injured			−3.812	0.515	−6.784	0.246
Witnessed serious injury			7.621	0.028*	7.244	0.034*
Witness people dying			−2.235	0.531	−3.271	0.354
Witnessed people dead			5.334	0.133	4.849	0.165
Post-trauma						
Post-event support					16.651	0.004**

Notes:
1. Anxiety symptoms measured by Spence Childhood Anxiety Scale (SCAS) score.
2. Those present in includes those in Omagh town centre when the bomb exploded or shortly after and/or witnessed related traumatic events.
3. Dummy variables for school included in model but excluded from the table.
4. Model 1 Adjusted R2 =0.096; Model 2 Adjusted R2 = 0.180; Model 3 Adjusted R2 = 0.210.
*5. P: *significant at the 0.05 level; **significant at the 0.01 level.*

2.4 DISCUSSION

The first aim of the present study was to assess the extent and nature of psychiatric morbidity among children (aged 8 to 13 years) 15 months after a car bomb explosion. The results suggest high levels of psychiatric morbidity, particularly probable PTSD, in the children. Even with the general reduction in the levels of PTSD reactions that tends to occur with time [20,35] and the relatively low numbers with direct exposure, the levels of probable PTSD reported in this study would appear to be high [6-8] and in line with rates found in warfare studies of children [2,22]. A number of factors may be relevant to this finding. First, the location of the incident was outside shops in the main street in the centre of a small market town and many school children will have continued to pass by the bombsite on a regular basis, providing a continual reminder of the incident and recurrent trigger of trauma memories. Secondly, the bombing was unexpected in the context of the ongoing political process, coming just four months after an agreement was signed between the British and Irish Governments that provided a basis for a political settlement and reform. In the preceding months the main paramilitary groups had declared ceasefires raising hopes and expectations that a period of peace had begun. After the explosion many children and young people reported that they thought the bomb alert was merely a hoax. Furthermore, telephone warnings of the explosion were provided which was an established practice during the Northern Ireland conflict to ensure the area under threat is evacuated. However on this occasion, ambiguous information about the location of the bomb misled the police who unintentionally moved some people towards the car containing the explosive device. After the incident, it was frequently reported that the sense of shock was intense because people believed they were standing in a safe place, not beside the car that contained the bomb. Many children and families were moved to streets nearby and were not directly exposed to the explosion but a theme that dominated the media reports afterwards was how many more might easily have been unintentionally diverted to stand beside the car bomb. In the days that followed the explosion these items about intentionality which has been linked with PTSD in younger age [21] and confusion about the location of the bomb were repeatedly discussed in the media and throughout the Omagh community.

Also, the group nature of the Omagh bombing may have contributed to higher rates of probable PTSD, which is consistent Trickey and colleague's meta-analysis [5] that found group trauma to be significant for younger children compared to individual trauma. It is also possible that a number of the children were subsequently re-exposed to distress in the days and months following the bomb in the 15 months prior to the data being collected. In addition to potential stressors linked to more normal life events, during the weeks that followed the Omagh bombing a repeated series of hoax phone calls to the local police led to the town centre being evacuated on a number of occasions. Some studies suggest that young people are vulnerable to relapse if exposed to such subsequent stressors [20,35].

TABLE 7: Predictors of depression 15 months after explosion among children present in Omagh (N= 241)

Variable	Model 1		Model 2		Model 3	
	β	p	β	p	β	p
Pre-trauma						
Age	−0.036	0.903	-.005	0.987	0.015	0.960
Male	−2.640	0.003**	−2.338	0.007**	−2.418	0.005**
Previous psychological treatment	2.488	0.066	2.308	0.083	1.932	0.152
Peri-trauma						
Perceived life threat			2.581	0.037*	2.225	0.077
Physically injured			1.516	0.311	1.040	0.495
Witnessed serious injury			1.842	0.042*	1.884	0.037*
Witness people dying			−2.151	0.019*	−2.348	0.011*
Witnessed people dead			1.591	0.085	1.544	0.093
Post-trauma						
Post-event support					2.084	0.135

Notes:
1. Anxiety symptoms measured by Birleson Depression Scale (BDS) score.
2. Those present in Omagh includes those in Omagh town centre when the bomb exploded or shortly after and/or witnessed related traumatic events.
3. Dummy variables for school included in model but excluded from the table.
4. Model 1 Adjusted R2 =0.032; Model 2 Adjusted R2 = 0.081; Model 3 Adjusted R2 = 0.086.
5. P: *significant at the 0.05 level; **significant at the 0.01 level.

Our second aim was to consider if type of exposure to a traumatic event increases PTSD symptoms in children to a greater extent than symptoms of general emotional distress. Our findings that children exposed to the bomb reported higher levels of probable PTSD and psychological distress than those not exposed (Figure 1) supports the findings from other studies [24,25]. Our study also indicates that direct exposure is more closely associated with an increase in PTSD symptoms than general psychiatric distress (Figure 1). Our finding that there is a trend, albeit non-significant, for an increase in PTSD and general psychiatric distress with increased exposure type (higher rates for "being present at the time" as opposed to "being present after" the explosion) provides some support for the finding from Foy and colleague's review [12] that temporal proximity is an important mediator of PTSD in children.

A novel consideration in our study is the concept of near miss which as far as we can discover has not been extensively researched in children. In this study the data suggests that the near miss group (those children who were in town but missed the explosion and the aftermath) differed significantly on the PTSD measure from those children directly exposed (p<0.05) but did not differ significantly on the PTSD measure from the loss group (p=0.630) or the no exposure group (p=0.174). Children in the near miss group, however, did not differ significantly from the direct exposure groups in their depression symptom levels (p= 0.432) or anxiety symptom levels (p=0.334), whereas those in the loss and no exposure groups had significantly lower levels of general psychiatric distress compared with those directly exposed. The mean IES score is higher in the loss group than the near miss group but the differences on all measures between the near miss and loss group were not statistically significant. However we have to be cautious about these findings because of the small number in the "near miss" category (N= 20) and the restricted statistical power to calculate differences with this group. These "near miss" findings are similar to the findings of a community study of adults after the Omagh bombing [46] which found that those in the "near miss" group did not differ in PTSD or general psychiatric measures from those who had no exposure.

Our third aim was to consider which individual and trauma characteristics predict chronic PTSD symptoms. In relation to pre-trauma factors, our finding that age was a predictor of probable anxiety but not a predictor specifically of probable PTSD supports the findings from a number of previous studies [29,31,35] but we accept that the age range in our analysis was restricted to children and did not include adolescents. Only a small effect was reported for younger age by Trickey and colleague's [5] and our finding supports their conclusion that younger age is largely unrelated to whether a young person develops PTSD. Female gender has been reported as a small but significant risk factor for PTSD in adults [47] and children [5]. However, in our study when co-morbidity and post trauma support are controlled for in the analysis, the association between gender and PTSD is no longer significant. As discussed earlier, Trickey and colleagues [5] reported that younger age has a moderating effect on gender as a risk factor for PTSD in children. In this study girls reported higher levels of probable depression and anxiety than boys and these associations remained significant after peri- and post trauma factors were added to the regression analysis (Models 2 and 3, Tables 6 and 7). Similar gender differences were reported in another study of school children in Belfast after a bomb had destroyed their school [48] as indicated earlier, recognised that negative affect is often externalised in boys in the form of behavioural symptoms [32].

Peri-traumatic factors that significantly predicted increased IES scores in this study were witnessing people injured and reporting a perceived life threat. However, when co-morbidity and post trauma support were controlled for, these peri-traumatic factors were no longer significant. Children who witnessed injured people were also at higher risk of depression. These findings are consistent with other studies [29] and both factors were reported as risk factors with large effect sizes in Trickey and colleagues' meta-analysis [5]. Of those children who witnessed the aftermath of the bomb, almost all saw people injured, almost half those exposed saw people they thought were dead and one in ten received psychological/psychiatric interventions post-event. This exposure to such gruesome scenes may contribute to the high rates of probable PTSD for the exposure groups as found in other conflict related studies where PTSD rates as high as 87% [30] and 90% [21] were reported. However, it is interesting that the only significant exposure predictor in our study was "seeing people injured" which was a significant predictor on all 3 outcome measures the IES, SCAS and BDS. In the Omagh bomb a large number of children and young people suffered burns and shrapnel injuries resulting for some in permanent disabilities including loss of sight and amputated limbs.

Post-trauma factors that were considered included "support received for difficulties experienced following the bomb" which was significantly associated with anxiety but not specifically PTSD or depression. Those receiving post-event interventions who were present in Omagh and exposed to the bomb had significantly higher depression and anxiety scores compared with those not receiving post-event support, however, no differences in PTSD scores were noted. Social support has been reported elsewhere as a risk factor for PTSD with a large effect size in both adults [47] and children [5]. Our finding is interesting because the Omagh bombing occurred in a changed political context, an early phase of peace-building with the main paramilitary groups on ceasefire, and so the social policy response was different to previous events. In the aftermath of the tragedy, political leaders and many celebrities visited the town and thousands of people attended vigils and memorial services. Government funding was made available specifically to provide supports for the bomb victims and to co-ordinate a response involving health, social, educational agencies and voluntary, faith and community groups. Despite these policy and community initiatives, whilst our study found that social support was linked to anxiety this factor did not appear to have had an effect specifically on traumatic symptoms in younger children.

Co-morbid psychological problems have been reported as risk factors with large effects in Trickey and colleagues' meta analysis [5]. In our study, of those children classified as reaching PTSD caseness, 10% also met probable caseness for anxiety and 9% probable caseness for depression. Over one third (38%) of those children reaching probable depression caseness also met probable caseness for anxiety. Co-morbid psychological problems had a moderating effect on pre-trauma characteristics and exposure factors in predicting probable PTSD. Our findings are consistent with other studies that have identified co-morbid symptoms as amongst the highest risk factors for chronic post trauma distress in children [49].

2.5 CONCLUSIONS

High rates of PTSD have been found in studies of children living in conflict areas [19,30,35]. Similar to patterns in adults [6] chronic post-trauma symptoms persist in a substantial sub-group of children and can severely interfere with functioning [20,50,51]. It is important that these children, whose needs may not be fully recognised and under-reported by parents [2], are identified as early as possible and offered effective therapies and support. Our study is one of a growing number of school-based studies that have been organised after single incident traumas for screening and assessing children [16,20] and providing early treatment responses [52]. Our findings that witnessing people injured and reporting a perceived life threat were significant risk factors and that co-morbid anxiety mediates the effect of exposure, age and gender as predictors of PTSD adds to the growing literature base identifying specific key factors for screening and assessing children after traumatic events.

2.5.1 LIMITATIONS

Our data was gathered 15 months after the bomb so it is likely that screening in the immediate aftermath of the bomb would have identified higher levels of PTSD symptomotology. Our questionnaire did not capture any traumas or significant life events that children may have been experienced in the intervening period that may have compounded an initial traumatic reaction to the bombing. Self-report questionnaires were used in the screening and we recognise these are only an indicator of probable psychiatric disorders and do not provide a complete accurate diagnosis. We were unable to collect multi-informant data from parents or teachers which would have provided confirmatory data to identify morbidity amongst the sample. While the overall sample size was large, the number of children who were directly exposed to the bombing was relatively small. This will have reduced the statistical power of the regression models. Finally, the study assessed psychological symptoms but did not measure the impact of symptoms on daily functioning.

REFERENCES

1.   National Institute for Clinical Evidence & National Collaborating Centre for Mental Health NICE: The Management of PTSD in Adults and Children in Primary and Secondary Care. London: Guideline 26; 2005.

2. Dyregrov A, Yule W: A review of PTSD in children. Child and Adolescent Mental Health 2006, 11:176-184.

3. American Psychiatric Association: Diagnostic and statistical manual of mental disorders. 4th edition. Washington, DC: APA; 1994.

4. Scheeringa M, Zeanah CH, Drell MJ, Larrieu JA: Two approaches to the diagnosis of posttraumatic stress disorder in infancy and early childhood. J Am Acad Child Psy 1995, 34:191-200.

5. Trickey D, Siddaway AP, Meiser-Stedman R, Serpell L, Field AP: A meta-analysis of risk factors for post-traumatic stress disorder in children and adolescents. Clin Psychol Rev 2012, 32:122-138.

6. Kessler RC, Sonnega A, Bromet E, Hughes M, Nelson CB: Posttraumatic stress disorder in the National Comorbidity Survey. Arch Gen Psychiatry 1995, 52:1048-1060.

7. Melzer H, Gatward R, Goodman R, Ford T: Mental health of children and adolescents in Great Britain. London: The Stationary Office; 2000.

8. Fletcher KE: Childhood posttraumatic stress disorder. In Child psychopathology, Edited by Mash EJ, Barkley R. New York: Guilford Press; 1996:248-276.

9. Dalgleish T, Meiser-Stedman R, Smith P: Cognitive aspects of posttraumatic stress reactions and their treatment in children and adolescents: an empirical review and some recommendations. Behav Cogn Psychoth 2005, 33:459-486.

10. La Greca AM, Prinstein MJ: Hurricanes and Earthquakes. In Helping children cope with disasters and terrorism. Edited by La Greca AM, Silverman WK, Vernberg EM, Roberts MC. Washington: American Psychological Association; 2002:107-138.

11. Stallard P, Salter E, Velleman R: Posttraumatic stress disorder following road traffic accidents. A second prospective study. Eur Child Adoles Psy 2004, 13:172-178.

12. Foy DW, Madvig BT, Pynoos RS, Camilleri AJ: Etiologic factors in the Development of Posttraumatic stress disorder in children and adolescents. J School Psychol 1996, 4:133-145.

13. La Greca AM, Silverman WK, Wasserstein SB: Children's predisater functioning as a predictor of posttraumatic stress following Hurricane Andrew. J Consul Clin Psych 1998, 66:883-892.

14. Lonigan CJ, Shannon MP, Finch AJ, Daugherty TK, Saylor CM: Children's reactions to natural disasters: Symptom severity and degree of exposure. Adv Behav Res Ther 1991, 13:135-154.

15. Vernberg EM, La Greca AM, Silverman WK, Prinstein M: Predictors of post-disaster functioning following Hurricane Andrew. J Abnorm Psychol 1996, 105:237-248.

16. Pfefferbaum B, Nixon SJ, Krug RS, Tivis RD, Moore VL, Brown JM, Pynoos RS, Foy D, Gurwich RH: Clinical needs assessment of middle and high school students following the 1995 Oklahoma Bombing. Am J Psych 1999, 156:1069-1074.

17. Nadir K, Pynoos R, Fairbanks L, Fredrick C: Children's PTSD reactions one year after a sniper attack at their school. Am J Psych 1990, 147:1526-1530.

18. Smith P, Perrin S, Yule W, Rabe-Hesketh S: War exposure and maternal reactions in the psychological adjustment of children from Bosnia-Herzegovina. J Child Psychol Psych 2001, 42:395-404.

19. Thabet AAM, Vostanis PV: Post traumatic stress disorder reactions in children of war. J Child Psychol Psych 1999, 24:291-298.

20. Thabet AAM, Vostanis PV: Post traumatic stress disorder reactions in children of war: a longitudinal study. Child Abuse Neglect 2000, 24:291-298.

21. Pynoos RS, Goenjian A, Tashjian M, Karakashian M, Manjikian R, Manoukian G, Steinberg AM, Fairbanks LA: Post-traumatic stress reaction in children following the 1988 Armenian earthquake. Brit J Psychiat 1993, 163:239-247.

22. Smith P, Perrin S, Yule W, Hacam B, Stuvland R: War exposure among children from Bosnia-Hercegovina: psychological adjustment to a community sample. J Trauma Stress 2002, 15:147-156.

23. Schwarzwald J, Weisenberg M, Waysman M, Solomon Z, Waysman M: Stress reaction of school-age children to the bombardment by SCUD missiles: a 1-year follow-up. J Trauma Stress 1994, 7:657-666.

24. Cox CM, Kenardy JA, Hendrikz JK: A meta-analysis of risk factors that predict psychopathology following accidental trauma. J Spec Pediatr Nurs 2008, 13:98-110.

25. Pine DS, Cohen JA: Trauma in children and adolescents: Risk and treatment of psychiatric sequelae. Biol Psychiatry 2002, 51:519-531.

26. Greiger TA, Waldrep DA, Lovasz MM, Ursano RJ: Follow up of Pentagon employees two years after the terrorist attack of September 11, 2001. Psychiatr Serv 2005, 56:1374-1378.

27. Pfefferbaum B, Seale TW, McDonald NB, Brandt EN Jr, Rainwater SM, Maynard BT, Meierhoefer B, Miller PD: Posttraumatic stress two years after the Oklahoma City bombing in youth geographically distant from the explosion. Psychiatry 2000, 63:358-370.

28. Ahern J, Galea S, Resnick H, Kilpatrick D, Bucuvalas M, Gold J, Vlahov D: Television images and psychological symptoms after the September 11 terrorist attacks. Psychiatry 2002, 65:289-300.

29. Green BL, Korol M, Grace MC: Children and disaster: age, gender and parental effects on PTSD symptoms. J Am Acad Child Psy 1991, 30:945-951.

30. Ahmad A, Sofi MA, Sundelin–Wahlsten AL: Posttraumatic stress disorder in children after the military operation "Anfal" in Iraqi Kurdistan. Eur Child Adolesc Psychiatry 2000, 9:235-243.

31. Aiko PJ, De Vries A, Kassam-Adams N, Cnaan A, Sherman-Slate E, Gallagher PR, Winston FK: Looking beyond the physical injury: Posttraumatic stress disorder in children and parents after paediatric traffic injury. Paediatrics 1999, 6:1292-1299.

32. Pfefferbaum B: Posttraumatic stress disorder in children: a review of the past 10 years. J Am Acad Child Psy 1997, 11:1503-1511.

33. Yule W, Perrin S, Smith P: Post-traumatic stress reactions in children and adolescents. In Post-traumatic stress disorders: Concepts and therapy. Edited by Yule W. Chichester: John Wiley & Sons; 1999:25-50.

34. Max JE, Castillo CS, Robin DA, Lindgren SD, Smith WL, Arndt S: Posttraumatic stress symptomatology after childhood traumatic brain injury. J Nerv Ment Dis 1998, 186:589-596.

35. Laor N, Wolmer L, Cohen DJ: Mothers' functioning and children's symptoms 5 years after a SCUD Missile Attack. Am J Psychiat 2001, 158:1020-1026.

36. Meiser-Stedman R: Toward a cognitive-behavioral model of PTSD in children and adolescents. Clin Child Fam Psychol Rev 2002, 5:217-232.

37. Perrin S, Smith P, Yule W: Practitioner review: the assessment and treatment of post-traumatic stress disorder in children and adolescents. J Child Psychol Psych 2000, 41:277-289.

38. Hubbard J, Realmuto GM, Northwood AK, Masten AS: Comorbidity of psychiatric diagnoses with posttraumatic stress disorder in survivors of childhood trauma. J Am Acad Child Psy 1995, 34:1167-1173.

39. Giaconia RM, Reinherz HZ, Silverman AB, Pakiz B, Frost AK, Cohen E: Traumas and posttraumatic stress disorder in a community population of older adolescents. J Am Acad Child Psy 1995, 34:1369-1380.

40. Breslau N, Davis GC, Andreski P, Peterson E: Traumatic events and posttraumatic stress disorder in an urban population of young people. Arch Gen Psychiatry 1991, 48:216-222.

41. McDermott M, Duffy M, McGuinness D: Addressing the psychological needs of children and young people in the aftermath of the Omagh Bomb. Child Care In Practice 2003, 10:141-154.

42. Horowitz MJ, Wilner N, Alvarez W: Impact of Event Scale: a measure of subjective distress. Psychometric Medicine 1979, 41:209-218.

43. Perrin S, Meiser-Stedman R, Smith P: The Childrens revised Impact of Event Scale (CRIES): validity as a screening instrument. Behav Cogn Psychother 2005, 33:487-498.

44. Birleson P: The validity of depressive disorder in childhood and the development of a self-rating scale: a research report. J Am Acad Child Psy 1981, 22:73-78.

45. Spence SH: A measure of anxiety symptoms among children. Behav Res Ther 1998, 36:545-566.

46. Duffy M, Bolton D, Gillespie K, Ehlers A, Clark DM: A community study of the psychological effects of the Omagh Car Bomb on Adults. PLoS ONE 2013, 8(9):e76618.

47. Brewin CR, Andrews B, Valentine JD: Meta-analysis of risk factors for post-traumatic stress disorder in trauma-exposed adults. J Consult Clin Psychol 2000, 68:748-766.

48. Joseph S, Cairns E, McCollam P: Political Violence, coping, and depressive symptomatology in Northern Irish children. Pers Indiv Differ 1993, 15:471-473.

49. Lai BS, La Greca AM, Auslander BA: Children's symptoms of postrtraumatic stress and depression after a natural disaster: Comorbidity and risk factors. Short J Affect Dis 2013, 146:71-78.

50. Yule W, Bolton D, Udwin O, Boyle S, O'Ryan D, Nurrish J: The long–term psychological effects of a disaster experienced in adolescence - I: The incidence and course of PTSD. Journal of Child Psychology and Psychiatry 2000, 4:503-511.

51. Morgan L, Scourfield J, Williams D, Jasper A, Lewis G: The Aberfan disaster: a thirty-three-year follow-up of the survivors. British Journal of Psychiatry 2003, 182:532-536.

52. Amaya-Jackson L, Reynolds V, Murray MC, McCarthy G, Nelson A, Cherney MS, Lee R, Foa E, March JS: Cognitive behavioural Treatment for pediatric posttraumatic stress disorder: protocol and application in schools and community settings. Cognitive and Behavioural Practice 2003, 10:204-213.

There is one supplemental table that is not available in this version of the article. To view this additional information, please use the citation on the first page of this chapter.

RELATIONSHIP BETWEEN THE COMT-VAL158MET AND BDNF-VAL66MET POLYMORPHISMS, CHILDHOOD TRAUMA, AND PSYCHOTIC EXPERIENCES IN AN ADOLESCENT GENERAL POPULATION SAMPLE

HUGH RAMSAY, IAN KELLEHER, PADRAIG FLANNERY, MARY C. CLARKE, FIONNUALA LYNCH, MICHELLE HARLEY, DEARBHLA CONNOR, CAROL FITZPATRICK, DEREK W. MORRIS, AND MARY CANNON

3.1 INTRODUCTION

Much research has established that psychotic experiences have a higher prevalence in the general population than psychotic disorders [1-4]. A meta-analysis of all community studies of psychotic experiences in children and adolescents found a median population prevalence of 17% in children aged 9-12 years and 7.5% in those aged 13-18 years [5]. Psychotic experiences in adolescence are associated with high risk for severe psychopathology, both in the immediate term and later into adulthood, including both psychotic

Relationship between the COMT-Val158Met and BDNF-Val66Met Polymorphisms, Childhood Trauma and Psychotic Experiences in an Adolescent General Population Sample. © *Ramsay H, Kelleher I, Flannery P, Clarke MC, Lynch F, Harley M, Connor D, Fitzpatrick C, Morris DW, and Cannon M. PLoS ONE, **8**,11 (2013), doi:10.1371/journal.pone.0079741. Licensed under Creative Commons Attribution License, http://creativecommons.org/licenses/by/3.0/.*

[6-8] and non-psychotic disorders [9-12]. Psychotic experiences share many important risk factors with schizophrenia [13]. For example, in the case of familial risk [14], there is covariation of psychotic experiences with maternal schizophrenia-spectrum disorder [15]. In common with psychotic disorders, psychotic experiences are more prevalent in adolescents who have had traumatic experiences, including physical abuse, exposure to domestic violence and unwanted sexual experiences [16-27].

We previously suggested that genes for psychosis may, in fact, be genes for the broader 'extended psychosis phenotype', made up not just of individuals with psychotic disorders but also a much larger population of individuals with psychotic experiences [28]. Investigation of the genetic aetiology of the extended psychosis phenotype may provide novel insights into the genetic underpinnings of psychosis.

Two frequently studied genetic variants in psychosis research are the catechol-o-methyl-transferase (COMT) Val158Met and the brain-derived neurotrophic factor (BDNF) Val66Met polymorphisms [29-32]. COMT plays an important role in the metabolism of catecholamines, such as dopamine and norepinephrine in the central nervous system. The Val158Met polymorphism is associated with a 3- to 4-fold variation in enzymatic activity [33,34] between the high activity Val/Val genotype and the low activity Met/Met genotype. Diverse gene-environment interactions have been reported in the case of COMT-Val158-Met, particularly in moderating risk for psychotic disorder [35,36], for example, in the case of cannabis use in adolescence [36,37]. Indeed, recent research has identified gene-gene-environment interaction involving COMT in the case of subjectively reported daily life stress using an experience sampling method in patients with schizophrenia [38]. Furthermore, the COMT-Val158Met Val allele has been associated with self-reported psychotic experiences in the context of stress and cannabis use in a Dutch adult population sample [39]and with the stress of army induction in a Greek male conscript sample [40]. COMT-Val158Met has also been associated with increased schizotypal personality trait scores in Val-Val individuals exposed to higher levels of self-reported childhood trauma [41]. BDNF, on the other hand, has an established role in neuronal development and cell survival in response to stress [42] and is abnormally expressed in schizophrenia [43]. Interestingly, the BDNF-Val66Met polymorphism has been shown to moderate the impact of childhood adversity on later expression of affective symptoms [44,45], suggesting the possibility of gene-environment interactions and one study reported an interaction between childhood abuse and Val66Met in predicting psychotic experiences [29].

3.1.1 AIMS OF THE STUDY

The present study aimed to explore the role of the COMT-Val158Met and BDNF-Val66Met polymorphisms in two community samples with psychotic experiences. We wished to test for a direct association between these polymorphisms and psychotic experiences in the general population. Furthermore, we aimed to test a gene-environment interaction, specifically, that those with the COMT-Val158Met Val-Val genotype or the BDNF-Val66Met Met-Met/Val-Met genotypes are more susceptible to psychotic experiences following trauma than those with other genotypes.

3.2 MATERIALS AND METHODS

3.2.1 SAMPLE

The sample consisted of 237 adolescents from two independent studies: 123 participants from the Adolescent Brain Development (ABD) study and 114 participants from the Challenging Times (CT) study. The study methodologies have been previously reported [46-48]. Details of recruitment for both studies are presented in Figure 1.

Briefly, the ABD study took place in counties Dublin and Kildare, Ireland, which contains a mixture of urban and suburban housing types of different socioeconomic status. Out of twenty seven schools approached, sixteen agreed to participate (59.3%), while four said they may participate at another time and seven (25.9%) declined participation. Participants were aged 11-13 years and were in the two most senior classes in the Irish primary school system. Written parental consent was required for adolescents to take part and 1,131 signed consent forms were returned out of 2,190 distributed (52%). Of the 1,131 adolescents who took part in the survey study, 656 (58%) indicated an interest in participating in the interview study and a random sample of 212 of these attended for interview. Those who participated at interview did not differ from those who did not on levels of psychopathology as determined by the SDQ. Those who attended for interview were also representative of the Irish population in terms of socioeconomic status and ethnic background. Among this sample, a total of 168 individuals provided genetic samples. There was insufficient DNA for analysis in 45 of these samples, resulting in sufficient DNA for genotyping the BDNF-Val66Met or COMT-Val158Met polymorphisms in 123 participants.

The CT study took place in the geographical catchment area of a child and adolescent mental health team in north Dublin, containing a population of 137,000. This urban area includes pockets of severe inner city deprivation, large suburban housing estates and more affluent areas of private housing. The population includes a higher proportion from lower socioeconomic backgrounds compared to the overall Irish population. A sample of 12 secondary schools in the catchment area was selected using stratified random sampling according to the approximate socio-economic status of the school. Of these 12 schools, eight participated in the study. One school did not participate owing to concerns that questioning pupils about suicidal thoughts might be harmful and three schools that agreed to participate were excluded because insufficient parental consent forms were received (<50%). The exclusion of these schools did not skew the socio-economic distribution of the remaining schools. A total of 743 students aged 12-15 years in eight schools were screened for psychopathology using the Strengths and Difficulties Questionnaire [49], which measures child psychopathology risk, and the Children's Depression Inventory [50][51], which measures the cognitive, affective and behavioural signs of depression. Written informed consent was obtained from the parent or guardian of participants. One hundred and forty pupils scored above the threshold on these instruments and all of these adolescents were invited to full psychiatric interview, of whom 117 (83.6%) agreed to attend. A comparison group of 173 healthy adolescents, matched for gender and school, were also invited to interview and 94 (54%) agreed to attend. Of these 211, 169 (80%) attended for further follow-up [52] and provided genetic samples. Of these, 55 did not contain sufficient DNA for analysis, resulting in genotyping of 114 participants for the COMT Val158Met or BDNF-Val66Met polymorphisms.

Figure 1: Recruitment to the Adolescent Brain Development and Challenging Times studies

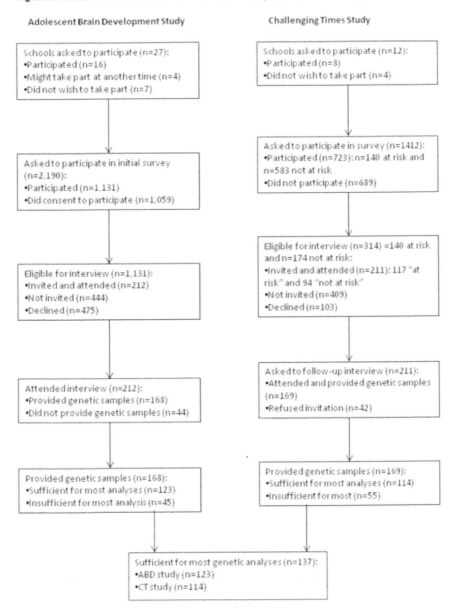

FIGURE 1: Recruitment to Adolescent Brain Development and Challenging Times studies.

In total, between both studies, there was sufficient DNA from 226 participants for analysis of COMT-Val158Met (115 from ABD study and 111 from CT study) and 222 participants for analysis of BDNF-Val66Met (116 from ABD study and 106 from CT study). This data is not available online owing to the necessity to protect the confidentiality of the young people who participated (and discussed the sensitive areas of psychotic experiences and history of abuse). However, we are happy to facilitate requests from researchers to freely review the data.

3.2.2 GENOTYPING

Genetic analysis was carried out using DNA extracted from saliva samples. The COMT (rs4680) and BDNF (rs6265) SNPs were genotyped using Taqman® SNP genotyping assays on a 7900HT sequence detection system (Applied Biosystems). The call rate for the Taqman genotyping was >95% and the samples were in Hardy-Weinberg equilibrium (p>0.05). Along with these samples, thirteen HapMap CEU DNA samples (www.hapmap.org) were genotyped for each SNP (rs4680 and rs6265) for quality control purposes and were found to be 100% concordant with available HapMap data for these SNPs.

3.2.3 EXPOSURE AND OUTCOME MEASURES

Participants and their parents were interviewed using the Schedule for Affective Disorders and Schizophrenia for School-aged Children (K-SADS), Present and Lifetime versions. The K-SADS is a well-validated semi-structured research diagnostic interview for the assessment of Axis I psychiatric disorders in children and adolescents [53]. Children and parents answered the same questions but were interviewed separately. All interviewers were trained psychiatrists or psychologists. The psychosis section of the K-SADS was used to assess the participants' psychotic experiences. They recorded extensive notes of potential psychotic phenomena in this section of the interview and a clinical consensus meeting, which took place following the interviews, classified these as a binary variable, having psychotic experiences or not, masked to diagnoses and all other information on the participants. This measure of psychotic experiences was our outcome variable. As part of the interview, a range of traumatic experiences were enquired about from both parent and child, including instances of physical abuse, sexual abuse and exposure to domestic (inter-parental) violence. A report from either child or parent of physical or sexual abuse or of witnessing domestic violence was classified as a childhood trauma [17]. Polymorphism genotypes were converted for ease of analysis into binary variables based on the findings of previous studies. The COMT-Val158Met polymorphism was categorised as either Val-Val or Met-Met/Val-Met, based on research supporting a gene-environment association in the case of Val-Val with transient psychotic experiences and schizotypal traits [40,41]. The BDNF-Val66Met polymorphism was categorised as either Val-Val or Met-Met/Val-Met, based on the previous finding of a gene environment association with psychotic experiences in Met carriers [29].

3.2.4 ETHICS STATEMENT

Ethical approval for the Adolescent Brain Development study was received from Beaumont Hospital medical ethics committee and for the Challenging Times study from the Mater Misericordiae University Hospital medical ethics committee. Informed assent was obtained from participants under age 16 years with informed consent received from their parents. In addition, all individual data was anonymised with no identifying characteristics present. Written informed consent was obtained from the parent or guardian of participants.

3.2.5 STATISTICAL ANALYSIS

Differences between those who participated in interview and provided genetic samples and those who participated but did not were assessed using chi-squared tests. Similarly, differences between those with versus without psychotic experiences were assessed using chi-squared tests or Fisher's exact test as appropriate.

Based on the hypothesis that COMT-Val158Met genotype can moderate the influence of childhood trauma on psychotic experiences, we tested whether psychotic experiences would be predicted by an interaction between a gene (either COMT-Val158Met or BDNF-Val66Met) and an environment (childhood trauma). To analyse the interaction between childhood trauma and genotypes (COMT-Val158Met and BDNF-Val66Met), logistic regression was used, with the absence or presence of psychotic experiences as the dependent variable. Sex, school grade and cannabis use were used as covariates. School grade was included as a categorical variable with four levels, one for each grade. The first two grades were recruited for the ABD study and the second two grades for the CT study. This variable therefore approximated the effects of both age and study group. Cannabis use and sex were also included to control for potential confounding by these factors. A logistic regression model was first fitted to determine the main effects of childhood trauma and COMT-Val158Met genotype (model 1). Second, to explore the interaction effect of COMT-Val158Met and childhood trauma, a second model was fitted with an interaction term (model 2) and this was compared to the first model using the likelihood ratio test. This process was also repeated separately in the same manner to examine the main effect of BDNF-Val66Met genotype and interaction between this genotype and childhood trauma.

Finally, where these tests indicated an important G X E interaction, the effect of this interaction was explored within each genotype group, using stratified tables with odds ratios and 95% confidence intervals. Statistical analyses were conducted using Stata version 11.0 for Windows. All statistical tests were two-sided; P-values less than 0.05 were judged statistically significant.

3.3 RESULTS

3.3.1 DESCRIPTIVE STATISTICS

Those who attended for interview and provided usable genetic samples were broadly similar to those who attended for clinical interview but did not provide usable genetic samples (details in table 1), though a higher proportion of males and a lower proportion of those in the highest school grade provided usable genetic samples.

TABLE 1: Comparison of those with genetic data to those without genetic data, according to demographic factors, trauma and psychotic experiences.

Variable		No (n=208) (%)	Yes (n=237) (%)	χ^2	D_f	P-value~
		Genetic data				
Sex	Male	78 (38)	128 (62)	11.25	1	0.001*
	Female	127 (54)	109 (46)			
School grade	5th grade	41 (43)	55 (57)	10.83	3	0.013*
	6th grade	64 (48)	68 (52)			
	7th grade	47 (37)	79 (63)			
	8th grade	49 (60)	33 (40)			
Social class	Class I/II	77 (39)	118 (61)	2.61	2	0.272
	Class IIIN	50 (38)	81 (62)			
	Class IIIN/IV/V	12 (27)	33 (73)			
Psychotic experiences	No	177 (47)	200 (53)	0.11	1	0.742
	Yes	30 (45)	37 (55)			
Traumatic experiences	No	193 (48)	216 (53)	1.01	1	0.315
	Yes	13 (37)	21 (62)			
Cannabis use	No	131 (37)	226 (63)	1.95	1	0.162
	Yes	10 (53)	9 (47)			

D_f= degrees of freedom; ~P-values based on Chi-squared test or Fisher's exact test as appropriate

The sample was 54% male and all participants were of white European origin. Psychotic experiences were present in 37/237 (17%) of the total sample. Childhood trauma was reported by 21/237 (9%). This included 13 (5.5%) who witnessed domestic violence, 9 (3.8%) who described child physical abuse and 4 (1.7%) who described child sexual abuse. There was overlap in these individuals, with some experiencing multiple forms of childhood trauma. Table 2 presents the group characteristics of those with psychotic experiences and those without psychotic experiences. The prevalence of psychotic experiences did not differ significantly according to sex, childhood trauma, cannabis use, COMT-Val158Met genotype or BDNF-Val66Met genotype. Consistent with previous research [5], psychotic experiences were less common in the older age groups, who were assessed as part of the CT study (age 12-13 years P<0.01; age 13-14 years P=0.03).

TABLE 2: Comparison of cases (with psychotic symptoms) and controls, according to demographic factors, trauma and COMT and BDNF genotype.

Risk factor (total number)		No (n=200) (%)	Yes (n=37) (%)	χ^2	D_f	P-value~
		Psychosis				
Sex (n=237)	Male	105 (82)	23 (18)	-	-	-
	Female	95 (87)	14 (13)	1.17	1	0.28
Age (n=235)	10-11 years	39 (71)	16 (29)	-	-	-
	11-12 years	53 (78)	15 (22)	0.79	1	0.37
	12-13 years	76 (96)	3 (4)	16.92	1	0.00
	13-14 years	30 (91)	3 (9)	4.82	1	0.03
Exposed to trauma (n=237)	No	184 (85)	32 (15)	-	-	-
	Yes	16 (76)	5 (24)	1.18	1	0.28
COMT genotype (n=226)	Met-Met/ Val-Met	134 (83)	27 (17)	-	-	-
	Val-Val	55 (85)	10 (15)	0.06	1	0.80
BDNF genotype (n=222)	Val-Val	130 (86)	22 (14)	-	-	-
	Met-Met/ Val-Met	59 (84)	11 (16)	0.06	1	0.81
Cannabis use (n=235)	No	191 (85)	35 (15)	-	-	-
	Yes	7 (78)	1 (22)	0.30	1	0.59

D_f= degrees of freedom; ~P-value is calculated based on Fisher's exact test.

3.3.2 GENE-ENVIRONMENT INTERACTION BETWEEN COMT-VAL158MET AND CHILDHOOD TRAUMA WITH RESPECT TO PSYCHOTIC EXPERIENCES

Table 3 presents the association between childhood trauma and genotype and psychotic experiences, controlling for sex, school grade (which adjusted for study sample and age group) and cannabis use. In the COMT and BDNF models without interaction, childhood trauma was borderline associated with psychotic experiences (COMT model: OR=3.17, P=0.086; BDNF model: OR=3.25, P=0.073). Neither COMT-Val158Met genotype nor BDNF-Val66Met genotype were associated with psychotic experiences in the models without interaction. Inclusion of an interaction term for gene (COMT-Val158Met genotype) X environment (childhood trauma) resulted in a model that was borderline superior (P=0.057) to the model without an interaction term, using the likelihood ratio test. The P-value for the interaction term between COMT-Val158Met X abuse was borderline significant (P=0.063), suggesting further stratified analysis would be appropriate. When interaction between BDNF-Val66Met and childhood trauma was examined in the same way, the model with interaction was not superior to the model without interaction on the likelihood ratio test ((χ^2<0.01, P=0.958).

TABLE 3: Results from hierarchical binary logistic regression models.

	Models without interaction		Models with interaction	
	OR (95% CI)	P-value	OR (95% CI)	P-value
COMT models (n=223)				
Abuse (0=absent, 1=present)	3.17 (0.85, 11.81)	0.874	1.33 (0.23, 7.62)	0.749
COMT Val158Met genotype (0=Val-Met/Met-Met, 1=Val-Val)	1.07 (0.45, 2.53)	0.086	0.79 (0.31, 2.06)	0.636
COMT Val158Met genotype X abuse	-	-	17.16 (0.86, 344.25)	0.063
BDNF models (n=218)				
Abuse	3.25 (0.90, 11.78)	0.073	3.18 (0.67, 15.00)	0.145
BDNF Val66Met genotype (0=Val-Val, 1=Met-Met/Val-Met)	0.98 (0.42, 2.31)	0.970	0.98 (0.39, 2.43)	0.958
BDNF Val66Met genotype X abuse	-	-	1.07 (0.08, 14.92)	0.958

All models controlled for sex, school year (categorical) and cannabis use. OR=odds ratios; 95% CI=95% confidence interval. *Statistically significant (p<0.05)

Following this, the effect of childhood trauma on psychotic experiences was examined in each COMT-genotype sub-group (see table 4). Childhood trauma was associated with psychotic experiences in those with the COMT-Val-Val genotype (OR=7.43, 95% CI: 1.12-49.11), but not the COMT-Met-Met/Val-Met genotype (OR=0.81, 95% CI: 0.17-3.88) (see table 2). There was borderline evidence in support of an interaction between genotype and childhood trauma (p=0.06) on the chi-squared test for interaction, which is a conservative measure of interaction.

TABLE 4: Association between childhood trauma and psychotic experiences, stratified by COMT-Val158Met group.

Gene	Variant	Frequency of psychotic experiences		OR	95% CI	P-value*	Test for homogeneity
		No trauma	Trauma				
COMT-Val158Met (n=226)	Val-Val (n=65)	7/59	3/6	7.43	1.12, 49.11	0.04	0.06
	Met-Met & Val-Met (n=161)	25/147	2/14	0.81	0.17, 3.88	1.00	

*OR, Odds ratio; 95% CI, 95% Confidence Interval *P-value is calculated based on Fisher's exact test.*

3.4 DISCUSSION

The current study provides, to our knowledge, the first evidence of a gene-environment interaction between the COMT-Val158Met genotype and childhood traumatic experiences in terms of risk for clinically assessed psychotic experiences in the population. When stratified by COMT-Val158Met genotype, individuals with the Val-Val genotype who experienced childhood trauma were borderline significantly more likely to report psychotic experiences compared to adolescents with the Val/Met and Met/Met genotypes.

3.4.1 COMT-VAL158MET POLYMORPHISM AND PSYCHOPATHOLOGY

Our findings should be viewed in the context of advances in understanding of the genetic underpinnings of the extended psychosis phenotype. Binbay et al examined familial risk for this extended phenotype, which ranges from psychotic experiences without impairment to clinical psychotic disorder. They suggested that there are high prevalence and low prevalence genetic risks, interacting with environmental risks, operating across the extended psychosis phenotype but particularly in those with clinical psychotic disorders [54]. In this context, the borderline interaction observed here between COMT-Val158Met genotype and childhood trauma in association with psychotic experiences, a relatively common outcome in this age group, may reflect the operation of a high prevalence genetic risk factor. However, the young age of our sample and differences in the clinical significance of psychotic experiences among different age groups suggest caution in approaching this interpretation.

Our results suggesting a possible interaction between the COMT-Val158Met Val-Val genotype and childhood trauma in association with psychotic experiences are consistent with previous findings regarding gene-environment interactions between COMT-Val158Met and stressful/traumatic experiences in association with the extended psychosis phenotype. Consistent with our findings, the Val allele was associated with self-reported psychotic symptoms interacting with cannabis use in an adult Dutch population sample [39] and with transient self-reported psychotic experiences during the stress of army induction in a sample (aged 18-24 years) of Greek male conscripts [40] and Val-Val genotype has been associated with more enduring schizotypal traits in those with the Val-Val genotype (but not other genotypes) in the context of self-reported childhood trauma in an adult sample [41]. It is interesting that we have found similar associations in adolescents, despite the changing clinical significance of psychotic experiences with increasing age [10].

It has become increasingly clear that psychotic experiences at this age mark risk for a wide range of psychopathology, not limited to psychotic disorders [1,2,11,55]. Indeed, a majority of young people in the population who report psychotic experiences have at least one non-psychotic Axis-

1 psychiatric disorder [10]. Furthermore, these symptoms are associated with high risk for suicidal behaviour [56-58]. In addition to psychosis outcomes, the COMT-Val158Met polymorphism has been associated with a diverse range of psychiatric outcomes, including rapid-cycling bipolar disorder [59], obsessive-compulsive disorder [60,61], attention deficit hyperactivity disorder [62-64] and suicidal behaviour [65]. COMT-Val158Met has also been shown to interact with environmental stress in association with major depression [66] and post-traumatic stress disorder [67] in the presence of the Met allele. These findings suggest that the COMT-Val-158Met polymorphism is associated with diverse outcomes in different environmental contexts. Given the common overlap between psychotic experiences and non-psychotic disorders, examination of COMT-Val-158Met and childhood trauma interaction with non-psychotic disorders and/or disorder severity in association with psychotic experiences may be fruitful.

The molecular mechanisms of gene-environment interactions remain obscure. Childhood trauma-dependent DNA-demethylation has been associated with increased stress-dependent gene transcription and long-term dysregulation of the stress hormone system in the case of FK506 binding protein 5, a regulator of the stress hormone system [68]. Childhood trauma has the potential to alter methylation at other sites on the genome. Methylation and functional studies would be necessary to further explore this in the case of COMT and childhood trauma.

3.4.2 BDNF AND PSYCHOTIC EXPERIENCES

We found no association between BDNF genotype and psychotic experiences. Furthermore, we found no gene-environment interaction between BDNF genotype and childhood trauma in predicting psychotic experiences. This differs from the findings of Alemany and colleagues [29], who found that BDNF-Val66Met genotype moderated the effect of childhood abuse on the positive dimension of psychotic-like experiences, measured using the Community Assessment of Psychic Experiences (CAPE) in a sample from a college campus. There are a number of possible reasons for this difference in findings. Firstly, our samples differed in terms of age and assessment methodology. Psychotic experiences are more common in early adolescence and their significance may therefore differ from that noted in an older sample. Questionnaire outcomes may also differ from those of clinical interview. A second possible explanation for the difference in findings is sample size, with the possibility of type II error in the context of our small sample.

3.4.3 STRENGTHS AND LIMITATIONS

Among the strengths of this study is the use of gold-standard clinical interviews of both parent and child to ascertain psychotic experiences and childhood trauma. Most studies on psychosis and childhood trauma have been conducted with adults, meaning there is a considerable temporal gap between events and symptom assessment, increasing the risk of recall bias. Given that our research was conducted in youths, however, the assessment of psychotic experiences took place relatively soon after the time of the traumatic events, reducing recall bias risk. A further strength of this study is that we have controlled for important potential confounders to a gene environment interaction, particularly cannabis use [69].

A number of limitations must be considered in evaluating the results of this study. Firstly, psychotic symptoms and trauma history were determined simultaneously, meaning that we cannot be certain that childhood trauma preceded psychotic experiences. Furthermore, the timing of trauma was not clearly ascertained. Recent findings suggest that the trauma is likely to be a causal risk factor for psychotic experiences but that this relationship is also partly bi-directional [26]. However, data on temporality is not available in this case. Secondly, the analysis was performed by pooling two samples in order to increase overall sample size and power. The study populations were broadly similar but differed in age and in the prevalence of psychotic experiences. This difference in psychotic experience prevalence was explained and adjusted for by including a variable for school grade (reflecting both study group and age) in analyses. Furthermore, the sampling methodology resulted in recruitment of a relatively small number from larger surveys, opening up a risk of selection bias. However, those who participated in the ABD study had similar psychopathology scores on the SDQ to those who did not participate, while the CT study included a representative control group without significant SDQ psychopathology. This indicates that the control group for this study (those without psychotic symptoms) differed slightly in their composition. However, this was controlled for in analysis. A further limitation of this study is the sample size. The borderline significance of the interaction between COMT genotype and childhood trauma indicates that chance should be considered a possible explanation of this interaction, though the test for interaction is conservative, making this less likely. The low study numbers limited the power of this study to detect weaker associations between genotype and psychotic experiences and between genotype interacting with non-Val-Val genotypes. However, this was not the main hypothesis of this study. Given the low study numbers, the findings presented must be regarded as preliminary and in need of replication.

3.5 CONCLUSION

Our findings provide evidence in support of a gene-environment interaction between the COMT-Val158Met polymorphism and traumatic experiences in terms of risk for psychotic experiences in the population. This adds to the existing literature suggesting increased paranoia [70], schizotypal traits [41] and psychotic experiences [39,40] among Val-Val individuals in the context of stress or trauma. Specifically, young people with childhood traumatic experiences who had the COMT Val-Val genotype were more likely to experience psychotic experiences compared with Val-Met and Met-Met individuals. Clinically, this may suggest that this group could benefit from closer monitoring and support in the context of childhood trauma. Further research investigating this interaction in larger populations over a longer period would assist in clarifying the clinical prognosis and whether this population could benefit from early clinical interventions. Furthermore, such research will help to confirm whether genetic risk is specific to psychotic disorder or is, in fact, related more broadly to a broader phenotype.

REFERENCES

1. Wigman JTW, van Nierop M, Vollebergh WAM, Lieb R, Beesdo-Baum K et al. (2012) Evidence that psychotic symptoms are prevalent in disorders of anxiety and depression, impacting on illness onset, risk, and severity--implications for diagnosis and ultra-high risk research. Schizophr Bull 38: 247-257. doi:10.1093/schbul/sbr196. PubMed: 22258882.

2. Scott J, Martin G, Bor W, Sawyer M, Clark J et al. (2009) The prevalence and correlates of hallucinations in australian adolescents: results from a national survey. Schizophr Res 107: 179-185. doi:10.1016/j.schres.2008.11.002. PubMed: 19046858.

3. Laurens KR, Hobbs MJ, Sunderland M, Green MJ, Mould GL (2012) Psychotic-like experiences in a community sample of 8000 children aged 9 to 11 years: an item response theory analysis. Psychol Med 42: 1495-1506. doi:10.1017/S0033291711002108. PubMed: 21999924.

4. van Os J, Linscott RJ, Myin-Germeys I, Delespaul P, Krabbendam L (2009) A systematic review and meta-analysis of the psychosis continuum: evidence for a psychosis proneness-persistence-impairment model of psychotic disorder. Psychol Med 39: 179-195. doi:10.1017/S0033291708003814. PubMed: 18606047.

5. Kelleher I, Connor D, Clarke MC, Devlin N, Harley M et al. (2012) Prevalence of psychotic symptoms in childhood and adolescence: a systematic review and meta-analysis of population-based studies. Psychol Med 42: 1857-1863. doi:10.1017/S0033291711002960. PubMed: 22225730.

6. Welham J, Scott J, Williams G, Najman J, Bor W et al. (2009) Emotional and behavioural antecedents of young adults who screen positive for non-affective psychosis: a 21-year birth cohort study. Psychol Med 39: 625-634. doi:10.1017/S0033291708003760. PubMed: 18606046.

7. Poulton R, Caspi A, Moffitt TE, Cannon M, Murray R et al. (2000) Children's self-reported psychotic symptoms and adult schizophreniform disorder: a 15-year longitudinal study. Arch Gen Psychiatry 57: 1053-1058. doi:10.1001/archpsyc.57.11.1053. PubMed: 11074871.

8. Linscott RJ, van Os J (2013) An updated and conservative systematic review and meta-analysis of epidemiological evidence on psychotic experiences in children and adults: on the pathway from proneness to persistence to dimensional expression across mental disorders. Psychol Med 43: 1133-1149. doi:10.1017/S0033291712001626. PubMed: 22850401.

9. Scott J, Martin G, Welham J, Bor W, Najman J et al. (2009) Psychopathology during childhood and adolescence predicts delusional-like experiences in adults: a 21-year birth cohort study. Am J Psychiatry 166: 567-574. doi:10.1176/appi.ajp.2008.08081182. PubMed: 19339357.

10. Kelleher I, Keeley H, Corcoran P, Lynch F, Fitzpatrick C et al. (2012) Clinicopathological significance of psychotic experiences in non-psychotic young people: evidence from four population-based studies. Br J Psychiatry 201: 26-32. doi:10.1192/bjp.bp.111.101543. PubMed: 22500011.

11. Werbeloff N, Drukker M, Dohrenwend BP, Levav I, Yoffe R et al. (2012) Self-reported attenuated psychotic symptoms as forerunners of severe mental disorders later in life. Arch Gen Psychiatry 69: 467-475. doi:10.1001/archgenpsychiatry.2011.1580. PubMed: 22213772.

12. Gale CK, Wells JE, McGee MA, Browne MAO (2011) A latent class analysis of psychosis-like experiences in the new zealand mental health survey. Acta Psychiatr Scand 124: 205-213. doi:10.1111/j.1600-0447.2011.01707.x. PubMed: 21495982.

13. Kelleher I, Cannon M (2011) Psychotic-like experiences in the general population: characterizing a high-risk group for psychosis. Psychol Med 41: 1-6. doi:10.1017/S003329171100047X. PubMed: 20624328.

14. Varghese D, Wray NR, Scott JG, Williams GM, Najman JM et al. (2013) The heritability of delusional-like experiences. Acta Psychiatr Scand 127: 48-52. doi:10.1111/acps.12038_1. PubMed: 22881212.

15. Polanczyk G, Moffitt TE, Arseneault L, Cannon M, Ambler A et al. (2010) Etiological and clinical features of childhood psychotic symptoms: results from a birth cohort. Arch Gen Psychiatry 67: 328-338. doi:10.1001/archgenpsychiatry.2010.14. PubMed: 20368509.

16. Lataster T, van Os J, Drukker M, Henquet C, Feron F et al. (2006) Childhood victimisation and developmental expression of non-clinical delusional ideation and hallucinatory experiences: victimisation and non-clinical psychotic experiences. Soc Psychiatry Psychiatr Epidemiol 41: 423-428. doi:10.1007/s00127-006-0060-4. PubMed: 16572272.

17. Kelleher I, Harley M, Lynch F, Arseneault L, Fitzpatrick C et al. (2008) Associations between childhood trauma, bullying and psychotic symptoms among a school-based adolescent sample. Br J Psychiatry 193: 378-382. doi:10.1192/bjp.bp.108.049536. PubMed: 18978317.

18. Read J, van Os J, Morrison AP, Ross CA (2005) Childhood trauma, psychosis and schizophrenia: a literature review with theoretical and clinical implications. Acta Psychiatr Scand 112: 330-350. doi:10.1111/j.1600-0447.2005.00634.x. PubMed: 16223421.

19. Morgan C, Fisher H (2007) Environment and schizophrenia: environmental factors in schizophrenia: childhood trauma--a critical review. Schizophr Bull 33: 3-10. PubMed: 17105965.

20. Spataro J, Mullen PE, Burgess PM, Wells DL, Moss SA (2004) Impact of child sexual abuse on mental health: prospective study in males and females. Br J Psychiatry 184: 416-421. doi:10.1192/bjp.184.5.416. PubMed: 15123505.

21. Bebbington PE, Bhugra D, Brugha T, Singleton N, Farrell M et al. (2004) Psychosis, victimisation and childhood disadvantage: evidence from the second british national survey of psychiatric morbidity. Br J Psychiatry 185: 220-226. doi:10.1192/bjp.185.3.220. PubMed: 15339826.

22. Janssen I, Krabbendam L, Bak M, Hanssen M, Vollebergh W et al. (2004) Childhood abuse as a risk factor for psychotic experiences. Acta Psychiatr Scand 109: 38-45. doi:10.1111/j.1600-0047.2004.00329.x. PubMed: 14674957.

23. Spauwen J, Krabbendam L, Lieb R, Wittchen HU, van Os J (2006) Impact of psychological trauma on the development of psychotic symptoms: relationship with

psychosis proneness. Br J Psychiatry 188: 527-533. doi:10.1192/bjp.bp.105.011346. PubMed: 16738342.

24. Varese F, Smeets F, Drukker M, Lieverse R, Lataster T et al. (2012) Childhood adversities increase the risk of psychosis: a meta-analysis of patient-control, prospective- and cross-sectional cohort studies. Schizophr Bull 38: 661-671. doi:10.1093/schbul/sbs050. PubMed: 22461484.

25. van Dam DS, van der Ven E, Velthorst E, Selten JP, Morgan C et al. (2012) Childhood bullying and the association with psychosis in non-clinical and clinical samples: a review and meta-analysis. Psychol Med 42: 2463-2474. doi:10.1017/S0033291712000360. PubMed: 22400714.

26. Kelleher I, Keeley H, Corcoran P, Ramsay H, Wasserman C et al. (2013) Childhood trauma and psychosis in a prospective cohort study: cause, effect, and directionality. Am J Psychiatry 170: 734-741. doi:10.1176/appi.ajp.2012.12091169. PubMed: 23599019.

27. Wigman JTW, van Winkel R, Ormel J, Verhulst FC, van Os J et al. (2012) Early trauma and familial risk in the development of the extended psychosis phenotype in adolescence. Acta Psychiatr Scand 126: 266-273. doi:10.1111/j.1600-0447.2012.01857.x. PubMed: 22486536.

28. Ian K, Jenner JA, Cannon M (2010) Psychotic symptoms in the general population - an evolutionary perspective. Br J Psychiatry 197: 167-169. doi:10.1192/bjp.bp.109.076018. PubMed: 20807956.

29. Alemany S, Arias B, Aguilera M, Villa H, Moya J et al. (2011) Childhood abuse, the bdnf-val66met polymorphism and adult psychotic-like experiences. Br J Psychiatry 199: 38-42. doi:10.1192/bjp.bp.110.083808. PubMed: 21719879.

30. Nurjono M, Lee J, Chong SA (2012) A review of brain-derived neurotrophic factor as a candidate biomarker in schizophrenia. Clin Psychopharmacol Neurosci 10: 61-70. doi:10.9758/cpn.2012.10.2.61. PubMed: 23431036.

31. Williams HJ, Owen MJ, O'Donovan MC (2007) Is comt a susceptibility gene for schizophrenia? Schizophr Bull 33: 635-641. doi:10.1093/schbul/sbm019. PubMed: 17412710.

32. Zammit S, Owen MJ, Evans J, Heron J, Lewis G (2011) Cannabis, comt and psychotic experiences. Br J Psychiatry 199: 380-385. doi:10.1192/bjp.bp.111.091421. PubMed: 21947654.

33. Weinshilboum RM, Otterness DM, Szumlanski CL (1999) Methylation pharmacogenetics: catechol o-methyltransferase, thiopurine methyltransferase, and histamine n-methyltransferase. Annu Rev Pharmacol Toxicol 39: 19-52. doi:10.1146/annurev.pharmtox.39.1.19. PubMed: 10331075.

34. Chen J, Lipska BK, Halim N, Ma QD, Matsumoto M et al. (2004) Functional analysis of genetic variation in catechol-o-methyltransferase (comt): effects on mrna, protein, and enzyme activity in postmortem human brain. Am J Hum Genet 75: 807-821. doi:10.1086/425589. PubMed: 15457404.

35. Tunbridge EM, Harrison PJ, Weinberger DR (2006) Catechol-o-methyltransferase, cognition, and psychosis: val158met and beyond. Biol Psychiatry 60: 141-151. doi:10.1016/j.biopsych.2005.10.024. PubMed: 16476412.

36. Henquet C, Rosa A, Delespaul P, Papiol S, Fananás L et al. (2009) Comt valmet moderation of cannabis-induced psychosis: a momentary assessment study of

'switching on' hallucinations in the flow of daily life. Acta Psychiatr Scand 119: 156-160. doi:10.1111/j.1600-0447.2008.01265.x. PubMed: 18808401.

37. Caspi A, Moffitt TE, Cannon M, McClay J, Murray R et al. (2005) Moderation of the effect of adolescent-onset cannabis use on adult psychosis by a functional polymorphism in the catechol-o-methyltransferase gene: longitudinal evidence of a gene x environment interaction. Biol Psychiatry 57: 1117-1127. doi:10.1016/j.biopsych.2005.01.026. PubMed: 15866551.

38. Peerbooms O, Rutten BPF, Collip D, Lardinois M, Lataster T et al. (2012) Evidence that interactive effects of comt and mthfr moderate psychotic response to environmental stress. Acta Psychiatr Scand 125: 247-256. doi:10.1111/j.1600-0447.2011.01806.x. PubMed: 22128864.

39. Vinkers CH, Van Gastel WA, Schubart CD, Van Eijk KR, Luykx JJ et al. (2013) The effect of childhood maltreatment and cannabis use on adult psychotic symptoms is modified by the comt val (158)met polymorphism. Schizophr Res : .

40. Stefanis NC, Henquet C, Avramopoulos D, Smyrnis N, Evdokimidis I et al. (2007) Comt val158met moderation of stress-induced psychosis. Psychol Med 37: 1651-1656. PubMed: 17640440.

41. Savitz J, van der Merwe L, Newman TK, Stein DJ, Ramesar R (2010) Catechol-o-methyltransferase genotype and childhood trauma may interact to impact schizotypal personality traits. Behav Genet 40: 415-423. doi:10.1007/s10519-009-9323-7. PubMed: 20033274.

42. Sofroniew MV, Howe CL, Mobley WC (2001) Nerve growth factor signaling, neuroprotection, and neural repair. Annu Rev Neurosci 24: 1217-1281. doi:10.1146/annurev.neuro.24.1.1217. PubMed: 11520933.

43. Shoval G, Weizman A (2005) The possible role of neurotrophins in the pathogenesis and therapy of schizophrenia. Eur Neuropsychopharmacol 15: 319-329. doi:10.1016/S0924-977X(05)80597-8. PubMed: 15820422.

44. Chen ZY, Jing D, Bath KG, Ieraci A, Khan T et al. (2006) Genetic variant bdnf (val66met) polymorphism alters anxiety-related behavior. Science 314: 140-143. doi:10.1126/science.1129663. PubMed: 17023662.

45. Aguilera M, Arias B, Wichers M, Barrantes-Vidal N, Moya J et al. (2009) Early adversity and 5-htt/bdnf genes: new evidence of gene-environment interactions on depressive symptoms in a general population. Psychol Med 39: 1425-1432. doi:10.1017/S0033291709005248. PubMed: 19215635.

46. Lynch F, Mills C, Daly I, Fitzpatrick C (2006) Challenging times: prevalence of psychiatric disorders and suicidal behaviours in irish adolescents. J Adolesc 29: 555-573. doi:10.1016/j.adolescence.2005.08.011. PubMed: 16202448.

47. Kelleher I, Murtagh A, Molloy C, Roddy S, Clarke MC et al. (2012) Identification and characterization of prodromal risk syndromes in young adolescents in the community: a population-based clinical interview study. Schizophr Bull 38: 239-246. doi:10.1093/schbul/sbr164. PubMed: 22101962.

48. Harley M, Kelleher I, Clarke M, Lynch F, Arseneault L et al. (2010) Cannabis use and childhood trauma interact additively to increase the risk of psychotic symptoms in adolescence. Psychol Med 40: 1627-1634. doi:10.1017/S0033291709991966. PubMed: 19995476.

49. Goodman R (2001) Psychometric properties of the strengths and difficulties questionnaire. J Am Acad Child Adolesc Psychiatry 40: 1337-1345. doi:10.1097/00004583-200111000-00015. PubMed: 11699809.

50. Kovacs M (1985) The children's depression, inventory (cdi). Psychopharmacol Bull 21: 995-998. PubMed: 4089116.

51. Kovacs M (1992) Children's Depression Inventory Manual. New York: Multi-Health Systems Inc.

52. Harley M, Connor D, Clarke M, Kelleher I, Coughlan H et al. (2013) The 'challenging times' study of mental health in young irish adults: an 8-year follow up cohort study.. Ir J Med Sci (In press.).

53. Kaufman J, Birmaher B, Brent D, Rao U, Ryan N (1996) The Schedule for Affective Disorders and Schizophrenia for School Aged Children: Present and Lifetime Version. Pittsburg: University of Pittsburgh: Western Psychiatric Institute and Clinic.

54. Binbay T, Drukker M, Elbi H, Tanık FA, Özkınay F et al. (2012) Testing the psychosis continuum: differential impact of genetic and nongenetic risk factors and co-morbid psychopathology across the entire spectrum of psychosis. Schizophr Bull 38: 992-1002. doi:10.1093/schbul/sbr003. PubMed: 21525167.

55. Fisher HL, Caspi A, Poulton R, Meier MH, Houts R et al. (2013) Specificity of childhood psychotic symptoms for predicting schizophrenia by 38 years of age: a birth cohort study. Psychol Med 43: 2077-2086. doi:10.1017/S0033291712003091. PubMed: 23302254.

56. Kelleher I, Lynch F, Harley M, Molloy C, Roddy S et al. (2012) Psychotic symptoms in adolescence index risk for suicidal behavior: findings from 2 population-based case-control clinical interview studies. Arch Gen Psychiatry 69: 1277-1283. doi:10.1001/archgenpsychiatry.2012.164. PubMed: 23108974.

57. Saha S, Scott JG, Johnston AK, Slade TN, Varghese D et al. (2011) The association between delusional-like experiences and suicidal thoughts and behaviour. Schizophr Res 132: 197-202. doi:10.1016/j.schres.2011.07.012. PubMed: 21813264.

58. Nishida A, Sasaki T, Nishimura Y, Tanii H, Hara N et al. (2010) Psychotic-like experiences are associated with suicidal feelings and deliberate self-harm behaviors in adolescents aged 12-15 years. Acta Psychiatr Scand 121: 301-307. doi:10.1111/j.1600-0447.2009.01439.x. PubMed: 19614622.

59. Kirov G, Murphy KC, Arranz MJ, Jones I, McCandles F et al. (1998) Low activity allele of catechol-o-methyltransferase gene associated with rapid cycling bipolar disorder. Mol Psychiatry 3: 342-345. doi:10.1038/sj.mp.4000385. PubMed: 9702744.

60. Karayiorgou M, Altemus M, Galke BL, Goldman D, Murphy DL et al. (1997) Genotype determining low catechol-o-methyltransferase activity as a risk factor for obsessive-compulsive disorder. Proc Natl Acad Sci U S A 94: 4572-4575. doi:10.1073/pnas.94.9.4572. PubMed: 9114031.

61. Meira-Lima I, Shavitt RG, Miguita K, Ikenaga E, Miguel EC et al. (2004) Association analysis of the catechol-o-methyltransferase (comt), serotonin transporter (5-htt) and serotonin 2a receptor (5ht2a) gene polymorphisms with obsessive-compulsive disorder. Genes Brain Behav 3: 75-79. doi:10.1046/j.1601-1848.2003.0042.x. PubMed: 15005715.

62. Caspi A, Langley K, Milne B, Moffitt TE, O'Donovan M et al. (2008) A replicated molecular genetic basis for subtyping antisocial behavior in children with attention-

deficit/hyperactivity disorder. Arch Gen Psychiatry 65: 203-210. doi:10.1001/arch-genpsychiatry.2007.24. PubMed: 18250258.

63. Eisenberg J, Mei-Tal G, Steinberg A, Tartakovsky E, Zohar A et al. (1999) Haplo-type relative risk study of catechol-o-methyltransferase (comt) and attention deficit hyperactivity disorder (adhd): association of the high-enzyme activity val allele with adhd impulsive-hyperactive phenotype. Am J Med Genet 88: 497-502. doi:10.1002/(SICI)1096-8628(19991015)88:5. PubMed: 10490706.

64. Hawi Z, Millar N, Daly G, Fitzgerald M, Gill M (2000) No association between catechol-o-methyltransferase (comt) gene polymorphism and attention deficit hyperactivity disorder (adhd) in an irish sample. Am J Med Genet 96: 282-284. doi:10.1002/1096-8628(20000612)96:3. PubMed: 10898900.

65. Kia-Keating BM, Glatt SJ, Tsuang MT (2007) Meta-analyses suggest association be-tween comt, but not htr1b, alleles, and suicidal behavior. Am J Med Genet B Neuro-psychiatr Genet 144B: 1048-1053. doi:10.1002/ajmg.b.30551. PubMed: 17525973.

66. Mandelli L, Serretti A (2013) Gene environment interaction studies in depression and suicidal behavior: an update. Neurosci Biobehav Rev (Epub ahead of print). PubMed: 23886513.

67. Kolassa I, Kolassa S, Ertl V, Papassotiropoulos A, De Quervain DJ (2010) The risk of posttraumatic stress disorder after trauma depends on traumatic load and the cate-chol-o-methyltransferase val (158)met polymorphism. Biol Psychiatry 67: 304-308.

68. Klengel T, Mehta D, Anacker C, Rex-Haffner M, Pruessner JC et al. (2013) Allele-specific fkbp5 dna demethylation mediates gene-childhood trauma interactions. Nat Neurosci 16: 33-41. PubMed: 23201972.

69. Bendall S, Jackson HJ, Hulbert CA, McGorry PD (2008) Childhood trauma and psychotic disorders: a systematic, critical review of the evidence. Schizophr Bull 34: 568-579. PubMed: 18003630.

70. Simons CJP, Wichers M, Derom C, Thiery E, Myin-Germeys I et al. (2009) Subtle gene-environment interactions driving paranoia in daily life. Genes Brain Behav 8: 5-12. doi:10.1111/j.1601-183X.2008.00434.x. PubMed: 18721261.

CHAPTER 4

CHILDHOOD MALTREATMENT AND COPING IN BIPOLAR DISORDER

LEDO DARUY-FILHO, ELISA BRIETZKE,
BRUNO KLUWE-SCHIAVON, CRISTIANE DA SILVA FABRES,
AND RODRIGO GRASSI-OLIVEIRA

4.1 INTRODUCTION

Adverse experiences in childhood have been recognized as common events in individuals with bipolar disorder (BD), with approximately half (49%) of these individuals reporting at least some form of abuse or neglect during childhood (Garno, Goldberg, Ramirez, & Ritzler, 2005; Leverich et al., 2002). Furthermore, a history of childhood maltreatment (CMT) has been associated with unfavorable characteristics of BD (Alvarez, Roura, Oses, Foguet, Sola, & Arrufat, 2011; Angst, Gamma, Rossler, Ajdacic, & Klein, 2011; Daruy-Filho, Brietzke, Lafer, & Grassi-Oliveira, 2011; Etain, Henry, Bellivier, Mathieu, & Leboyer, 2008; McIntyre et al., 2008), including early age of onset (Carballo et al., 2008; Dienes, Hammen, Henry, Cohen, & Daley, 2006), recurrence (Brown, McBride, Bauer, & Williford, 2005), and decreased response to treatment (Marchand, Wirth, & Simon, 2005).

Childhood maltreatment has also been significantly associated with impairment in cognitive performance among various samples. The neurocognitive domains that appear to be associated with childhood trauma are memory (Grassi-Oliveira, Ashy, & Stein, 2008; Grassi-Oliveira, Stein, Lopes, Teixeira, & Bauer, 2008; Ritchie et al., 2011), attention, and executive function (Viola, Tractenberg, Pezzi, Kristensen, & Grassi-Oliveira, 2013). Neuroimaging findings indicate that CMT is associated with a reduction of the volume of the hippocampus, amygdala, and anterior cingulate cortex in clinical and nonclinical human samples (Bremner et al., 1997; Stein, Koverola, Hanna, Torchia, & McClarty, 1997; Treadway, Grant, Ding, Hollon, Gore, & Shelton, 2009; Vythilingam et al., 2002; Woon, & Hedges, 2008).

A robust body of evidence indicates that exposure to early life stressors, not just CMT, can negatively impact the clinical course of BD (Kapczinski et al., 2008; Post, & Leverich, 2006). This impact is exerted through changes in immune, endocrine, and molecular mechanisms that modulate neuroplasticity (Bender, Alloy, Sylvia, Urosevic, & Abramson, 2010; Grassi-Oliveira et al., 2008; Kauer-Sant'Anna et al., 2007) and may be influenced by individual differences in stress responsivity. Interestingly, one's coping abilities may be impaired by the effects of stress mediators on relevant neurofunctional circuits (Grassi-Oliveira, Daruy-Filho, & Brietzke, 2010).

According to Folkman, Lazarus, Gruen, & DeLongis (1986), coping comprises the set of mechanisms used by an individual to deal with a stressful situation. These mechanisms include direct strategies for problem resolution, cognitive reevaluation, acceptance, and social support rather than nonadaptive alternatives, such as avoidance, denial, or emotional thought (Compas, 2006). Coping functions include the ability to evaluate situations and make the best decision possible, and some authors have postulated that coping is intimately linked to cognitive function. Because of this, impairments in the integrity of these mechanisms could be associated with inefficient coping strategies.

Childhood abuse and neglect create neurodevelopmental toxicity that consequently impairs all of the functions that are intimately linked to cognitive function, including coping skills (Barker-Collo, Read, & Cowie, 2012; Del-Ben, Vilela, de Crippa, Hallak, Labati, & Zuardi, 2001; Grassi-Oliveira et al., 2008; Sesar, Simic, & Barisic, 2010). As a result, abused and neglected individuals may likely use less adaptive forms of stress management and infrequently use coping strategies that focus on the resolution of problems. The objective of the present study was to verify the existence of an association between CMT and coping strategies in individuals with BD Type 1.

4.2 METHODS

Female outpatients with BD Type 1 and without psychiatric comorbidities, aged 18 to 65 years, were selected from an outpatient unit of the Hospital Materno-Infantil Presidente Vargas in Porto Alegre, Brazil. The diagnoses of BD and comorbidities were confirmed using the Semi-structured Clinical Interview for the *Diagnostic and Statistical Manual of Mental Disorders,* 4th edition (SCID-I-CV; Del-Ben et al., 2001). Only euthymic patients were included, with euthymia defined as a score on the Young Mania Rating Scale < 7 and 21-item Hamilton Depression Rating Scale < 7 (Hamilton, 1960; Vilela, Crippa, Del-Ben, & Loureiro, 2005; Young, Biggs, Ziegler, & Meyer, 1978). Patients with diagnoses of other mental disorders, such as psychotic disorders, substance abuse or dependence, and dementia, or other organic disorders were excluded. All of the procedures for data collection were conducted by a trained research team.

Before their inclusion in the study, all of the participants provided written informed consent. The investigation protocol was approved by the ethics committee of the Hospital Materno-Infantil Presidente Vargas and Pontifícia Universidade Católica do Rio Grande do Sul.

4.2.1 COPING

Coping was evaluated using the Brazilian-Portuguese versions of the Ways of Coping Questionnaire (WCQ) and Brief COPE, two self-applied instruments. The WCQ (Folkman et al., 1986; Seidl, Tróccoli, & Zannon, 2001) comprises a set of 45 items, in which the frequency of use of different coping strategies is presented in a Likert format, from 1 ("I never do this") to 5 ("I always do this"). The participant is asked to select a specific stress-related event and, using a 5-point Likert-type scale, indicate how he would respond to that event. The coping strategies are divided into four factors: (i) focused on the problem (i.e., active efforts to manage, cope, solve, or reappraise the problem), (ii) focused on emotion (i.e., efforts to regulate the emotional states associated with the stressor as a way to reduce emotional discomfort without the objective to solve the problem; these include emotional reactions, such as rage, anxiety, guilt, avoidance, and passive behavior), (iii) religious or fantastic thought (i.e., religious behavior, thoughts, and faith that help when coping with problems), and (iv) search for social support (i.e., actively search for information or emotional support).

The Brief COPE (Carver, 1997; Ribeiro, & Rodrigues, 2004) is a Likert-type, self-applied instrument that consists of a set of 28 items that cover 14 coping styles in the original version and nine coping styles in the Portuguese version. The individual is asked how he managed stressful situations, from 0 ("I never did this") through 3 ("I did this a lot"). After factorial analysis with samples of BD patients, the original 14 factors were grouped into three independent factors: (i) focused on the problem (i.e., active coping, planning positive reinterpretation, acceptance, and search for instrumental, emotional, and social support), (ii) adaptive and focused on emotion (i.e., apply strategies of self-distraction, humor, expression of feelings, and religious thought), and (iii) maladaptive and focused on emotion (i.e., use of a substance to cope, such as alcohol and drugs, denial, guilt, and behavioral disengagement). This factorial regrouping was performed with the objective of adapting the results to the theoretical constructs proposed by the authors.

4.2.2 CHILDHOOD MALTREATMENT

A history of CMT was evaluated using the Childhood Trauma Questionnaire (CTQ; Grassi-Oliveira, Stein, & Pezzi, 2006). The CTQ is a self-applied instrument that evaluates the impact of traumatic events that occurred in adolescents and adults during childhood. The CTQ is a set of 28 items that investigates five subscales of traumatic events: (i) physical abuse, (ii) emotional abuse, (iii) sexual abuse, (iv) physical neglect, and (v) emotional neglect. The questionnaire is presented in a Likert-type scale that represents how frequently events were experienced, from 1 (Never) to 5 (significant frequency or always), resulting in a score from 5 to 25 for each of the five components of the scale. In the present study, the cutoffs adopted were proposed by the cohort of Walker et al. (1999) and intended to identify a positive history of CMT (Walker et al., 1999).

4.2.3 STATISTICAL ANALYSES

A one-sample Kolmogorov-Smirnov test was used to test the normality of all of the continuous variables. To elucidate relationships between the factors of the WCQ and Brief COPE and between coping factors and CMT, Pearson's correlation test was performed. In this case, for variables that were not normally distributed, Spearman's correlation test was performed. Stepwise linear regression analysis was performed to identify independent factors (CMT) associated with the coping factors of the WCQ and Brief COPE. Values of $p < .05$ were considered statistically significant. The data were analyzed using SPSS version 16.0 (SPSS, Chicago, IL, USA).

4.3 RESULTS

Thirty female patients with BD Type 1 were included in the study. The demographic variables, distributions of coping strategies, and histories of CMT are presented in Table 1.

Emotional abuse was the most frequent type of CMT in the sample, with 80% of the individuals reporting some type of emotional abuse during childhood. More than half of the females reported a history of physical abuse and neglect (66.7% and 63.3%, respectively). Emotional neglect during childhood was found in 43.3% of the individuals, and 26.7% of those individuals reported sexual abuse.

The use of fantastic thoughts or religious practices was the most frequent modality for coping measured by the WCQ, which was used by two-thirds of the sample. A similar frequency was found using the Brief COPE. The WCQ indicated that 56% of the patients frequently used strategies that were focused on managing specific stressors, with responses including "sometimes," "frequently," and "always." Forty-six percent of the strategies were focused on emotion, and 33% were focused on the search for social support. The most frequently used coping styles (i.e., "sometimes" and "frequently") for individuals assessed using the Brief COPE were focusing on the problem (66.7%) and focusing on emotion (46.7%). The maladaptive style of coping was the least frequent style used by the individuals in the sample (16.7%).

4.3.1 COPING MEASUREMENTS: WAYS OF COPING QUESTIONNAIRE AND BRIEF COPE

The regrouping of the 14 original factors of the Brief COPE into three factors that showed similarities to the WCQ is presented in Table 2. The two factors of the Brief COPE that suggested styles of adaptive coping were found to be associated with the WCQ factors "focused on the problem""and "religious or fantastic thought." The search for social support was inversely associated with coping strategies focused on emotion. Maladaptive coping and a focus on emotion were directly related to the WCQ's coping strategy that focused on emotion.

4.3.2 COPING AND CHILDHOOD MALTREATMENT

Table 3 shows the results of the associations between coping and a history of CMT in the exploratory correlational analysis. The severity of CMT was generally positively associated with a preference for coping styles and strategies that centered on emotional reactions rather than on more adaptive coping strategies that focused on problem-solving.

The pattern of coping as related to emotional expression was associated with a history of neglect (physical or emotional) and physical and emotional abuse. Additionally, the use of coping strategies that focused on the problem was negatively associated with the presence of physical abuse and emotional neglect.

The use of adaptive coping strategies and styles that focus on emotion, evaluated by the Brief COPE, showed an inverse relationship with the severity of CMT, especially with regard to emotional neglect, whereas the presence of childhood sexual abuse was related only to an adaptive pattern of coping measured by the WCQ.

TABLE 1: Description of the sample (n = 30)

	Mean (SD)
Clinical and Demographic Data	
Age (years)	42.77 (12.36)
Age of first mood episode (years)	26.07 (11.23)
Duration of illness (years)	17.6 (10.65)
Number of episodes	4.75 (3.32)
Number of hospitalizations	3.77 (4.5)
HAM-D (21-item) score	5.57 (4.7)
YMRS score	2.9 (3.88)
Childhood Maltreatment	
Components of CTQ (5-25)	
Emotional abuse	14.96 (5.67)
Emotional neglect	13.63 (5.24)
Physical abuse	10.86 (5.53)
Physical neglect	9.63 (4.46)
Sexual abuse	7.43 (4.40)
Total CTQ (25-125)	56.53 (18.67)
Coping	
WCG (1–5)	
Religious or fantastic thoughts	3.32 (0.76)
Focused on the problem	3.06 (0.67)
Focused on emotion	2.98 (0.77)
Search for emotional support	2.76 (0.84)
Brief COPE (0–3)	
Focused on the problem	1.65 (0.59)
Adaptive and focused on emotion	1.51 (0.58)
Maladaptive	1.12 (0.49)

HAM-D, Hamilton Depression Rating Sale; YMRS, Young Mania Ratin Scale; CTQ, Childhood Trauma Questionnaire; WCQ, Ways of Copy Questionnaire.

The objective of the present study was to evaluate the impact of the severity of CMT on the preference for coping styles and strategies, for which a multiple regression analysis was conducted (Tables 4 and 5). Each factor of the WCQ and Brief Cope was included as a dependent variable in the equations, and the subscales of the CTQ were considered independent variables. The subscale for sexual abuse was excluded from the regression because its results did not present a normal distribution with strong positive asymmetry.

TABLE 2: Correlation between factors of the WCQ and Brief COPE

		Brief COPE		
		Focused on the problem	Adaptive and focused on emotion	Maladaptive and focused on emotion
WCQ	Focused on the problem	0.39[b]	0.49[a]	−0.39[b]
	Focused on emotion	−0.53[a]	−0.07	0.54[b]
	Fantastic or religious thoughts	0.13	0.41[b]	−0.00
	Search for social support	0.38[b]	0.20	0.03

WCQ, Ways of Coping Questionnaire; Pearson Correlation, [a]p ≤ 0.01, [b]p ≤ 0.05

TABLE 3: Exploratory analysis between coping and childhood maltreatment (n = 30).

	Total CTQ*	Physical neglect*	Emotional neglect*	Sexual abuse**	Physical abuse*	Emotional abuse*
WCQ						
Focused on the problem	−0.31	−0.02	−0.53[a]	0.40[b]	−0.45[b]	−0.34
Focused on emotion	0.42[b]	0.20	0.28	0.19	0.33	0.44[b]
Fantastic and religious thoughts	−0.13	−0.09	−0.29	0.43[b]	−0.18	−0.10
Search for social support	−0.01	0.04	−0.29	0.38[b]	−0.06	−0.70
Brief COPE						
Focused on the problem	−0.42[b]	−0.25	−0.49[a]	0.02	−0.29	−0.34
Adaptive and focused on emotion	−0.52[b]	−0.46[a]	−0.48[a]	0.02	−0.48[a]	−0.44[b]
Maladaptive and focused on emotion	0.60[a]	0.36[b]	0.55[a]	0.13	0.64[a]	0.44[b]

*WCQ, Ways of Coping Questionnaire, *Pearson Correlation, **Spearman Correlation, [a]p ≤ 0.01, [b]p ≤ 0.05*

TABLE 4: Linear regression: Coping (WCQ) and history of childhood maltreatment

	R	DR^2	DF	df	B	p
Focused on the problem	0.53	0.25	11.08	1.29		0.002
Physical neglect					0.33	ns
Emotional neglect					−0.53	0.002
Physical abuse					−0.14	ns
Emotional abuse					0.07	ns
Focused on emotion	0.44	0.16	6.78	1.29		0.01
Physical neglect					−0.00	ns
Emotional neglect					−0.05	ns
Physical abuse					0.10	ns
Emotional abuse					0.44	0.01
Fantastic or religious thought	0.33	−0.03	0.78	4.29		ns
Search for social support	0.42	0.05	1.40	4.29		ns

ns, nonsignificant; WCQ, Ways of Coping Questionnaire

TABLE 5: Linear regression: Coping (Brief COPE) and history of childhood maltreatment

	R	DR^2	DF	df	B	p
Focused on the problem	0.49	0.22	9.20	1.29		0.005
Physical neglect					0.00	ns
Emotional neglect					−0.49	0.005
Physical abuse					0.12	ns
Emotional abuse					0.02	ns
Adaptive and focused on emotion	0.48	0.21	8.76	1.29		0.006
Physical neglect					−0.31	ns
Emotional neglect					−0.28	ns
Physical abuse					−0.48	0.006
Emotional abuse					−0.23	ns
Maladaptive and focused on emotion	0.64	0.39	19.6	1.29		0.000
Physical neglect					0.11	ns
Emotional neglect					0.18	ns
Physical abuse					0.64	0.004
Emotional abuse					0.08	ns

A linear relationship was found between the frequency of emotional neglect during childhood on the WCQ and the decreased use of coping strategies that focused on the problem (Table 4). Similarly, the severity of emotional abuse was directly proportional to the frequency of using coping strategies that focused on emotional control.

Table 5 shows a similar linear regression, but it included data from the Brief COPE. Participants who were subjected to physical abuse during childhood presented coping styles that appeared to depend on the frequency and severity of abuse. More importantly, however, is the frequent use of maladaptive coping strategies that focused on emotion rather than the less frequent use of adaptive coping strategies. The frequency of emotional neglect during childhood was found to be a significant predictor of the infrequent implementation of coping strategies that focused on a resolution of the stressor.

4.4 DISCUSSION

The present results support the hypothesis that traumatic events during childhood negatively interfere with the way adult individuals with BD cope with stress. Bipolar patients who were subjected to CMT, especially physical neglect and physical abuse, presented a preference for coping styles and strategies that focused on emotional control and were associated with the cognitive reappraisal of problems and strategies that included the avoidance of coping. Furthermore, the styles and strategies associated with problem solving were less frequently used in the subpopulation of BD patients who were subjected to CMT.

One important aspect is the difference between the instruments used to evaluate coping. The WCQ is applied based on one specific stressor chosen by the patient and is based on the model of coping strategies by Folkman et al. (1986). This approach provides evidence of the way the individual copes with a specific problem. The Brief COPE, in contrast, evaluates the most common way to cope with stressors (coping traits or coping styles; Carver, Scheier, & Weintraub, 1989). Both concepts evaluated by WCQ and Brief COPE were similar showing association with CMT.

Reducing the 14 factors from the Brief COPE to three factors was compatible with the theoretical framework presented by the WCQ. An important difference between these two instruments is the strategies that focus on emotion. In the WCQ, the factor "coping focused on emotion" combines several strategies related to emotional control, such as anger and guilt, which imply intrinsic dysfunctional characteristics. In the Brief COPE, the authors considered that certain strategies (e.g., humor and distraction) are functional despite being focused on emotion. Because of this, a specific factor for those strategies was created.

Furthermore, the present results are consistent with previous studies that used samples from different populations. Walsh, Fortier, & Dilillo (2010) found a relationship between a history of childhood sexual abuse and emotion-oriented coping in a sample of university students. Sesar et al. (2010) and Hager, & Runtz (2012) studied coping and CMT and proposed that avoidance and denial strategies are among the more common coping strategies.

Our hypothesis suggests that the relationship between CMT and dysfunctional coping is based on the impact of severe stress during critical periods of neurodevelopment, thus impairing cognition and affecting memory (Ritchie et al., 2011), attention, and executive function (Savitz et al., 2008). These findings are corroborated by structural neuroimaging studies that described volumetric reductions of the hippocampus (Bremner et al., 1997; Stein et al., 1997; Woon, & Hedges, 2008). Additionally, the size of the amygdala was found to be reduced in a pediatric population who was exposed to maltreatment (Weniger, Lange, Sachsse, & Irle, 2008), although increased activity was also found (Grant, Cannistraci, Hollon, Gore, & Shelton, 2011). Moreover, the mediators involved in the neuroprogression of BD, such as inflammation and neurotrophic factors (Berk, 2011), may be influenced by CMT. In fact, CMT is associated with reprogramming of the release of brain-derived neurotrophic factor (Elzinga, 2011), and a close association between traumatic events and cytokines has been observed (Guo, 2012).

The hippocampus, amygdala, and prefrontal cortex may be vulnerable to several secondary mechanisms that may compromise their functionality, resulting in difficulties in coping strategies that depend on the integrity of certain cognitive processes, particularly processes related to the resolution of problems and cognitive reappraisal, such as executive function, all of which depend on the prefrontal cortex (Clark, Rogers, Armstrong, Rakowski, & Kviz, 2008; Funahashi, 2001; Tanji, & Hoshi, 2008). Additionally, the activity of the amygdala, in association with the prefrontal cortex, can modulate emotional responses to ensure that behavior is appropriate for the specific context (Bachevalier, & Malkova, 2006; Garrett, & Chang, 2008). Individuals with amygdalar hyperactivity associated with prefrontal dysfunction exhibited behaviors and thoughts with more emotional characteristics when dealing with stressors (i.e., coping strategies focused on emotion).

Based on the present results, we hypothesize that BP patients exposed to CMT will use less adaptive strategies in a probable stressful situation in the future, rather than strategies directed toward problem resolution. Therefore, BP patients will tend to experience the stressor in a more negative way, thereby increasing the activation of the mediators involved in the neurophysiology of stress (Bender et al., 2010; Kauer-Sant'Anna et al., 2007) and allostatic load (Kapczinski et al., 2008). These facts can postulate the moderating role of coping between CMT and bipolar disorder progression. Thus, a new focus for psychosocial interventions may be introduced that is centered on readjusting coping strategies to relieve the progressive nature of the disorder (Grassi-Oliveira, Daruy-Filho, & Brietzke, 2010).

The limitations of the present study include the small sample size and its restriction to women. More studies are necessary to elucidate specific cognitive processes associated with coping strategies and the neuroanatomical and neurofunctional processes impacted by traumatic events during childhood that affect the strategies chosen by an individual. Moreover, prospective studies may investigate individual differences in coping abilities as a modulator of the impact of CMT on BD.

REFERENCES

1.   Alvarez, M. J., Roura, P., Oses, A., Foguet, Q., Sola, J., & Arrufat, F. X. (2011). Prevalence and clinical impact of childhood trauma in patients with severe mental disorders. Journal of Nervous and Mental Disease, 199(3), 156-161.

2. Angst, J., Gamma, A., Rossler, W., Ajdacic, V., & Klein, D. N. (2011). Childhood adversity and chronicity of mood disorders. European Archives of Psychiatry and Clinical Neuroscience, 261(1), 21-27.
3. Bachevalier, J., & Malkova, L. (2006). The amygdala and development of social cognition: theoretical comment on Bauman, Toscano, Mason, Lavenex, and Amaral (2006). Behavioral Neuroscience, 120(4), 989-991.
4. Barker-Collo, S., Read, J., & Cowie, S. (2012). Coping strategies in female survivors of childhood sexual abuse from two Canadian and two New Zealand cultural groups. Journal of Trauma & Dissociation, 13(4), 435-447.
5. Bender, R. E., Alloy, L. B., Sylvia, L. G., Urosevic, S., & Abramson, L. Y. (2010). Generation of life events in bipolar spectrum disorders: a re-examination and extension of the stress generation theory. Journal of Clinical Psychology, 66(9), 907-926.
6. Berk, M., Kapczinski, F., Andreazza, A. C., Dean, O. M., Giorlando, F., Maes, M., Yücel, M., Gama, C. S., Dodd, B., Magalhães, P. V. S., Amminger, P., McGorry, P., & Malhi, G. S. (2011). Pathways underlying neuroprogression in bipolar disorder: Focus on inflammation, oxidative stress and neurothophic factors. Neurosciendce and Biobehavioral Reviews, 35(3), 804-17.
7. Bremner, J. D., Randall, P., Vermetten, E., Staib, L., Bronen, R. A., Mazure, C., Capelli, S., McCarthy, G., Innis, R. B., & Charney, D. S. (1997). Magnetic resonance imaging-based measurement of hippocampal volume in posttraumatic stress disorder related to childhood physical and sexual abuse: a preliminary report. Biological Psychiatry, 41(1), 23-32.
8. Brown, G. R., McBride, L., Bauer, M. S., & Williford, W. O. (2005). Impact of childhood abuse on the course of bipolar disorder: a replication study in U.S. veterans. Journal of Affective Disorders, 89(1-3), 57-67.
9. Carballo, J. J., Harkavy-Friedman, J., Burke, A. K., Sher, L., Baca-Garcia, E., Sullivan, G. M., Grunebaum, M. F., Parsey, R. V., Mann, J. J., & Oquendo, M. A. (2008). Family history of suicidal behavior and early traumatic experiences: additive effect on suicidality and course of bipolar illness? Journal of Affective Disorders, 109(1-2), 57-63.
10. Carver, C. S. (1997). You want to measure coping but your protocol's too long: consider the brief COPE. International Journal of Behavioral Medicine, 4(1), 92-100.
11. Carver, C. S., Scheier, M. F., & Weintraub, J. K. (1989). Assessing coping strategies: a theoretically based approach. Journal of Personality and Social Psychology, 56(2), 267-283.
12. Clark, M. A., Rogers, M. L., Armstrong, G. F., Rakowski, W., & Kviz, F. J. (2008). Differential response effects of data collection mode in a cancer screening study of unmarried women ages 40-75 years: a randomized trial. BMC Medical Research Methodology, 8, 10.
13. Compas, B. E. (2006). Psychobiological processes of stress and coping: implications for resilience in children and adolescents-comments on the papers of Romeo & McEwen and Fisher et al. Annals of the New York Academy of Science, 1094, 226-234.
14. Daruy-Filho, L., Brietzke, E., Lafer, B., & Grassi-Oliveira, R. (2011). Childhood maltreatment and clinical outcomes of bipolar disorder. Acta Psychiatrica Scandinavica, 124, 427-434.

15. Del-Ben, C. M., Vilela, J. A., de Crippa, J. A. S., Hallak, J. E. C., Labati, C. M., & Zuardi, A. W. (2001). Reliability of the Structured Clinical Interview for DSM-IV-Clinical Version translated into Portuguese. Brazilian Journal of Psychiatry, 23(3), 156-159.

16. Dienes, K. A., Hammen, C., Henry, R. M., Cohen, A. N., & Daley, S. E. (2006). The stress sensitization hypothesis: understanding the course of bipolar disorder. Journal of Affective Disorders, 95(1-3), 43-49.

17. Elzinga, B. M., Molendijk, M. L., Voshaar, R. C. O., Bus, B. A. A., Prickaerts, J., Spinhoven, P., & Penninx, B. J. W. H. (2011). The impact of childhood abuse and recent stress on serum brain-derived neurotrophic factor and the moderating role of BDNF Val66Met. Psychopharmacology, 214(1), 319-28.

18. Etain, B., Henry, C., Bellivier, F., Mathieu, F., & Leboyer, M. (2008). Beyond genetics: childhood affective trauma in bipolar disorder. Bipolar Disorder, 10(8), 867-876.

19. Folkman, S., Lazarus, R. S., Gruen, R. J., & DeLongis, A. (1986). Appraisal, coping, health status, and psychological symptoms. Journal of Personality and Social Psychology, 50(3), 571-579.

20. Funahashi, S. (2001). Neuronal mechanisms of executive control by the prefrontal cortex. Neuroscience Research, 39(2), 147-165.

21. Garno, J. L., Goldberg, J. F., Ramirez, P. M., & Ritzler, B. A. (2005). Impact of childhood abuse on the clinical course of bipolar disorder. British Journal of Psychiatry, 186, 121-125.

22. Garrett, A., & Chang, K. (2008). The role of the amygdala in bipolar disorder development. Development and Psychopathology, 20(4), 1285-1296.

23. Grant, M. M., Cannistraci, C., Hollon, S. D., Gore, J., & Shelton, R. (2011). Childhood trauma history differentiates amygdala response to sad faces within MDD. Journal of Psychiatric Research, 45(7), 886-895.

24. Grassi-Oliveira, R., Ashy, M., & Stein, L. M. (2008). Psychobiology of childhood maltreatment: effects of allostatic load? Revista Brasileira de Psiquiatria, 30(1), 60-68.

25. Grassi-Oliveira, R., Daruy-Filho, L., & Brietzke, E. (2010). New perspectives on coping in bipolar disorder. Psychology & Neuroscience, 3(2), 161-165.

26. Grassi-Oliveira, R., Stein, L. M., Lopes, R. P., Teixeira, A. L., & Bauer, M. E. (2008). Low plasma brain-derived neurotrophic factor and childhood physical neglect are associated with verbal memory impairment in major depression: a preliminary report. Biological Psychiatry, 64(4), 281-285.

27. Grassi-Oliveira, R., Stein, L. M., & Pezzi, J. C. (2006). [Translation and content validation of the Childhood Trauma Questionnaire into Portuguese language]. Revista de Sa úde Pública, 40(2), 249-255.

28. Guo, M., Liu, T., Guo, J. C., Jiang, X. L., Chen, F., & Gao, Y. S. (2012). Study on serum cytokine levels in posttraumatic stress disorder patients. Asian Pacific Journal of Tropical Medicine, 5(4), 323-5.

29. Hager, A. D., & Runtz, M. G. (2012). Physical and psychological maltreatment in childhood and later health problems in women: an exploratory investigation of the roles of perceived stress and coping strategies. Child Abuse & Neglect, 36(5), 393-403.

30. Hamilton, M. (1960). A rating scale for depression. Journal of Neurology, Neurosurgery and Psychiatry, 23, 56-62.

31. Kapczinski, F., Vieta, E., Andreazza, A. C., Frey, B. N., Gomes, F. A., Tramontina, J., Kauer-Sant'anna, M., Grassi-Oliveira, R., & Post, R. M. (2008). Allostatic load in bipolar disorder: implications for pathophysiology and treatment. Neuroscience and Biobehavioral Reviews, 32(4), 675-692.

32. Kauer-Sant'Anna, M., Tramontina, J., Andreazza, A. C., Cereser, K., da Costa, S., Santin, A., Yathan, L. N. & Kapczinski, F. (2007). Traumatic life events in bipolar disorder: impact on BDNF levels and psychopathology. Bipolar Disorder, 9(Suppl. 1), 128-135.

33. Leverich, G. S., McElroy, S. L., Suppes, T., Keck, P. E., Jr., Denicoff, K. D., Nolen, W. A., Altshuler, L. L., Rush, A. J., Kupka, R., Frye, M. A., Autio, K. A., &. Post, R. M. (2002). Early physical and sexual abuse associated with an adverse course of bipolar illness. Biological Psychiatry, 51(4), 288-297.

34. Marchand, W. R., Wirth, L., & Simon, C. (2005). Adverse life events and pediatric bipolar disorder in a community mental health setting. Community Mental Health Journal, 41(1), 67-75.

35. McIntyre, R. S., Soczynska, J. K., Mancini, D., Lam, C., Woldeyohannes, H. O., Moon, S., Konarski, J. Z., & Kennedy, S. H. (2008). The relationship between childhood abuse and suicidality in adult bipolar disorder. Violence and Victims, 23(3), 361-372.

36. Post, R. M., & Leverich, G. S. (2006). The role of psychosocial stress in the onset and progression of bipolar disorder and its comorbidities: the need for earlier and alternative modes of therapeutic intervention. Development and Psychopathology, 18(4), 1181-1211.

37. Ribeiro, J., & Rodrigues, A. (2004). Questões acerca do coping: a propósito do estudo de adaptação do BRIEF Cope. Psicologia, Saúde e Doenças, 5(1), 3-15.

38. Ritchie, K., Jaussent, I., Stewart, R., Dupuy, A. M., Courtet, P., Malafosse, A., & Ancelin, M. L. (2011). Adverse childhood environment and late-life cognitive functioning. International Journal of Geriatric Psychiatry, 26, 503-510.

39. Savitz, J. B., van der Merwe, L., Stein, D. J., Solms, M., & Ramesar, R. S. (2008). Neuropsychological task performance in bipolar spectrum illness: genetics, alcohol abuse, medication and childhood trauma. Bipolar Disorder, 10(4), 479-494.

40. Seidl, E. M. F., Tróccoli, B. T., & Zannon, C. M. L. C. (2001). Análise fatorial de uma medida de estratégias de enfrentamento. Psicologia: Teoria e Pesquisa, 17(3), 225-234.

41. Sesar, K., Simic, N., & Barisic, M. (2010). Multi-type childhood abuse, strategies of coping, and psychological adaptations in young adults. Croatian Medical Journal, 51(5), 406-416.

42. Stein, M. B., Koverola, C., Hanna, C., Torchia, M. G., & McClarty, B. (1997). Hippocampal volume in women victimized by childhood sexual abuse. Psychological Medicine, 27(4), 951-959.

43. Tanji, J., & Hoshi, E. (2008). Role of the lateral prefrontal cortex in executive behavioral control. Physiological Review, 88(1), 37-57.

44. Treadway, M. T., Grant, M. M., Ding, Z., Hollon, S. D., Gore, J. C., & Shelton, R. C. (2009). Early adverse events, HPA activity and rostral anterior cingulate volume in MDD. PLoS One, 4(3), e4887.

45. Vilela, J. A., Crippa, J. A., Del-Ben, C. M., & Loureiro, S. R. (2005). Reliability and validity of a Portuguese version of the Young Mania Rating Scale. Brazilian Journal of Medical and Biological Research, 38(9), 1429-1439.

46. Viola, T. W., Tractenberg, S. G., Pezzi, J. C., Kristensen, C. H., & Grassi-Oliveira, R. (2013) Childhood physical neglect associated with executive functions impairments in crack cocaine-dependent women. Drug Alcohol Depend, 132(1-2), 271-276. doi: 10.1016/j.drugalcdep.2013.02.014

47. Vythilingam, M., Heim, C., Newport, J., Miller, A. H., Anderson, E., Bronen, R., Brummer, M., Staib, L., Vermetten, E., Charney, D. S., Nemeroff, C. B., & Bremner, J. D. (2002). Childhood trauma associated with smaller hippocampal volume in women with major depression. American Journal of Psychiatry, 159(12), 2072-2080.

48. Walker, E. A., Gelfand, A., Katon, W. J., Koss, M. P., Von Korff, M., Bernstein, D., & Russo, J. (1999). Adult health status of women with histories of childhood abuse and neglect. American Journal of Medicine, 107(4), 332-339.

49. Walsh, K., Fortier, M. A., & Dilillo, D. (2010). Adult coping with childhood sexual abuse: a theoretical and empirical review. Aggression and Violent Behavior, 15(1), 1-13.

50. Weniger, G., Lange, C., Sachsse, U., & Irle, E. (2008). Amygdala and hippocampal volumes and cognition in adult survivors of childhood abuse with dissociative disorders. Acta Psychiatrica Scandinavica, 118(4), 281-290.

51. Woon, F. L., & Hedges, D. W. (2008). Hippocampal and amygdala volumes in children and adults with childhood maltreatment-related posttraumatic stress disorder: a meta-analysis. Hippocampus, 18(8), 729-736.

52. Young, R. C., Biggs, J. T., Ziegler, V. E., & Meyer, D. A. (1978). A rating scale for mania: reliability, validity and sensitivity. British Journal of Psychiatry, 133, 429-435.

PART II

EARLY TRAUMA AND THE BRAIN

CHAPTER 5

A PRELIMINARY STUDY OF THE INFLUENCE OF AGE OF ONSET AND CHILDHOOD TRAUMA ON CORTICAL THICKNESS IN MAJOR DEPRESSIVE DISORDER

NATALIA JAWORSKA, FRANK P. MACMASTER, ISMAEL GAXIOLA, FILOMENO CORTESE, BRADLEY GOODYEAR, AND RAJAMANNAR RAMASUBBU

5.1 INTRODUCTION

Major depressive disorder (MDD) is a common psychiatric disorder with a high burden of disease, yet its neural underpinnings remain elusive. A handful of studies have assessed spatial patterns of cortical thickness in MDD. Interestingly, the regional patterns of cortical thinning in MDD do not perfectly reflect what would be expected from the neuroimaging literature (i.e., cortical thinning is not confined to cognitive and emotive centers; fronto-cortico-limbic structures) [1]. Further, extant literature is not consistent with respect to which cortical regions are typically thicker/thinner in the disorder. This indicates a need for further study with careful

A Preliminary Study of the Influence of Age of Onset and Childhood Trauma on Cortical Thickness in Major Depressive Disorder. © *Jaworska N, MacMaster FP, Gaxiola I, Cortese F, Goodyear B, and Ramasubbu R.* BioMed Research International, **2014** (2014), http://dx.doi.org/10.1155/2014/410472. Licensed under a Creative Commons Attribution 3.0 Unported License, http://creativecommons.org/licenses/by/3.0/.

attention to factors that may influence cortical thickness in MDD, such as age of disorder onset and past trauma and neglect.

The majority of research on cortical thickness in MDD has focused on assessing elderly depressed individuals (typically defined as >60 years of age; late-life MDD). For instance, one group found no cortical thickness differences between older depressed females and controls [2]. Similarly, Colloby et al. [3] noted no cortical thickness differences in frontal lobe structures between older individuals with MDD and controls. However, they found a tendency for decreased cortical thickness in MDD in the left frontal pole/pars orbitalis and the right medial orbitofrontal region. In yet another study, Kumar et al. [4] noted a thinner right isthmus in an elderly depressed cohort compared with controls. Another group found that elderly depressed individuals demonstrated thinner cortices in frontal (medial/superior), superior parietal, and inferior temporal regions [5]. Further, treatment nonresponders (versus responders) demonstrated thinner cortices in bilateral posterior cingulate and parahippocampal regions, the left paracentral, pre/cuneus and insular cortices as well as the right medial orbitofrontal, lateral occipital, and superior postcentral cortices [5]. Yet others reported a thinner bilateral dorsolateral prefrontal cortex (DLPFC) and thinner postcentral region in elderly depressed individuals relative to controls. Cortical thinning was also found in the left prefrontal (orbito-frontal, pars triangularis), rostral anterior cingulate, medial/superior temporal, and parietal cortices as well as in the pre/paracentral gyri. Right hemisphere thinning was noted in the pars opercularis, rostral middle frontal, precuneus, and isthmus cortices in elderly individuals with MDD [6]. Finally, cortical thickness in the frontal pole, superior/middle frontal gyrus, orbitofrontal gyrus, and anterior cingulate gyrus was thinner in elderly depressed patients relative to controls [7]. In sum, while some have noted no cortical thickness differences in elderly depressed versus control individuals, others have. Research points to decreased cortical thickness in prefrontal regions, particularly in the orbitofrontal area, in superior/middle frontal aspects (including the DLPFC) as well as para/postcentral regions and the cuneus/isthmus regions in elderly individuals with MDD.

Assessments of cortical thickness in nonelderly adults (i.e., those younger than 60 years of age) with MDD are sparse. Järnum et al. [8] found thinner cortices in MDD patients (middleaged) compared with controls in the orbitofrontal cortex, superior temporal lobe, and insula. Further, depressed nonremitters exhibited a thinner posterior cingulate cortex compared to those in remission. Similarly, another group noted thinner cortices in nonelderly (18–60 years of age) individuals with MDD in the left parahippocampal gyrus, orbitofrontal cortex as well as in the right middle/superior frontal gyri (DLPFC), middle temporal gyrus, and insula [9]. Yet another group noted that depressed adults showed cortical thinning in the bilateral superior/middle frontal gyri, right precentral gyrus (i.e., DLPFC), and right orbitofrontal gyrus. Smaller clusters of cortical thinning existed in the parietal (bilateral inferior parietal regions and left post-central gyrus), temporal (left entorhinal and middle temporal cortex), and occipital lobes (left lateral occipital and lingual gyrus). Regions that were thicker in MDD were the left anterior insula and lateral orbitofrontal gyrus [10]. Recently, van Eijndhoven et al. [11] assessed medication-naïve patients during their first major depressive episode (MDE) or after their first MDE. The medial orbitofrontal cortex was thinner in the MDD patients than in controls. Conversely, the temporal pole and the caudal anterior and posterior cingulate cortices were thicker. This was evident in both currently depressed and recovered patients, suggesting trait-versus state-specific abnormalities. Thus, assessments of nonelderly adults with MDD suggest cortical thinning in the medial orbitofrontal cortex (though lateral regions may be associated with thickening), insula, DLPFC, and the middle temporal cortex—which partially overlap with findings in elderly depressed individuals.

Finally, a handful of groups have assessed cortical thickness in pediatric MDD and found thinner cortices in the right pericalcarine, postcentral, and superior parietal gyri as well as the left supramarginal gyrus. The pediatric MDD cohort (≤18 years of age) exhibited thicker bilateral temporal pole cortices [12], consistent with the results in adults [11]. Additionally, our group observed thicker bilateral middle frontal gyri and left caudal cingulate gyrus in MDD adolescents compared to controls [13].

Potential factors that may contribute to the inconsistency in cortical thickness findings in MDD include the age of the sample examined, medication status, illness severity, sex, and age of MDD onset. The latter is perhaps the most pertinent as later childhood/adolescence is marked by extensive brain changes [14–16]. As such, early MDD onset (i.e., pediat-

ric/adolescent onset) may interfere with normal neurodevelopmental trajectories and manifest as structural abnormalities in adulthood. Further,
early MDD onset appears to be associated with increased risk for disorder
recurrence, illness burden, and psychiatric comorbidities [14]. This suggests that early MDD onset may be associated with specific neurobiological features. However, few studies have assessed the effect of age of onset
on cortical thickness in MDD. A recent study examined the association
between age of MDD onset (in this case, early: <24 years; late: >25 years)
and cortical thickness [1]. Reductions were found in the DLPFC, pre/post-
central gyri and the lingual gyrus in the early MDD onset group versus
controls. Further analyses revealed thicker cortices in the early versus late
MDD onset groups in the bilateral posterior cingulate cortex. Conversely,
the left parahippocampal, right lingual, right fusiform, and right precuneus
gyri were thinner in the early versus the late onset MDD group. Another
group assessed elderly depressed patients with earlier (<60 years) and late-
life (>60 years) MDD onset and found that the left anterior cingulate was
thinner in the late-life onset group [17]. Though preliminary, such data
suggest that age of onset may play a role in the spatial distribution of cortical thickness findings in MDD.

Early adverse events increase the possibility of MDD development
later in life [18]. Early trauma/maltreatment may interfere with normal
brain development. Previous work has reported cortical thickness reductions in maltreated versus nonmaltreated children in the anterior cingulate, superior frontal gyrus, orbitofrontal cortex, left middle temporal regions, and lingual gyrus [19]. Heim et al. [20] also reported widespread
cortical thinning as a function of childhood adversity (assessed by the
Childhood Trauma Questionnaire (CTQ)). CTQ scores were specifically
associated with anterior cingulate gyrus, precuneus, and parahippocampal gyrus cortical thinning. These studies parallel morphometric work
that has noted grey matter density and volumetric reductions in medial/
prefrontal regions and cingulate in adults and children with a history of
maltreatment/trauma (e.g., physical neglect) [21–26]. These structures
have been implicated in emotion regulation and memory processing and
tend to exhibit morphometric and functional changes in MDD. Further,
this research suggests that maltreatment/trauma is associated with structural modulations persisting into adulthood. To our knowledge, the inter-

action between age of MDD onset and trauma history on cortical thickness in depression has not been assessed.

As such, this pilot study examined cortical thickness in nonelderly adults (i.e., <60 years of age) with MDD to expand on the relatively scant and inconsistent literature on the matter. Second, we sought to assess whether differences existed in pediatric (<24 years of age) compared with adult MDD onset (>25 years of age) on cortical thickness, in an effort to replicate and expand on previous work. Third, we examined whether differences existed in cortical thickness in depressed adults with childhood sexual and/or physical abuse (sexual + physical abuse group—referred to simply as the abuse group) versus those who experienced no sexual and/or physical abuse but experienced emotional neglect/abuse and/or physical neglect (no sexual + physical abuse group (referred to simply as the non-abuse group); the abuse group also experienced emotional and physical neglect; Table 1). The interaction between age of MDD onset and trauma was also explored.

TABLE 1: Characteristics of MDD onset groups (pediatric/adult MDD onset) and controls.

Characteristics	MDD (overall)	Pediatric MDD onset	Adult MDD onset	HC
N	36	20	16	18
Sex (F/M)	22/14	12/8	10/6	10/8
Age (yrs.)	37.1 ± 11.2	31.5 ± 10.5	44.1 ± 7.7	31.9 ± 9.2
Baseline HAMD17	22.1 ± 4.1	20.7 ± 4.1	23.9 ± 3.4	—
Duration of current MDE (yrs.)	5.1 ± 5.4	4.9 ± 5.7	5.4 ± 5.3	—
Time since MDD onset (yrs.)	12.3 ± 9.2	14.4 ± 10.7	9.8 ± 6.3	—
MDD onset (yrs.)	24.8 ± 10.1	17.1 ± 4.8	34.3 ± 5.6	—

HC: healthy controls; HAMD17: Hamilton Depression Rating Scale; MDD: major depressive disorder; MDE: major depressive episode; means ± SDs presented.

We expected thinner cortices in orbitofrontal, DLPFC, para-/postcentral, and insular cortices in MDD (versus controls) as well as greater reductions in the pediatric (versus adult) MDD onset group in the DLPFC and posterior inferior temporal regions relative to the adult onset group. Finally, we expected greater thinning in the MDD group with a history of childhood abuse in cortical regions comprising the frontal-limbic network. No directional hypotheses existed regarding age of MDD onset and trauma history due to lack of precedent literature.

TABLE 2: Characteristics of childhood abuse and nonabuse MDD groups.

Characteristics	Nonabuse MDD group	Abuse MDD group
N	19	12
Sex (F/M)	10/9	8/4
Age (yrs.)	36.4 ± 12.6	40.0 ± 9.5
Baseline HAMD17	21.2 ± 4.0	24.7 ± 3.9
Duration of current MDE (yrs.)	4.7 ± 5.2	5.4 ± 6.1
Time since MDD onset (yrs.)	11.1 ± 9.4	11.9 ± 9.1
MDD onset (yrs.)	25.3 ± 11.1	28.1 ± 7.3
CTQTotal	51.0 ± 6.1	62.9 ± 11.1
CTQ "neglect" score	39.8 ± 5.6	42.5 ± 6.5
CTQ "abuse" score	11.2 ± 1.1	21.3 ± 6.7

CTQ: Childhood Trauma Questionnaire; CTQ "neglect" score: emotional neglect + physical neglect + emotional abuse; CTQ "abuse" score: physical abuse + sexual abuse; $HAMD_{17}$: Hamilton Depression Rating Scale; MDD: major depressive disorder; MDE: major depressive episode; means ± SDs presented.

5.2 METHODS

5.2.1 PARTICIPANTS

Thirty-six adults (age range: 19–58 years) with a primary diagnosis of MDD were tested. Clinical diagnoses were made by the study psychiatrist (R.R.) according to the Structured Clinical Interview for DSM (Diagnostic and Statistical Manual of Mental Disorders) IV-TR Diagnoses, Axis

I, Patient Version (SCID-IV-I/P) criteria. The Hamilton Rating Scale for Depression (HAMD$_{17}$) was used to assess symptom severity [27], with patients being included if they had an HAMD$_{17}$ score of $\geq$18. All participants were free of psychotropic medications for a minimum of three weeks at time of neuroimaging. Exclusion criteria included bipolar disorder (BP-I/II or NOS), psychosis history, a clinically significant anxiety disorder, current (<6 months) substance abuse/dependence, neurological disorders, eating disorders, unstable medical condition, and significant suicide risk. Participants with magnetic resonance imaging (MRI) contraindications (e.g., pregnancy, metal implants, and claustrophobia) were also excluded. Twenty patients had MDD onset at <24 years of age (pediatric onset) and 16 patients had MDD onset at >25 years of age (adult onset). Childhood traumatic events were assessed with the Childhood Trauma Questionnaire-Short Form (CTQ-SF) [28]. The CTQ-SF (referred to simply as the CTQ) consists of five subscales with five questions each (range: 1–5): emotional abuse, physical abuse, sexual abuse, emotional neglect, and physical neglect as well as a total score (CTQ$_{Total}$). For this study, cut-off scores of a minimum of 8 on the physical abuse, physical neglect and sexual abuse subscales, 10 on the emotional abuse subscale, and 15 on the emotional neglect subscale were used. These thresholds are linked with moderate-to-severe levels of abuse and neglect [29]. MDD patients were divided into two groups based on early exposure to physical or sexual abuse: group 1 (abuse group): sexual and/or physical abuse (N = 12); group 2 (nonabuse group): no sexual and/or physical abuse (but presence of emotional neglect/abuse and/or physical neglect) (N = 19). Most patients who had a history of physical or sexual abuse also experienced some form of emotional maltreatment. Five MDD subjects did not complete the CTQ and were not included in the analyses pertaining to trauma.

Eighteen healthy controls (HCs) without any psychiatric history were also tested; HCs were not included in the cortical thickness analyses regarding trauma history. Informed consent was obtained prior to study initiation in compliance with the Conjoint Health Research Ethics Board at the University of Calgary. Participant characteristics are presented in Tables 1 and 2.

5.2.2 MAGNETIC RESONANCE IMAGING (MRI): 3D IMAGE ACQUISITION

Images were collected at the Seaman Family MR Centre (Foothills Hospital, University of Calgary) with a 3 T General Electric scanner (Signa LX, Waukesha, WI, USA) using a receive-only eight-channel RF head coil. A 3D T1-weighted magnetization prepared rapid acquisition gradient echo (MPRAGE) image was acquired (TR = 8.3 ms; TE = 1.8 ms; flip angle = 20°; voxel size = 0.5 × 0.5 × 1 mm; 1 mm slice thickness; 176 slices).

5.2.3 CORTICAL THICKNESS ANALYSES

Cortical thickness analyses were carried out using FreeSurfer software (http://surfer.nmr.mgh.harvard.edu/). Detailed procedures on cortical thickness analyses using FreeSurfer have been published [30–33]. In brief, T1-weighted images were intensity-normalized (correcting for magnetic field inconsistencies) and then a skull-stripping procedure was applied to remove extracerebral voxels. A researcher (F.M.-blind to identity/diagnoses) then carried out manual edits to the skull-stripped images. Scans subsequently underwent a segmentation procedure using an estimation of the structure of the grey-white interface. In order to create a smooth spherical representation of the grey-white interface and pial surface, each scan was covered with a triangular tessellation and inflated. Inflated scans were then aligned to FreeSurfer's default reference template via a 2D warp based on cortical folding patterns. Once smoothed using a circularly symmetric Gaussian kernel, sulci and gyri curvature patterns were aligned and the average cortical thickness was measured at each surface point. A uniform surface-based spherical coordinate system was created by transforming the reconstructed surfaces into parameterizable surfaces. An averaging procedure (50 iterations) was applied to smooth the surface and the reconstructed pial surface refined with a deformable surface algorithm. Data was again aligned on a common spherical coordinate system. Cortical thickness was determined by measuring and averaging the distance between the grey-white and pial surfaces [30–33].

5.2.4 STATISTICAL ANALYSES

Groups were compared on demographic and clinical indices using one-way analyses of variance (ANOVAs). These analyses were first carried out between the MDD versus HC groups; subsequently, assessments were conducted with three levels (MDD, pediatric onset, adult onset) comprising the group variable. Clinical and demographic features were also compared with one-way ANOVAs between the pediatric and adult onset groups (Table 1).

A multivariate ANOVA (MANOVA) was carried out to assess cortical thickness differences across regions (see Table 1 in Supplementary Material available online at http://dx.doi.org/10.1155/2014/410472) between the MDD and HC groups. The MANOVA was followed by exploratory repeated-measures ANOVAs (rmANOVA; hemisphere as the within- and group (MDD, HC) as the between-subject factor) for each of regional cortical thickness measures (significance set at $P < 0.01$).

Subsequently, a multivariate analysis of covariance (MANCOVA) was carried out to assess cortical thickness measure differences across regions between the three groups (HC, pediatric onset, and adult onset); age was used as a covariate since it differed in the HC versus the adult and pediatric onset groups (Section 3.2). The MANCOVA was followed by exploratory rmANCOVAs (age as covariate; hemisphere as the within-subject factor; group (HC, pediatric onset, and adult onset) as the between-subject factor) assessing thickness in each cortical region; significance was set at $P < 0.01$.

One-way ANOVAs were carried out to compare the MDD groups with childhood abuse + neglect versus nonabuse + neglect (i.e., abuse and non-abuse groups, resp.) on pertinent demographic and clinical variables. A MANCOVA was carried out to assess cortical thickness measure differences across regions between the two groups (nonabuse, abuse); $HAMD_{17}$ scores were used as a covariate as they differed between the abuse and nonabuse groups (Section 3.3). This was followed by exploratory rmANCOVAs ($HAMD_{17}$ as covariate; hemisphere as the within-subject factor; group (abuse, non-abuse) as the between-subject factor) assessing thickness in each cortical region; significance was set at $P < 0.01$.

MANCOVAs ($HAMD_{17}$ scores and age as covariates) were carried out with the two MDD onset (adult, pediatric) and two trauma groups (abuse, nonabuse) as independent variables on cortical thickness measures across regions. Exploratory rmANCOVAs (HAMD$_{17}$ scores and age as covariates; hemisphere as within and groups as between-subject factors) were then carried out for thickness in each cortical region; significance was set at $P < 0.01$.

Finally, exploratory Spearman's correlations were carried out (for the MDD group) between CTQTotal scores, abuse scores (physical + sexual abuse CTQ scores) and neglect scores (emotional abuse + emotional neglect + physical neglect CTQ scores), and all regional cortical thickness measures; significance was set at $P < 0.005$. Similarly, correlations were carried out between HAMD$_{17}$ and all regional cortical thickness measures (MDD group only); significance was set at $P < 0.005$. Unless stated otherwise, means and standard deviations (SDs) are presented for all results. All cortical thickness measures are expressed as mm.

5.3 RESULTS

5.3.1 TWO GROUP ANALYSES (HC AND MDD)

One-way ANOVAs (group (MDD and HCs) as the between-subject factor) revealed no main effect of group on age (Table 1).

The MANOVA (group (HC, MDD) as fixed factor and regions as dependent variables) yielded no main effects of group on cortical thickness. In an effort to replicate previous research, the MANOVA was followed up by exploratory rmANOVA, with hemisphere as the within- and group (MDD, HC) as between-subject factors on cortical thickness (per region). Significance was set at $P < 0.01$ to minimize false positives and control for multiple comparisons. The main effects of hemisphere, as found by the rmANOVAs, on cortical thickness in various brain regions are listed in Table 3. The rmANOVA revealed a main effect of group (MDD, HC) on frontal pole thickness ($F[1,52] = 7.05$, $P = 0.01$), with a thicker cortex in the MDD ($3.19 \pm .38$) versus the HC group ($2.95 \pm .30$; Figure 1). A trend for a main effect of group was noted on transverse temporal thickness ($F[1,52] = 6.49$, $P = 0.014$), with a thinner cortex in the MDD ($3.12 \pm .17$) versus the HC group ($3.22 \pm .15$).

TABLE 3: Cortical thickness hemispheric differences.

Region (cortical thickness)	Hemisphere effect	value
Caudal middle frontal cortex	L > R	.003
Entorhinal cortex	L > R	.001
Fusiform cortex	L > R	<.001
Inferior parietal cortex	L > R	.005
Inferior temporal cortex	L > R	<.001
Isthmus cingulate cortex	L > R	<.001
Lingual cortex	R > L	.005
Pars orbitalis cortex	R > L	<.001
Pericalcarine cortex	L > R	.006
Precentral cortex	L > R	.003
Precuneus cortex	R > L	.001
Rostral anterior cingulate cortex	R > L	<.001
Rostral middle frontal cortex	R > L	<.001
Superior frontal cortex	L > R	<.001
Superior parietal cortex	R > L	.001
Superior temporal cortex	L > R	<.001

L: left hemisphere; R: right hemisphere.

5.3.2 THREE GROUP ANALYSES (ADULT MDD ONSET, PEDIATRIC MDD ONSET, AND HCS) AND TWO GROUP ANALYSES (ADULT AND PEDIATRIC MDD ONSET)

One-way ANOVAs were carried out with group as the independent variable (3 groups: pediatric onset: onset <24 yrs; adult onset: onset >25 yrs; HCs) and age as the dependent variable. A main effect of group existed (F[2,51] = 9.97, P < 0.001); follow-up comparisons indicated a difference between the adult MDD onset and both the HC (P < 0.001) and pediatric MDD onset groups (P < 0.001), with the adult onset group being the oldest (Table 1).

Further one-way ANOVAs were carried out between the pediatric versus adult MDD onset groups on other pertinent variables (i.e., HAMD$_{17}$ scores, duration of current MDE, and time since diagnoses). A main effect of group was noted for $HAMD_{17}$ scores ($F[1,34] = 6.50$, $P = 0.015$), with higher scores in the adult verses the pediatric MDD onset group.

The MANCOVA (age as a covariate) yielded no main effect of group (3 groups: HC, adult onset, and pediatric onset) on cortical thickness. However, given the pilot nature of this work, the MANCOVA was followed up with exploratory rmANCOVAs (age as a covariate; hemisphere as the within- and group as the between-subject factor) assessing thickness in each cortical region. Significance was set at $P < 0.01$. A main effect of hemisphere was noted on cortical thickness in the rostral middle frontal cortex ($F[1,50] = 7.38$, $P = 0.009$; right > left). A trend for a main effect of group (3 groups) on frontal pole cortex thickness was noted ($F[2,50] = 4.64$, $P = 0.014$), with a thinner cortex in the HC group ($2.93 \pm .30$) versus the pediatric MDD onset group ($3.22 \pm .39$; $P = 0.005$).

5.3.3 TWO GROUP ANALYSES (MDD GROUPS: CHILDHOOD ABUSE GROUP AND NONABUSE GROUP)

One-way ANOVAs were conducted to compare the childhood abuse (N = 12) versus non-abuse (N = 19) groups on pertinent demographic and clinical variables (i.e., time since MDD diagnosis, $HAMD_{17}$ scores, current age, age of MDD onset, and duration of current MDE). A main effect of group (neglect, abuse) was found on $HAMD_{17}$ scores ($F[1,29] = 4.21$, $P = 0.049$), with higher scores in the abuse group. The abuse group also had higher CTQTotal ($F[1,29] = 27.38$, $P < 0.001$) and, expectedly, abuse scores ($F[1,29] = 42.04$, $P < 0.001$) than the nonabuse group (Table 2).

The MANCOVA, with group (nonabuse, abuse) as the independent variable, was carried out on cortical thickness measures ($HAMD_{17}$ scores were the covariate)—no main group effect on cortical thickness existed. Exploratory rmANOVAs (group as between- and hemisphere as within-subject factors, $HAMD_{17}$ as the covariate) yielded no significant results, apart from a weak trend for a main effect of group ($F[1,28] = 3.26$, $P = 0.082$) on frontal pole cortical thickness. This trend was followed up with univariate ANOVAs assessing frontal pole thickness in each hemisphere ($HAMD_{17}$ as covariate).

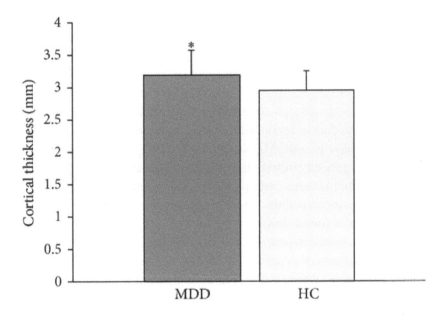

FIGURE 1: The group with major depressive disorder (MDD) had a thicker frontal pole compared with the healthy control (HC) group (*P = .01).

A trend for main effect of group on right frontal pole cortex thickness ($F[1,28] = 4.20$, $P = 0.05$) was found, with a thicker cortex in the abuse ($3.36 \pm .37$) versus the nonabuse group ($3.10 \pm .25$).

An inverse correlation was found between left precuneus cortex thickness ($r = -0.57$, $P < 0.001$, $N = 31$) as well as right middle temporal cortex thickness ($r = -0.59$, $P < 0.001$, $N = 31$) and CTQ_{Total} scores. Similarly, an inverse correlation existed between both left ($r = -0.51$, $P = 0.003$, $N = 31$) and right ($r = -0.54$, $P = 0.002$, $N = 31$) frontal pole cortex thickness and CTQ_{Total} scores. An inverse correlation also existed between right frontal pole cortex thickness and "abuse" scores ($r = -0.50$, $P = 0.004$, $N = 31$). An inverse relation existed between left inferior parietal cortex thickness (-0.59, $P < 0.001$, $N = 31$) as well as right superior parietal cortex thickness (-0.53, $P = 0.002$, $N = 31$) and "neglect" scores.

Finally, Chi-square tests revealed no significant difference in the proportion of the pediatric versus adult MDD onset individuals in either the abuse or nonabuse groups. MANCOVAs ($HAMD_{17}$ scores and age as covariates) were carried out with the two MDD onset (adult and pediatric) and two childhood trauma groups (abuse and nonabuse) as the independent variables on cortical thickness measures. No main group effects or interactions were found. Exploratory rmANOVAs ($HAMD_{17}$ scores and age as covariates; hemisphere as with- and groups as between-subject factors; significance was set at $P < 0.01$) yielded a trend for an onset group × childhood trauma group interaction for superior temporal cortex thickness ($F[1,25] = 5.98$, $P = 0.022$), with pairwise comparisons indicating a trend for a difference in cortical thickness between the pediatric ($N = 8$; $2.59 \pm .14$) and adult ($N = 8$; $2.77 \pm .14$) MDD onset groups with childhood abuse ($P = 0.02$). For frontal pole thickness, an onset group × childhood trauma group × hemisphere interaction trend existed ($F[1,25] = 5.07$, $P = 0.033$). Pairwise comparisons indicated a trend for a difference ($P = 0.026$) in right frontal pole cortical thickness between the abuse ($N = 8$; $3.01 \pm .15$) and nonabuse ($N = 8$; $3.42 \pm .41$) groups in the adult MDD onset cohort.

5.4 DISCUSSION

In brief, this pilot study aimed to contribute to existing literature on cortical thickness in depressed adults in two ways. First, we sought to clarify the effect of age of MDD onset on spatial cortical thickness patterns. Second, we investigated the role of childhood trauma, in the form of abuse or nonabuse history (though both groups experienced neglect), on cortical thickness in MDD and its interaction with age of disorder onset. We found thicker frontal pole cortices in the MDD versus HC group. Conversely, a tendency for a thinner transverse temporal cortex existed in MDD. With respect to age of onset, clinically, the adult versus pediatric onset group exhibited higher $HAMD_{17}$ scores. The pediatric onset group had a thicker frontal pole cortex than HCs. In comparisons of MDD groups with childhood abuse versus nonabuse history (the abused group also exhibited neglect and had higher CTQ_{Total} scores), the abuse group had greater $HAMD_{17}$ scores. A tendency for a thicker cortex was noted in the abuse versus nonabuse group in the right frontal pole. Inverse correlations existed between the left precuneus, right middle temporal as well as bilateral frontal pole cortical thickness, and CTQ_{Total} scores. Inverse relations were also noted between right frontal pole cortex thickness and CTQ abuse scores as well as between the left inferior and right superior parietal cortex thickness and CTQ neglect scores. Finally, the superior temporal cortex tended to be thinner in the pediatric versus adult onset groups with childhood abuse. Additionally, the right frontal pole cortex tended to be thinner in the abuse versus nonabuse groups in the adult onset group.

The role of the frontal poles in MDD (and outside the context of the disorder) is not well understood. Neuroimaging studies suggest that frontal poles play a role in "cognitive branching" (i.e., flexibility) as they are activated when performing several subgoals while keeping in mind another (main) goal. Though a handful of functional MRI (fMRI) studies have implicated frontal pole activity in response to antidepressant interventions [34, 35], few morphometric studies of the frontal poles in MDD exist. Much of the work linking the frontal poles with depression stems from stroke research, where greater depression severity has been associated with increased lesion proximity to the frontal poles [36].

Unlike Sheline et al. [7], we noted thicker frontal pole cortices in MDD versus HCs. However, since their sample consisted of late-life depressed individuals while ours was comprised of relatively young-to-middle aged adults, the results may not be directly comparable. They also found that thinner frontal pole cortices existed in patients who did not achieve remission compared to those who did. Given that greater frontal pole cortical thickness in our study was driven by the pediatric onset group, it is feasible that these individuals may have been more likely to be treatment responders (versus the adult onset group). However, as response was not assessed in the current study (though this represents a worthy future direction), this interpretation is speculative. Further, because the pediatric onset group was characterized by lower $HAMD_{17}$ scores than the adult onset group, thicker frontal pole cortices may reflect a neurocompensatory/adaptive mechanism in the disorder. Neurocompensatory mechanisms are more likely during adolescence, which is a period associated with extensive brain plasticity [14–16]. Additionally, the right frontal pole cortex tended to be thinner in the abuse versus nonabuse groups in the adult MDD onset group suggesting that more pronounced trauma might make the brain susceptible to the neural consequences associated with a psychiatric condition in adulthood.

Few studies have assessed (or reported on) the significance of the transverse temporal cortex (Heschl's gyrus) in MDD. One fMRI study found greater right Heschl's gyrus activation during an emotive processing/attention control task in individuals with a family history of depression versus those without a family history [37], suggesting that the region may play some role in MDD. However, postmortem examinations yielded no differences in neural or glial cell density or cortical thickness in the Heschl's gyrus between individuals with MDD and HCs [38]. Another group found volumetric reductions in the superior temporal gyrus (not Heschl's gyrus specifically) in recovered depressed participants [39], which is somewhat consistent with our observed trend for a thinner transverse temporal cortex in MDD. We also noted a tendency for a thinner superior temporal cortex in the pediatric versus adult onset groups with childhood abuse suggesting that abuse during critical neurodevelopmental periods may influence cytoarchitecture within this region.

Childhood maltreatment is strongly associated with increased risk for psychiatric disorder development [22]. By extension, neural abnormalities associated with trauma/maltreatment may increase psychiatric disorder vulnerability. Previous work has reported reduced cortical thickness and volume in maltreated versus nonmaltreated children in the anterior cingulate, superior frontal gyrus, and orbitofrontal cortex [19, 25]. Similarly, childhood emotional maltreatment and physical neglect were associated with reductions in medial prefrontal cortex volumes in adults [21, 22]. Nondepressed subjects with a family history of MDD and a history of emotional abuse exhibited smaller DLPFC, medial prefrontal, and anterior cingulate cortices than controls [23]. Yet another group found that decreased cingulate volume in individuals with MDD was related to abuse history [5]. Finally, Dannlowski et al. [24] reported reduced grey matter volumes in regions including the orbitofrontal cortex and anterior cingulate gyrus in adults with high CTQ scores. The above indicates that prefrontal, anterior cingulate cortex, and lateral temporal regions (areas implicated in MDD) tend to be rather consistently affected by maltreatment/ trauma. Further, research suggests that maltreatment/trauma is associated with structural damage that persists into adulthood. These results mimic our findings of an inverse relation between cortical thickness in the frontal poles, precuneus, and middle temporal regions and CTQ_{Total} scores as well as inverse relations between abuse scores and left inferior and right superior parietal cortical thickness.

The primary limitation of this study was its exploratory nature as well as the small sample size, specifically when groups were split by abuse/ nonabuse history. Assessments of interactions between age of MDD onset and childhood trauma (i.e., 2×2 group comparisons) on cortical thickness—though highly novel—were statistically underpowered. Further, in an effort to correct for multiple comparisons and decrease false positive rates, we included covariates when appropriate; inclusion of covariates further decreases power. Due to these limitations, it was not feasible to meaningfully explore the influence of sex on cortical thickness in this study, which may have been informative. In a similar vein, although we attempted to correct for multiple comparisons by adjusting our significance level, true corrections (e.g., Bonferroni) were not applied, though this should be done in comparable future work. As such, our findings and conclusions should be treated as preliminary and with caution, warranting further replication and expansion with a larger sample size.

Briefly, the focal future direction of this work is to assess cortical thickness in a large sample of well-characterized depressed individuals in terms of their trauma history and MDD onset age in order to disambiguate the contributions of these factors in influencing cortical cytoarchitecture in MDD. Greater clarity is needed to better understand the multiple, likely interacting, factors that contribute to altered cortical thickness patterns in MDD. Assessments of such well-characterized samples over time (versus cross-sectionally) would also allow us to gain better insight regarding the neurodevelopmental processes across the lifespan in the context of depression.

REFERENCES

1. W. Truong, L. Minuzzi, C. N. Soares et al., "Changes in cortical thickness across the lifespan in major depressive disorder," Psychiatry Research, vol. 214, no. 3, pp. 204–211, 2013.
2. P. C. M. P. Koolschijn, N. E. M. van Haren, H. G. Schnack, J. Janssen, H. E. Hulshoff Pol, and R. S. Kahn, "Cortical thickness and voxel-based morphometry in depressed elderly," European Neuropsychopharmacology, vol. 20, no. 6, pp. 398–404, 2010.
3. S. J. Colloby, M. J. Firbank, A. Vasudev, S. W. Parry, A. J. Thomas, and J. T. O'Brien, "Cortical thickness and VBM-DARTEL in late-life depression," Journal of Affective Disorders, vol. 133, no. 1-2, pp. 158–164, 2011.
4. A. Kumar, O. Ajilore, A. Zhang, D. Pham, and V. Elderkin-Thompson, "Cortical thinning in patients with late-life minor depression," The American Journal of Geriatric Psychiatry, 2013.
5. R. S. Mackin, D. Tosun, S. G. Mueller et al., "Patterns of reduced cortical thickness in late-life depression and relationship to psychotherapeutic response," The American Journal of Geriatric Psychiatry, vol. 21, pp. 794–802, 2013.
6. H. K. Lim, W. S. Jung, K. J. Ahn et al., "Regional cortical thickness and subcortical volume changes are associated with cognitive impairments in the drug-naive patients with late-onset depression," Neuropsychopharmacology, vol. 37, no. 3, pp. 838–849, 2012.
7. Y. I. Sheline, B. M. Disabato, J. Hranilovich et al., "Treatment course with antidepressant therapy in late-life depression," The American Journal of Psychiatry, vol. 169, pp. 1185–1193, 2012.

8. H. Järnum, S. F. Eskildsen, E. G. Steffensen et al., "Longitudinal MRI study of cortical thickness, perfusion, and metabolite levels in major depressive disorder," Acta Psychiatrica Scandinavica, vol. 124, no. 6, pp. 435–446, 2011.

9. G. Wagner, C. C. Schultz, K. Koch, C. Schachtzabel, H. Sauer, and R. G. Schlosser, "Prefrontal cortical thickness in depressed patients with high-risk for suicidal behavior," Journal of Psychiatric Research, vol. 46, pp. 1449–1455, 2012.

10. P.-C. Tu, L.-F. Chen, J.-C. Hsieh, Y.-M. Bai, C.-T. Li, and T.-P. Su, "Regional cortical thinning in patients with major depressive disorder: a surface-based morphometry study," Psychiatry Research, vol. 202, no. 3, pp. 206–213, 2012.

11. P. van Eijndhoven, G. van Wingen, M. Katzenbauer et al., "Paralimbic cortical thickness in first-episode depression: evidence for trait-related differences in mood regulation," The American Journal of Psychiatry, vol. 170, no. 12, pp. 1477–1486, 2013.

12. E. Fallucca, F. P. MacMaster, J. Haddad et al., "Distinguishing between major depressive disorder and obsessive-compulsive disorder in children by measuring regional cortical thickness," Archives of General Psychiatry, vol. 68, no. 5, pp. 527–533, 2011.

13. S. C. N. Reynolds, N. Jaworska, L. M. Langevin, and F. P. MacMaster, Cortical Thickness in Youth with Major Depressive Disorder, 2013.

14. J. Kaufman, A. Martin, R. A. King, and D. Charney, "Are child-, adolescent-, and adult-onset depression one and the same disorder?" Biological Psychiatry, vol. 49, no. 12, pp. 980–1001, 2001.

15. T. Paus, "Structural maturation of neural pathways in children and adolescents: in vivo study," Science, vol. 283, no. 5409, pp. 1908–1911, 1999.

16. M. Arain, M. Haque, L. Johal et al., "Maturation of the adolescent brain," Neuropsychiatric Disease and Treatment, vol. 9, pp. 449–461, 2013.

17. B. M. Disabato, C. Morris, J. Hranilovich et al., "Comparison of brain structural variables, neuropsychological factors, and treatment outcome in early-onset versus late-onset late-life depression," The American Journal of Geriatric Psychiatry, 2013.

18. K. M. Scott, D. R. Smith, and P. M. Ellis, "Prospectively ascertained child maltreatment and its association with DSM-IV mental disorders in young adults," Archives of General Psychiatry, vol. 67, no. 7, pp. 712–719, 2010.

19. P. A. Kelly, E. Viding, G. L. Wallace et al., "Cortical thickness, surface area, and gyrification abnormalities in children exposed to maltreatment: neural markers of vulnerability?" Biological Psychiatry, vol. 74, no. 11, pp. 845–852, 2013.

20. C. M. Heim, H. S. Mayberg, T. Mletzko, C. B. Nemeroff, and J. C. Pruessner, "Decreased cortical representation of genital somatosensory field after childhood sexual abuse," The American Journal of Psychiatry, vol. 170, pp. 616–623, 2013.

21. A.-L. van Harmelen, M.-J. van Tol, N. J. A. van der Wee et al., "Reduced medial prefrontal cortex volume in adults reporting childhood emotional maltreatment," Biological Psychiatry, vol. 68, no. 9, pp. 832–838, 2010.

22. T. Frodl, E. Reinhold, N. Koutsouleris, M. Reiser, and E. M. Meisenzahl, "Interaction of childhood stress with hippocampus and prefrontal cortex volume reduction in major depression," Journal of Psychiatric Research, vol. 44, no. 13, pp. 799–807, 2010.

23. A. Carballedo, D. Lisiecka, A. Fagan et al., "Early life adversity is associated with brain changes in subjects at family risk for depression," The World Journal of Biological Psychiatry, vol. 13, pp. 569–578, 2012.

24. U. Dannlowski, A. Stuhrmann, V. Beutelmann et al., "Limbic scars: long-term consequences of childhood maltreatment revealed by functional and structural magnetic resonance imaging," Biological Psychiatry, vol. 71, no. 4, pp. 286–293, 2012.

25. S. A. de Brito, E. Viding, C. L. Sebastian et al., "Reduced orbitofrontal and temporal grey matter in a community sample of maltreated children," Journal of Child Psychology and Psychiatry, vol. 54, pp. 105–112, 2013.

26. N. V. Malykhin, R. Carter, K. M. Hegadoren, P. Seres, and N. J. Coupland, "Fronto-limbic volumetric changes in major depressive disorder," Journal of Affective Disorders, vol. 136, no. 3, pp. 1104–1113, 2012.

27. M. Hamilton, "Development of a rating scale for primary depressive illness," The British Journal of Social and Clinical Psychology, vol. 6, no. 4, pp. 278–296, 1967.

28. D. P. Bernstein, J. A. Stein, M. D. Newcomb et al., "Development and validation of a brief screening version of the Childhood Trauma Questionnaire," Child Abuse and Neglect, vol. 27, no. 2, pp. 169–190, 2003.

29. M. M. Grant, C. Cannistraci, S. D. Hollon, J. Gore, and R. Shelton, "Childhood trauma history differentiates amygdala response to sad faces within MDD," Journal of Psychiatric Research, vol. 45, no. 7, pp. 886–895, 2011.

30. A. M. Dale, B. Fischl, and M. I. Sereno, "Cortical surface-based analysis: I. Segmentation and surface reconstruction," NeuroImage, vol. 9, no. 2, pp. 179–194, 1999.

31. B. Fischl and A. M. Dale, "Measuring the thickness of the human cerebral cortex from magnetic resonance images," Proceedings of the National Academy of Sciences of the United States of America, vol. 97, no. 20, pp. 11050–11055, 2000.

32. B. Fischl, M. I. Sereno, and A. M. Dale, "Cortical surface-based analysis: II. Inflation, flattening, and a surface-based coordinate system," NeuroImage, vol. 9, no. 2, pp. 195–207, 1999.

33. B. Fischl, M. I. Sereno, R. B. Tootell, and A. M. Dale, "High-resolution intersubject averaging and a coordinate system for the cortical surface," Human Brain Mapping, vol. 8, pp. 272–284, 1999.

34. G. S. Dichter, J. N. Felder, and M. J. Smoski, "The effects of brief behavioral activation therapy for depression on cognitive control in affective contexts: an fMRI investigation," Journal of Affective Disorders, vol. 126, no. 1-2, pp. 236–244, 2010.

35. J. Wu, M. S. Buchsbaum, J. C. Gillin et al., "Prediction of antidepressant effects of sleep deprivation by metabolic rates in the ventral anterior cingulate and medial prefrontal cortex," American Journal of Psychiatry, vol. 156, no. 8, pp. 1149–1158, 1999.

36. K. Narushima, J. T. Kosier, and R. G. Robinson, "A reappraisal of poststroke depression, intra- and inter-hemispheric lesion location using meta-analysis," Journal of Neuropsychiatry and Clinical Neurosciences, vol. 15, no. 4, pp. 422–430, 2003.

37. F. Amico, A. Carballedo, D. Lisiecka, A. J. Fagan, G. Boyle, and T. Frodl, "Functional anomalies in healthy individuals with a first degree family history of major depressive disorder," Biology of Mood & Anxiety Disorders, vol. 2, article 1, 2012.

38. D. Cotter, D. Mackay, S. Frangou, L. Hudson, and S. Landau, "Cell density and cortical thickness in Heschl's gyrus in schizophrenia, major depression and bipolar disorder," British Journal of Psychiatry, vol. 185, pp. 258–259, 2004.

39. T. Takahashi, M. Yücel, V. Lorenzetti et al., "An MRI study of the superior temporal subregions in patients with current and past major depression," Progress in Neuro-Psychopharmacology and Biological Psychiatry, vol. 34, no. 1, pp. 98–103, 2010.

CHAPTER 6

CHILDHOOD EMOTIONAL MALTREATMENT SEVERITY IS ASSOCIATED WITH DORSAL MEDIAL PREFRONTAL CORTEX RESPONSIVITY TO SOCIAL EXCLUSION IN YOUNG ADULTS

ANNE-LAURA VAN HARMELEN, KIRSTEN HAUBER, BREGTJE GUNTHER MOOR, PHILIP SPINHOVEN, ALBERT E. BOON, EVELINE A. CRONE, AND BERNET M. ELZINGA

6.1 INTRODUCTION

Chronic parental rejection can be considered a core aspect of Childhood Emotional Maltreatment (CEM; emotional abuse and/or emotional neglect) [1]. For instance, during episodes of CEM, children may be ignored, isolated, or siblings may be favored. CEM has severe and persistent adverse effects on behavior and emotion in adulthood [2], and CEM is a potent predictor of depressive and anxiety disorders in later life [3], [4]. Social rejection, ranging from active isolation to ignoring basic emotional needs, may enhance sensitivity towards future rejection [5]. Along these

Childhood Emotional Maltreatment Severity Is Associated with Dorsal Medial Prefrontal Cortex Responsivity to Social Exclusion in Young Adults. © van Harmelen A-L, Hauber K, Moor BG, Spinhoven P, Boon AE, Crone EA, and Elzinga BM. PLoS ONE, 9,1 (2014), doi:10.1371/journal.pone.0085107. Licensed under Creative Commons Attribution 4.0 International License, http://creativecommons.org/licenses/by/4.0/.

lines, individuals reporting CEM may be especially sensitive to (perceived) social rejection. Individuals high in rejection sensitivity have a tendency to expect, perceive, and overreact to social rejection, and show enhanced distress and related neural responses to social rejection in the lab [5]. Furthermore, rejection sensitivity (both behaviourally and in terms of brain responses) is positively related to the development and maintenance of depression, social anxiety, and borderline personality disorder symptoms [6], [7]. Therefore, enhanced distress and neural responses to (perceived) social rejection may be one of the mechanisms through which a history of CEM may predispose individuals to the development of depressive and anxiety disorders in later life. However, the subjective and neural responses to social rejection in individuals reporting CEM are currently unknown.

Social rejection in the lab has been examined most frequently with the Cyberball task [8], [9]. During an fMRI compatible variation of the Cyberball task, participants play two games of virtual toss with two other players (computer controlled confederates). In the first (inclusion) game, participants are thrown the ball an equal number of throws as compared to the other players. However, in the second (rejection/exclusion) game they may receive the ball once or twice in the beginning of the game, but thereafter never receive it again. Social exclusion during the Cyberball task induces a cascade of negative emotions, including anxiety, depression, reduced sense of belonging and meaningful existence, and a reduced sense of control, and lowered self-esteem [5], [10]–[13].

Neuroimaging studies have revealed a set of brain regions that are typically activated during social exclusion in the Cyberball task, primarily in cortical midline structures; the anterior cingulate cortex (ACC)/medial prefrontal cortex (mPFC), and Insula [14], [15]. The ACC and mPFC are vital for expectancy-violation, error-detection, the processing of cognitive conflict, and self- and other referential processing [16]–[18]. In line, a recent meta-analysis suggested that activation in these regions during social exclusion might be related with enhanced social uncertainty, social distress, and social rumination [15]. Activation in the dorsal ACC/mPFC and Insula have been related to self-reported distress during exclusion in the Cyberball game, however, not all studies found dorsal ACC/mPFC responsivity to social exclusion [14], [15], [19], [20], or only found it in the first trials of the exclusion game [11]. Furthermore, studies investigating adolescents and children found ventral ACC/mPFC responses to distress during social exclusion [11], [21]–[23]. Increased dorsal ACC/mPFC to exclusion may be dependent on individual differences. As dorsal mPFC activity is especially pronounced in individuals sensitive to interpersonal rejection [24], [25], anxiously attached [26], and/or having low self-esteem [27], [28]. Therefore, dorsal ACC/mPFC responsivity to social rejection may also be evident in individuals with CEM. However, CEM related brain functioning during social exclusion has not yet been examined.

We examined the impact of a history of CEM on brain functioning and emotional distress to social exclusion. We compared young adult patients reporting a moderate to extreme history of CEM (N = 26) with healthy controls (N = 20) reporting low to moderate CEM. We examined whole brain responses while specifying the mPFC, ACC and Insula as regions of interest (ROIs) because of their important role in social exclusion [14], [15]. We hypothesized that individuals reporting a history of CEM would show enhanced brain responses and emotional distress to social exclusion. Therefore, we hypothesized that the severity of CEM would show a dose-response relationship with self-reported distress and brain responsivity.

6.2 METHODS

6.2.1 ETHICS STATEMENT

All participants 18 years of age or older provided written informed consent. For participants that were under 18 years of age at the time of scanning, parental/legal guardian written consent was obtained. This study was conducted according to the principles expressed in the Declaration of Helsinki, and was approved by the Leiden University Medical Center Medical Ethics committee. All participants had uncompromised capacity to consent (i.e. exclusion criteria for our study included difficulty understanding the Dutch language, or a IQ < 80).

6.2.2 SAMPLE

We included a total of 26 out- and inpatients reporting moderate to extreme CEM ('CEM group') who were in treatment at a center for youth specialized mental health care in the Hague, the Netherlands (mean age = 18.31 years, SD = 1.23; 6 males) and 20 healthy controls reporting low to moderate CEM (mean age = 18.85, SD = 1.95; 6 males). The CEM and control groups were matched in terms of age ($F(1,44) = 1.38$, $P = .25$), gender ($X^2(1) = .28$, $P = .74$), and IQ ($F(1,44) = 2.76$, $P = .10$) (see Table 1). In the CEM group, 11 patients reported regular use of anti-depressant and anti-anxiogenic medication (n = 8 used SSRI's, n = 1 used the tricyclic antidepressant (TCA) = amitrypteline, and n = 3 used benzodiazepam).

TABLE 1: Demographics for the Control (n = 20) and CEM (n = 26) groups.

	Controls (n = 20)		CEM				
	Mean	SD	Mean	SD	Chi-Square	F	P
Gender M/F	6/14		6/20		0.281		0.74
IQ	111.5	9.54	107.0	8.76		2.76	0.10
Age	18.85	1.90	18.31	1.23		1.38	0.25
Emotional Abuse	5.2	0.89	11.81	4.20		47.70	0.00
Emotional Neglect	6.85	1.76	17.65	3.60		151.81	0.00
Physical Abuse	5.00	0.00	6.38	2.65		5.41	0.03
Physical Neglect	4.05	0.22	6.77	3.90		9.64	0.00
Sexual Abuse	5.45	1.00	9.15	2.66		34.75	0.00

Patients in the CEM group were excluded when they had a comorbid pervasive developmental disorder or psychosis (as measured with the SCID-I [29]). In addition, current substance abuse was also set as an exclusion criterion. Current substance abuse was measured through random urine samples that are mandatory for individuals admitted at the center.

Fifteen participants from the control group had participated earlier in a study on developmental differences in neural responses during social exclusion [11]). Twenty-six participants who were >15 years of age at the time of scanning in the Gunther Moor et al. study, and who had indicated that they could be approached for future research were contacted. Twenty-one participants agreed to participate and completed the Childhood Trauma Questionnaire (CTQ [30]). Five participants were excluded based on CTQ scores indicating a history of childhood abuse; two reported moderate to severe physical abuse (both scored 12), two reported severe emotional neglect (both scored 19), and one participant reported borderline moderate/severe emotional neglect (14). To further obtain a good match with the CEM group, five control participants were recruited from the general public through an recruitment website, and through adevertisements. All control participants included in this study indicated no history of psychiatric disorder, were not taking any psychotropic drugs and had scores of low-moderate emotional abuse (<12), emotional neglect (<14), and physical neglect (<10), and no physical abuse (<6), and sexual abuse (<6), on the CTQ, according to the cut offs [30] for low severity of abuse: emotional abuse: ≥ 9; emotional neglect: ≥ 10; physical neglect: ≥ 8; physical abuse: ≥ 8; and sexual abuse: ≥ 6.

Finally, exclusion criteria for all participants were left-handedness, or general contra-indications for MRI, such as metal implants, heart arrhythmia, and claustrophobia, difficulty understanding the Dutch language, or a IQ< 80 (all participants completed the WAIS, or if <18 years the WISC intelligence subscales similarities and block design [31], [32]).

6.2.3 ASSESMENT OF PSYCHOPATHOLOGY

In all patients with a history of CEM, DSM-IV axis I (psychiatric disorders) and DSM-IV axis II disorders (personality disorders) were assesed using the Structured Clinical Interview for DSM Disorders (SCID-I & SCID-II [29], [33]; please note that two patients in the CEM group had no SCID-I data). All patients in the CEM group had at least one axis I disorder (18 participants had multiple axis I disorders), and 19 participants had a concurrent axis II personality disorder (see Table 2 for all axis I and II diagnoses). Control participants over the age of 18 at the time of scanning reported no history of neurological or psychiatric disorders.

TABLE 2: Clinical characteristics of the CEM group.

SCID I	Depression	Alcohol abuse	Social phobia	Obesession	Generalized anxiety	PTSD	
# current	16		10	2	1	10	
# lifetime	9	3	4	1		3	
Total	24	3	14	3	1	13	
SCID II*	Avoidant	Dependent	Obsessive	Depressive	Passive Aggressive	Paranoid	Border-line
	11	2	3	10	1	5	7

Note: SCID II data for two participants is missing.

Control participants who were under the age of 18 at the time of scanning were screened for psychiatric disorders using the Child Behavioural Checklist (CBCL [34]) that was filled in by their parents. Control participants were only included in this study if they scored in the normal range of the CBCL (see Achenbach; 34). Control participants over the age of 18 at the time of scanning were screened for DSM-IV axis II personality disorders with the Dutch Questionnaire for Personality Characteristics (VKP [35]; Vragenlijst voor Kenmerken van de Persoonlijkheid). Because the VKP is know to be overly inclusive [35], controls with a score that indicated a 'probable' personality disorder on the VKP (n = 8) were also assessed with a SCID-II interview by a trained clinical psychologist (K.H.). All controls that were followed up with the SCID-II were free from personality disorder diagnoses.

6.2.4 CHILDHOOD EMOTIONAL MALTREATMENT

History of childhood emotional maltreatment was assessed using the Dutch version of the Childhood Trauma Questionnaire (CTQ [30], [36]). In the Dutch version of CTQ, a total of 24 items are scored on a 5-point scale, ranging from 1 = never true to 5 = very often true. The CTQ retrospectively assessed five subtypes of childhood abuse: emotional abuse, sexual abuse, physical abuse, emotional neglect and physical neglect. The CTQ is a sensitive and reliable screening questionnaire with Cronbach's alpha for the CTQ subscales varying between. 63-.91 [37].

In line with the American Professional Society on the Abuse of Children [1] and our previous studies on CEM [38], [39], emotional maltreatment in childhood was defined as a history of emotional neglect and/or emotional abuse before the age of 16 years. In line with the American Professional Society on the Abuse of Children [1] definition of emotional abuse that specifies that emotional abuse consists of parental isolating, intimidating, terrorizing, blaming, belittling, degrading, denying emotional responsibility or otherwise behaviour that is insensitive to the child's developmental needs, or can potentially damage the child emotionally, or psychologically and our previous studies on CEM [38], [39], CEM was defined as a history of emotional neglect and/or emotional abuse before the age of 16 years. In line with the idea that emotional abuse rarely occurs alone [40], in our sample, there was a significant correlation between emotional abuse and emotional neglect scores ($r = .54$, $p<.001$), and only three participants reported emotional abuse in isolation (i.e. they reported emotional abuse that was in the moderate to extreme range (CTQ scores>12) together with only emotional neglect that was in the moderate range (CTQ scores of 11,12, and 13 on emotional neglect). As only 3 individuals reported emotional abuse in isolation we were unable to perform separate analyses for the different emotional maltreatment types.

For the entire sample, overall CEM score was defined as the highest score on the emotional abuse or emotional neglect subscale of the CTQ (e.g., if emotional abuse score was 19, and emotional neglect score was 14, overall CEM score was 19). In our study, Cronbach's alpha for the emotional abuse subscale was 88, for the emotional neglect subscale 94, and for the combined emotional abuse and neglect subscales 83. The CEM group reported significantly higher levels of childhood abuse compared to controls on all subscales of the CTQ (all F's>5.41, P's<.03), see Table 1. Self-reported CEM ranged from low to extreme CEM across participants (see Figure 1). In the control group self-reported severity of CEM ranged from low to moderate, whereas in the CEM group severity of CEM ranged from moderate to extreme [30].

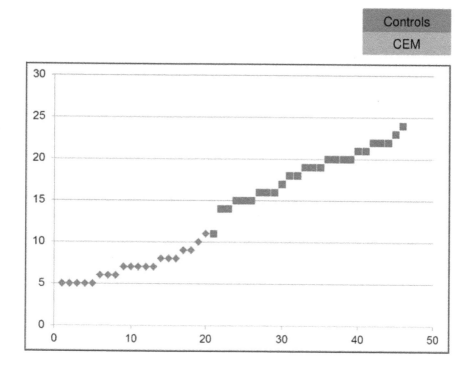

FIGURE 1: Distribution CEM severity across participants.

6.2.5 THE CYBERBALL GAME

In the Cyberball game [8], [9] participants played a game of virtual toss with two other players (computer controlled confederates), depicted using animated avatars. Participants were led to believe that the other players (one female, one male) played the game online on the internet. Fictitious names of the players (common Dutch names, counterbalanced between participants) were displayed on the screen just above their avatars (i.e. in the left and right hand corners of the screen). The participant's self was displayed on the screen as an animated hand, with the participant's name displayed just below the hand. In the Cyberball game, participants first played the inclusion game, followed by the exclusion game. During inclusion, participants threw the ball one-third of the total amount of throws (thus, achieving an equal number of throws as compared to the other players). During social exclusion, they received the ball once in the beginning of the game, but thereafter never received it again. Immediately after inclusion, and after exclusion, participants filled in two questionnaires that assessed their distress during the game (see below for specifics on the questionnaires). All instructions, and questionnaires were presented on the screen, and all instructions were read out loud (through the intercom) by the experimenter. Finally, and before starting the Cyberball game, participants were questioned whether they understood the instructions of the game.

Both Cyberball games consisted of a total of 30 ball tosses, and each game was administered in a separate run that lasted circa 5 minutes. The duration of each ball toss was fixed to 2 seconds. We added a random jitter interval (100–4000 ms.) in order to account for the reaction time of a real player. To further increase credibility of the Cyberball game, both games started with a loading screen that notified that 'the computer is trying to connect with the other players'.

6.2.6 DISTRESS: NEED SATISFACTION AND MOOD RATINGS

To assess distress after inclusion, exclusion, and after scanning (just before the debriefing; 'post scanning'), all participants completed the Need

Threat Scale [41], and a mood questionnaire [42]. The Need Threat Scale consists of eight items that measure self-esteem, belonging, meaningful existence, and control (each was measured with two questions). A high score on this scale indicates that the basic needs are threatened (i.e., low self-esteem, low sense of belonging to others, low sense of meaningful existence, and low sense of control). The mood questionnaire consisted of eight items that (two of each) measured feeling good/bad, relaxed/tense, happy/sad, and friendly/unfriendly. All items on the questionnaires were rated from 1 ('not at all') to 5 ('very much'), and a high score on this questionnaires indicates good mood). To enhance the readability of this paper, we inverted the need threat scores (in the original scale a high need threat score indicated low need threat), which explains the negative need threat scores in Figures 2 and S2.

After inclusion and exclusion, participants were instructed to describe their mood and need threat feelings during the inclusion and exclusion game. At post-scanning, participants were instructed to assess their current mood and need threat feelings.

6.2.7 FMRI DATA ACQUISITION

Upon arrival to the lab, we first familiarized the participants with the scanning environment and sounds, using a mock scanner, and recorded scanner sounds. Actual scanning was performed on a 3.0 Tesla Philips fMRI scanner in the Leiden University Medical Center. To restrict head motion, we inserted foam cushions between the coil and the head. Functional data were acquired using T2*-weighted Echo-Planar Images (EPI) (TR = 2.2 s, TE = 30 ms, slice-matrix = 80×80, slice- thickness = 2.75 mm, slice gap = 0.28 mm, field of view = 220). The two first volumes were discarded to allow for equilibration of T1 saturation effects. After the functional run, high-resolution T2-weighted images and high- resolution T1-weighted anatomical images were obtained.

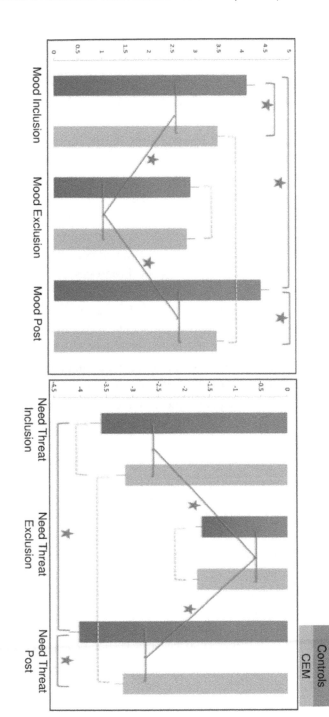

FIGURE 2: Self-reported Mood and Need threat for the Control and CEM groups. Note. Significant differences are indicated with an asterisk, whereas dotted lines depict non-significant differences. A high score on the Mood scale indicates high mood, whereas a high score on the Need Threat Scale indicates high need threat.

6.2.8 FMRI DATA ANALYSIS

Data were analyzed using Statistical Parametric Mapping (SPM8; Well-come Department of Cognitive Neurology, London), version 8, and MATLAB 12.b. Images were corrected for differences in timing of slice acquisition, followed by rigid body motion correction. Preprocessing further included normalization to reorientation of the functional images to the anterior commissure and spatial smoothing with an 8-mm full-width half- maximum Gaussian kernel. The normalization algorithm used a 12-parameter affine transformation together with a nonlinear transformation involving cosine basic functions, and resampled the volumes to 3 mm cubic voxels. Movement parameters never exceeded 1 voxel (<3 mm) in any direction for any subject or scan. Preprocessing of the fMRI time series data used a series of events convolved with a canonical hemodynamic response function (HRF) model. In line with Gunther Moor et al. [11] BOLD responses were distinguished for events on which participants received (inclusion), or did not receive the ball (exclusion). We divided the inclusion game in three conditions; 'receiving ('Ball inclusion game'), not receiving and playing the ball'. During the exclusion game, the first two trials where participants received and played the ball once were not analyzed, and all other throws were set as 'not receiving the ball ('No-ball exclusion game')'.

First level models were assessed using general linear model, with modeled events, and a basic set of cosine functions (to high pass filter the data) as covariates. The least-squares parameter estimates of height of the best-fitting canonical HRF for each condition were used in pair-wise contrasts. For all participants, contrasts between conditions were computed by performing one-tailed t-tests, treating participants as a random effect. To examine the effect of social exclusion and inclusion, for all analyses, we compared brain responses using the t contrast: 'No-ball exclusion game-Ball inclusion game'. This contrast has previously been used Gunter Moor et al [11], where it was associated with activations in regions commonly associated with Cyberball (i.e. Insula, the ACC, and mPFC). This analysis was also performed as a t-sample t-test to examine differences between the CEM group and the control group.

Next, individual differences were added as predictors in regression analyses. First, we examined whether activation in the contrast 'No-ball exclusion game-Ball inclusion game' was associated with the self-report measurements, using whole brain regression analyses with mood, or need threat scoresII after exclusion (i.e. a higher score indicates a better mood, or high needs threat) as regressors of interest.

In order to examine whether the severity of CEM (see Figure 1) was related to activation in the contrast 'No-ball exclusion game-Ball inclusion game', we performed whole brain multiple regression analyses with CEM score as regressor of interest, and physical abuse, physical neglect, and sexual abuse scores as regressors of no interest. We were unable to add diagnosis (yes/no) as regressor of interest in this model, as we only had SCID II data for $n = 7$ controls, and no SCID II data was available for all controls. When we calculated a binary presence vs. absence variable while setting all controls at 0, there was a very high correlation between CEM score and this binary variable ($r = .90$). Therefore, we choose to examine the impact of Axis I and Axis II diagnosis separately within the CEM group (see Text S1), while focussing on those disorders that are known to impact responses to social exclusion (Current Depression, and Borderline Personality Disorder). Activations related to other types of maltreatment (e.g. sexual/physical abuse) during exclusion were examined with a similar whole brain multiple regression analysis, while specifying a specific type of abuse as regressor of interest, and CEM and the other types of abuse as regressors of no interest. There was multicollinearity between CEM, physical neglect, physical abuse and sexual abuse (r's$>.31$, P,.04), however, when we repeated the regression analyses while only specifying CEM as predictor the main effects of CEM on brain activations remained unchanged.

For these analyses, brain activations were first examined at whole brain level with a threshold of $P<.005$ uncorrected, with a spatial extent $K>25$ voxels because this threshold and cluster extent have been suggested to provide a good balance between type 1 and type 2 errors [43]. Because of their presumed role during social exclusion, we then set the entire ACC, mPFC and Insula as Regions of interest (ROIs) (see also [14], [44]). If peak voxel activations fell within these predetermined ROIs, to further protect against Type 1 errors, we also report whether these activations were significant after small volume correction (SVC) for the spatial extent of the activated region (family wise error at the cluster level). For this SVC we used the automatic anatomical labeling (AAL) toolbox within the Wakeforest-pickatlas toolbox [45]. Brain activations where peak voxel activations fell outside our predetermined ROIs were examined at P<.05 FWE corrected at the whole brain level. All brain coordinates are reported in MNI atlas space. For illustration purposes, we extracted cluster activations (for the main effect of task) using the Marsbar region of interest toolbox [46].

6.2.9 BEHAVIORAL ANALYSES

Behavioral responses for the mood and need threat scales were analyzed using Group (CEM, Controls) by measurement moment (Inclusion, Exclusion, Post Scanning) Repeated Measures Analyses of Variances (ANOVAs) in IBM SPSS statistics 19. In addition, the relationship between severity of CEM across participants, and distress (mood and need threat scores) after inclusion, exclusion, and post scanning was assessed using correlational analyses. All analyses were Bonferroni corrected for multiple testing, and significance was set at P<.05 two-sided.

6.3 RESULTS

6.3.1 IMPACT OF SOCIAL EXCLUSION ON SELF-REPORTED MOOD AND NEED THREAT

A Group (CEM, Controls) by measurement moment (Inclusion, Exclusion, Post Scanning) rmANOVA on mood revealed a main effect of measurement moment on mood score (F(2,86) = 67.47, P<.001), and post-hoc t-tests showed that for both groups mood scores significantly decreased from inclusion to exclusion (t's> 5.58, Ps<.001), and significantly increased from exclusion to post scanning (t's<–4.53, P's<.001).

In addition, there was a main effect of group ($F(1,43) = 6.19$, $P = .02$), and there was a significant mood × group interaction ($F(2,86) = 9.52$, $P<.001$). Figure 2 shows that after inclusion, the CEM group reported significantly lower mood scores when compared to controls ($F(1,43) = 6.83$, $P = .012$), however after exclusion, this difference disappeared ($F(1,43) = .09$, $P = .77$). At post scanning, the CEM group again reported lower mood feelings compared to controls ($F(1,43) = 15.54$, $P = <.001$).

A Group (CEM, Controls) by measurement moment (Inclusion, Exclusion, Post Scanning) rmANOVA on need threat revealed a main effect of measurement moment on need threat scores ($F(2,88) = 162.80$, $P<.001$), and post-hoc t-tests indicated that need threat scores significantly increased from inclusion to exclusion in both groups (t's>9.08, P's<.001), and significantly decreased from exclusion to post scanning (t's>−7.80, P's<.001), suggesting that exclusion in the Cyberball task significantly increased threat related feelings across participants. There was a marginal main effect of group ($F(1,44) = 3.80$, $P = .06$), and a significant need threat × group interaction ($F(2,88) = 8.33$, $P<.001$). Post-hoc tests showed that after inclusion, the CEM group reported similar need threat when compared to controls ($F(1,44) = 2.62$, $P = .11$), which remained after exclusion ($F(1,44) = .24$, $P = .62$). However, at post scanning, the CEM group reported increased need threat feelings when compared to controls ($F(1,44) = 9.72$, $P = <.005$), see Figure 2.

6.3.2 RELATIONSHIP BETWEEN SEVERITY OF CEM AND SELF-REPORTED DISTRESS (MOOD AND NEED THREAT)

Across participants, correlation analyses revealed that the severity of the CEM score was negatively related to mood ($r = −.45$, $P <.001$) and positively with feelings of need threat ($r = .29$, $P<.05$) after inclusion. However, after exclusion, no relationships with CEM score and mood, nor need threat were found (r's<−.02, P's>.29). Finally, post scanning, CEM score was again significantly negatively related to mood ($r = −.49$, $P <.001$) and positively with need threat scores ($r = .58$, $P<.001$).

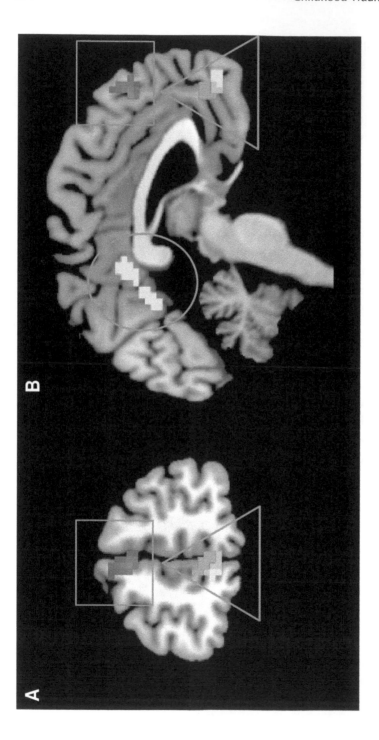

FIGURE 3: Brain responses to social exclusion ('No-ball exclusion game-Ball inclusion game') at y = −51 (A), x = 3 (B). Note. The green blobs depict the posterior cingulate (circle), and ventral mPFC cluster (triangle) that were related to social exclusion ('No-ball exclusion game-Ball inclusion game') across participants. The violet blob (triangle) depicts the ventral mPFC that was activated in response to need threat at exclusion across participants. The red blob depicts the dorsal mPFC cluster that was related to CEM across participants.

TABLE 3: Activations for the 'No-ball exclusion game - Ball inclusion game' contrast at P<.005, K>25.

				peak					ROI
			K	P_{FWE}	T	Z	P	x,y,z (mm)	P_{SVC}
Main effect across participants		Ventral mPFC	44	0.93	3.79	3.51	0.000	−3 57 −12	1.00
				1.00	3.15	2.98	0.001	6 57 −9	
				1.00	2.97	2.82	0.002	−9 45 −9	
		Posterior ACC	61	0.97	3.69	3.43	0.000	0 −36 36	0.09
				0.99	3.52	3.29	0.000	−6 −54 18	
		Inferior frontal gyrus	36	0.98	3.61	3.37	0.000	−42 27 15	
				1.00	3.31	3.11	0.001	−57 24 15	
				1.00	2.98	2.83	0.002	−54 27 6	
Mood exclusion	positive relationship	No signficant clusters							
	negative relationship	Frontal inferior Opperculum	35	1.00	3.31	3.11	0.001	54 9 27	
Need treat exclusion	positive relationship	ventral mPFC	31	0.92	3.81	3.53	0.000	−3 51 −6	ns
	negative relationship	No significant clusters							
CEM vs. Controls	CEM > Controls	Superior frontal gyrus	51	0.78	4.04	3.71	0.000	−24 24 51	
				1.00	2.84	2.70	0.003	−36 15 51	
		Angular gyrus	64	0.99	3.53	3.29	0.000	−51 −69 27	
				1.00	3.09	2.93	0.002	−42 −69 36	
				1.00	2.87	2.74	0.003	−33 −78 42	
	Controls > CEM	No significant clusters							
CEM severity	Negative	Superior Frontal Gyrus	56	0.71	4.15	3.77	0.000	−18 30 51	
		Dorsal Medial PreFrontal cortex	80	0.92	3.85	3.53	0.000	−3 48 33	0.05
				0.98	3.62	3.35	0.000	−12 48 42	
				1.00	2.97	2.81	0.002	6 60 30	

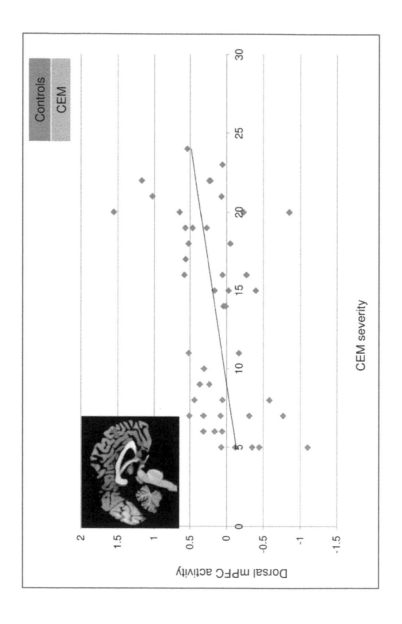

FIGURE 4: Relationship dorsal mPFC and CEM severity.

6.3.3 FMRI ANALYSES; MAIN EFFECT OF EXCLUSION>INCLUSION

Across participants, the contrast 'No-ball exclusion game-Ball inclusion game' resulted in activations in the posterior ACC ($x = 0$, $y = -36$, $z = 36$, $K = 61$, $Z = 3.43$, P<.001, ($P_{SVC} = .09$), and the ventral mPFC ($x = -3$, $y = 57$, $z = -12$, $K = 44$, $Z = 3.51$, P<.001, Figure 3). The activation in posterior ACC marginally survived SVC, but the ventral mPFC area did not survive SVC. All brain regions that were active at the reported threshold (P<.005, K>25) are presented in Table 3. An independent (CEM vs. Controls,) t-test in the same and the reversed contrast revealed no significant group differences.

6.3.4 IMPACT OF CEM SEVERITY ON BRAIN ACTIVATIONS DURING EXCLUSION ACROSS PARTICIPANTS

A whole brain regression analysis across all participants indicated that in the contrast 'No-ball exclusion game-Ball inclusion game' the severity of CEM score had a positive association with dorsal mPFC activation ($x = -3$, $y = 48$, $z = 33$, $K = 80$, $Z = 3.53$, P<.001, (PSVC <.05) (see Figures 3 and Figure 4). Interestingly, both within the control and CEM groups, dorsal mPFC activity in the same cluster was related with CEM severity (see Table S1, Figure S1). There were no significant negative brain activations (see Table 3), nor were there any brain activations related to physical abuse, physical neglect, nor sexual abuse for the contrast 'No-ball exclusion game-Ball inclusion game'.

6.3.5 CORRELATIONAL ANALYSES BETWEEN DISTRESS AND DORSAL MPFC ACTIVATION

Correlational analyses between activations in the dorsal mPFC cluster ($x = -3$, $y = 48$, $z = 33$), and self-reported Need Threat revealed a marginal positive relationships after inclusion ($r = .26$, P = .08), but not after exclusion, nor post measurement (r's<.17, P's>.25). Similar correlational analyses revealed that the dorsal mPFC activation was not related to self-reported mood at any of the measurement moments (r's<–.23, P's>.14).

6.3.6 BRAIN ACTIVATIONS RELATED TO DISTRESS ACROSS PARTICIPANTS

A whole brain regression analysis indicated that need threat scores after exclusion were related to activation in the ventral mPFC contrast 'No-ball exclusion game-Ball inclusion game' (x = –3, y = 51, z = –6, K = 31, P<.001), however, this did not survive SVC (P_{SVC} = 1) (Figure 3). The reversed contrast did not result in any significant differences in brain activation. Additionally, self-reported mood scores after exclusion were not associated with significant brain activations (positively, nor negatively) in the contrast 'No-ball exclusion game-Ball inclusion game'.

6.4 DISCUSSION

We examined whether individuals reporting CEM showed enhanced neural responses and emotional distress to social exclusion. We found a dose-response relationship between the severity of CEM and dorsal mPFC responsivity to social exclusion across participants, both in individuals reporting CEM and healthy Controls. Contrary to our expectations, we did not find differences in neural responses to social exclusion when comparing patients reporting moderate to extreme CEM with Controls reporting low to moderate CEM.

Across participants, we found that social exclusion was associated with increases in posterior ACC and ventral mPFC. Although the ventral mPFC response was not significant after small volume correction, ventral mPFC/ACC responsivity to exclusion is reported by numerous studies in adolescents and children [11], [21], [23], [47]. Interestingly, the ventral mPFC and posterior ACC have been implicated in a model for self-referential processing [48]; the posterior ACC is involved in the integration of autobiographical memory with emotional information about the self [48]. Whereas, the ventral mPFC is assumed to play a role in the more affective components of self-referential processing, through emotional appraisal of self-relevant information and the coupling of emotional and cognitive processing during self-referential processing [48]. In line with the more affective role of the ventral mPFC, we found that increases in self-reported needs threat after social exclusion (i.e. reduced self-esteem, sense of belonging, meaningful existence, and control) were positively related with ventral mPFC responsivity, albeit at sub-threshold level. Taken together, our findings of posterior ACC and ventral mPFC response during social exclusion suggest that social exclusion led to negative self- and other referential processing in our sample.

Social exclusion was related to decreases in mood, and increases in needs threat in our sample, which is in line with the idea of enhanced negative self-referential processing related to social exclusion in our participants. The CEM group reported lower mood after inclusion, and at post measurement, yet after exclusion there was no significant difference between the CEM and Control group. In line, the severity of a history of CEM was negatively related with mood after inclusion; however this relationship disappeared after exclusion. These findings may be due to a floor effect in self-reported mood scores, i.e. participants could only rate their distress on a 1–5 scale, and the CEM group already reported lower mood at inclusion, leaving them little space for further reductions. The CEM group also reported higher needs threat at post-measurement, whereas the need threat scores were not significantly different from the control group during in- or exclusion, even though both groups reported an increase in need threat after exclusion. Apparently, need threat feelings were restored at post measurement in the control group, whereas in the CEM group need threat remained relatively high. These findings suggest that, at least for needs threat, the control group seems to recover quicker in the aftermath of social exclusion compared individuals with CEM. Indeed, the severity of CEM was positively related with needs threat after inclusion and at post-measurement. These findings suggest that the CEM group may show persistent negative self- and other- referential processing at post-measurement level, which was also evident after inclusion, suggesting chronic negative self-referential processing in the CEM group. This is in line with findings of our research group that CEM is associated with more negative self-cognitions [38], and more frequent self and other referential processing (i.e. more intrusions of au-tobiographical interpersonal memories) [39].

We found that the severity of CEM was positively related with dorsal mPFC responsivity to social exclusion. CEM related dorsal mPFC respon-sivity may reflect a further increase in negative self-and other-referential processing in these individuals, since the mPFC is pivotal in self-refer-ential processing [20], [48]–[53]. And a recent meta-analysis suggested that dorsal mPFC responsivity to social exclusion is related with enhanced social uncertainty, social distress, and social rumination [15]. Dorsal mPFC in the self-referential processing model [48] has been suggested to be important for the evaluation and decision making of self-and other referential information (the evaluation whether information is relevant to the self). Therefore, our findings suggest that severity of CEM may be as-sociated with a further increase in negative self-and other referential think-ing during social exclusion. Perhaps individuals reporting CEM perceive social exclusion as especially relevant to themselves. Moreover, negative self- referential processing enhances (negative) bias and recall, resulting in more frequent, and more intense negative experiences, which in its turn enhances the negative self-referential cognitions [54]. This is consistent with the slower recovery in the CEM group, and with our previous find-ings of more negative and more frequent self and other referential process-ing in CEM [38], [39].

The finding of CEM related dorsal mPFC activity is of interest since animal studies utilizing paradigms that closely resemble CEM (e.g. mater-nal isolation/separation or isolation rearing) show that the mPFC is par-ticularly affected by early life emotional stress [55]–[60]. In line, patients and healthy controls reporting CEM show a reduction in dorsal mPFC volume [61]–[63], and dorsal mPFC hypo-activity during higher order cognitive processing [unpublished data]. Therefore, our findings that in-dividuals reporting CEM show enhanced dorsal mPFC responsivity dur-ing interpersonally stressful situations, suggest altered regulation/fluctua-tions of dorsal mPFC activity in individuals reporting CEM. Perhaps these findings resemble attenuation (mPFC hypo-activity) or increases (mPFC hyperactivity) in negative self- and other-referential processing in these individuals. Future studies should examine this.

Dorsal mPFC responsivity to social stress has been found to be predictive of current, and future depressive symptoms in healthy young adolescents aged 12–14 years old [7]. However, in our study we did not find that the CEM related dorsal mPFC responsivity was more prominent in our patient sample, nor was it related to a diagnosis of current depression. Across participants, mPFC responsivity was not related with self-reported mood or needs threat (although mPFC responsivity was only related with needs threat in the CEM group). Thus, our findings of CEM related enhanced mPFC responsivity in individuals with CEM may not be related with current (psychiatric) distress. Rather, these findings are more in line with the idea that increased negative self-and other referential thinking (dorsal mPFC) constitutes a vulnerability or sensitivity factor, that may underlie the emotional and behavioral vulnerabilities that have been reported in these individuals [64], [65]. And, only in interaction with other risk factors such as exposure to more recent adverse events, genetic make-up, or low social support, will this vulnerability eventually lead to psychopathology in later life [66].

The main effects of brain activations related to social exclusion in our sample were relatively weak. This may be related to the fact that we used the contrast 'No-ball exclusion game-Ball inclusion game' in order to calculate brain activations for social exclusion. The CEM group already reported lower mood at inclusion, and we found no reduction in self-reported needs threat, nor mood in the CEM group when compared to Controls after social exclusion. This suggests that social exclusion in our sample predominantly seemed to cause distress in the control group. In addition, because the CEM group already reported relatively low mood after inclusion, the social exclusion appeared to have a relatively little further impact on self-reported distress within the CEM group. In other words, even though the CEM group may be highly sensitive to social exclusion, they may also be chronically stressed. In that sense, additional social stress may therefore not further increase brain activations related to distress during social exclusion in these individuals. Therefore, including the CEM group when examining overall brain responses related to social exclusion ('No-ball exclusion game- Ball inclusion game') in our sample may have led to a reduction in those brain responses. This may also have blurred the overall brain responses to social exclusion.

Finally, contrary to our expectations, we found no group effects on brain activations to social exclusion when comparing the CEM group with healthy Controls. This may be explained by the fact that the CEM group reported moderate to extreme CEM, and the healthy Controls reported low to moderate CEM. Whereas, we found that the severity of CEM showed a positive association with dorsal mPFC responsivity. Therefore, low-moderate CEM in the control group may have reduced our chances of finding group differences, at least in dorsal mPFC responsivity. Moreover, the CEM and Control groups did not show subjective differences in self-reported distress during exclusion, which may have further reduced our chances of finding group differences in brain functioning.

There are some limitations that need to be addressed. First of all, although current Axis I depressive diagnosis, was not related to activations in the dorsal mPFC, we could not disentangle the effect of current depression from that of history of CEM in our analyses due to high multicollinearity. Although, the findings of CEM related dorsal mPFC responses to exclusion were found across participants, and were even apparent in the Control group, suggesting that an Axis I depressive diagnosis might not confound our findings. However, to better disentangle the impact of CEM from the impact of depressive diagnosis on brain functioning during social exclusion, future studies examining patients with depression with and without CEM, and controls with and without a history of CEM are needed.

Second, in our study we assessed CEM retrospectively, and we have to stress the relative subjectivity of self-reported CEM. Furthermore, self-reported CEM may be subject to biased recall, even though a review of studies in both patients and healthy controls showed that CEM is more likely to be under-reported than over-reported [67]. And it should be noted that the test-retest reliability of the CTQ subscales for emotional abuse and emotional neglect has been found satisfactory across different ranges of samples (i.e. college students, psychiatric patients, and convenience samples) [68]. Furthermore, in a large sample of patients and controls, it was found that retrospective recall of CEM was not affected by current mood state [4].

Third, although we assessed whether controls over the age of 18 had a history of psychiatric illnesses, they were not formally screened for DSM-IV axis I disorders. However, we found that DSM-IV axis I Current Depression, which is known to impact brain responses to social exclusion, was not related with activation in the CEM related mPFC cluster during social exclusion. Therefore, it is not very likely that unidentified DSM-IV axis I Current Depression in the control group may have confounded the results.

6.5 CONCLUSIONS

Taken together, we show that severity of CEM is positively related to dorsal mPFC responsivity to social exclusion in both patients with psychiatric disorders and healthy controls. The dorsal mPFC is vital for self and other-referential processing [48], [69]. Together with findings of more negative and more frequent self-referential processing in CEM [38], [39] and slower recovery in terms of need threat after the social exclusion task, our findings suggest increased dorsal mPFC activity during social exclusion may be related to more negative self-and-other-reflective thinking in individuals reporting CEM. Increased negative self-and-other referential thinking (dorsal mPFC) enhances vulnerability to the development of psychiatric disorders [54]. Therefore, our findings may be important in understanding the emotional and behavioral problems that has been reported in these individuals in adulthood [64], [65].

REFERENCES

1. APSAC APS on the A of C (1995) Guidelines for psychosocial evaluation of suspected psychological maltreatment in children and adolescents.
2. Hart H, Rubia K (2012) Neuroimaging of child abuse: a critical review. Front Hum Neurosci 6: 52. Available: http://www.pubmedcentral.nih.gov/articlerender.fcgi?artid=3307045&tool=pmcentrez&rendertype=abstract. Accessed 11 March 2013.
3. Iffland B, Sansen LM, Catani C, Neuner F (2012) Emotional but not physical maltreatment is independently related to psychopathology in subjects with various degrees of social anxiety: a web-based internet survey. BMC Psychiatry 12: 49. Available: http://www.pubmedcentral.nih.gov/articlerender.fcgi?artid=3528417&tool=pmcentrez&rendertype=abstract. Accessed 20 May 2013.
4. Spinhoven P, Elzinga BM, Hovens JGFM, Roelofs K, Zitman FG, et al. (2010) The specificity of childhood adversities and negative life events across the life span to

anxiety and depressive disorders. J Affect Disord 126: 103–112. Available: http://www.ncbi.nlm.nih.gov/pubmed/20304501. Accessed 11 March 2013.

5. DeWall CN, Bushman BJ (2011) Social Acceptance and Rejection: The Sweet and the Bitter. Curr Dir Psychol Sci 20: 256–260. Available: http://cdp.sagepub.com/content/20/4/256.full. Accessed 5 March 2013.

6. Rosenbach C, Renneberg B (2011) Abgelehnt, ausgeschlossen, ignoriert: Die Wahrnehmung sozialer Zurückweisung und psychische Störungen – eine Übersicht. Verhaltenstherapie 21: 87–98. Available: http://www.karger.com/doi/10.1159/000328 839. Accessed 20 May 2013.

7. Masten CL, Eisenberger NI, Borofsky L a, McNealy K, Pfeifer JH, et al. (2011) Subgenual anterior cingulate responses to peer rejection: a marker of adolescents' risk for depression. Dev Psychopathol 23: 283–292. Available: http://www.pubmedcentral. nih.gov/articlerender.fcgi?artid=3229829&tool=pmcentrez&rendertype=abstract. Accessed 19 March 2013.

8. Williams KD, Cheung CK, Choi W (2000) Cyberostracism: effects of being ignored over the Internet. J Pers Soc Psychol 79: 748–762 Available: http://www.ncbi.nlm. nih.gov/pubmed/11079239.

9. Williams KD, Jarvis B (2006) Cyberball: a program for use in research on interpersonal ostracism and acceptance. Behav Res Methods 38: 174–180 Available: http://www.ncbi.nlm.nih.gov/pubmed/16817529.

10. Boyes ME, French DJ (2009) Having a Cyberball: Using a ball-throwing game as an experimental social stressor to examine the relationship between neuroticism and coping. Pers Individ Dif 47: 396–401. Available: http://linkinghub.elsevier.com/retrieve/pii/S0191886909001615. Accessed 27 March 2013.

11. Moor BG, Güroğlu B, Op de Macks Z a, Rombouts S a RB, Van der Molen MW, et al. (2012) Social exclusion and punishment of excluders: neural correlates and developmental trajectories. Neuroimage 59: 708–717. Available: http://www.ncbi. nlm.nih.gov/pubmed/21791248. Accessed 28 February 2013.

12. Themanson JR, Khatcherian SM, Ball AB, Rosen PJ (2012) An event-related examination of neural activity during social interactions. Soc Cogn Affect Neurosci: 1–7. Available: http://www.ncbi.nlm.nih.gov/pubmed/22577169. Accessed 20 May 2013.

13. Zadro L, Williams KD, Richardson R (2004) How low can you go? Ostracism by a computer is sufficient to lower self-reported levels of belonging, control, self-esteem, and meaningful existence. J Exp Soc Psychol 40: 560–567. Available: http://linkinghub.elsevier.com/retrieve/pii/S0022103103001823. Accessed 6 March 2013.

14. Eisenberger NI (2012) The pain of social disconnection: examining the shared neural underpinnings of physical and social pain. Nat Rev Neurosci 13: 421–434. Available: http://www.ncbi.nlm.nih.gov/pubmed/22551663. Accessed 3 March 2013.

15. Cacioppo S, Frum C, Asp E, Weiss RM, Lewis JW, et al. (2013) A Quantitative Meta-Analysis of Rejection: 10–12. doi:10.1038/srep02027.

16. Etkin A, Egner T, Kalisch R (2011) Emotional processing in anterior cingulate and medial prefrontal cortex. Trends Cogn Sci 15: 85–93. Available: http://www.pubmedcentral.nih.gov/articlerender.fcgi?artid=3035157&tool=pmcentrez &rendertype=abstract. Accessed 1 March 2013.

17. Ridderinkhof KR, Ullsperger M, Crone E a, Nieuwenhuis S (2004) The role of the medial frontal cortex in cognitive control. Science 306: 443–447 Available: http://www.ncbi.nlm.nih.gov/pubmed/15486290.

18. Somerville LH, Heatherton TF, Kelley WM (2006) Anterior cingulate cortex responds differentially to expectancy violation and social rejection. Nat Neurosci 9: 1007–1008. Available: http://www.ncbi.nlm.nih.gov/pubmed/16819523. Accessed 10 March 2013.

19. Masten CL, Eisenberger NI, Pfeifer JH, Dapretto M (2010) Witnessing peer rejection during early adolescence: neural correlates of empathy for experiences of social exclusion. Soc Neurosci 5: 496–507. Available: http://www.pubmedcentral.nih.gov/articlerender.fcgi?artid=2957502&tool=pmcentrez&rendertype=abstract. Accessed 28 February 2013.

20. Yoshimura S, Ueda K, Suzuki S, Onoda K, Okamoto Y, et al. (2009) Self-referential processing of negative stimuli within the ventral anterior cingulate gyrus and right amygdala. Brain Cogn 69: 218–225. Available: http://www.ncbi.nlm.nih.gov/pubmed/18723260. Accessed 21 May 2013.

21. Masten CL, Eisenberger NI, Borofsky L a, Pfeifer JH, McNealy K, et al. (2009) Neural correlates of social exclusion during adolescence: understanding the distress of peer rejection. Soc Cogn Affect Neurosci 4: 143–157. Available: http://www.pubmedcentral.nih.gov/articlerender.fcgi?artid=2686232&tool=pmcentrez&rendertype=abstract. Accessed 28 February 2013.

22. Sebastian CL, Tan GCY, Roiser JP, Viding E, Dumontheil I, et al. (2011) Developmental influences on the neural bases of responses to social rejection: implications of social neuroscience for education. Neuroimage 57: 686–694. Available: http://www.ncbi.nlm.nih.gov/pubmed/20923708. Accessed 28 February 2013.

23. Bolling DZ, Pitskel NB, Deen B, Crowley MJ, McPartland JC, et al. (2011) Enhanced neural responses to rule violation in children with autism: a comparison to social exclusion. Dev Cogn Neurosci 1: 280–294. Available: http://www.pubmedcentral.nih.gov/articlerender.fcgi?artid=3129780&tool=pmcentrez&rendertype=abstract. Accessed 22 May 2013.

24. Eisenberger NI, Way BM, Taylor SE, Welch WT, Lieberman MD (2007) Understanding genetic risk for aggression: clues from the brain's response to social exclusion. Biol Psychiatry 61: 1100–1108. Available: http://www.ncbi.nlm.nih.gov/pubmed/17137563. Accessed 22 May 2013.

25. Burklund LJ, Eisenberger NI, Lieberman MD (2007) The face of rejection: rejection sensitivity moderates dorsal anterior cingulate activity to disapproving facial expressions. Soc Neurosci 2: 238–253. Available: http://www.pubmedcentral.nih.gov/articlerender.fcgi?artid=2373282&tool=pmcentrez&rendertype=abstract. Accessed 22 May 2013.

26. DeWall CN, Masten CL, Powell C, Combs D, Schurtz DR, et al. (2012) Do neural responses to rejection depend on attachment style? An fMRI study. Soc Cogn Affect Neurosci 7: 184–192. Available: http://scan.oxfordjournals.org/content/7/2/184.short. Accessed 28 February 2013.

27. Onoda K, Okamoto Y, Nakashima K, Nittono H, Yoshimura S, et al. (2010) Does low self-esteem enhance social pain? The relationship between trait self-esteem and anterior cingulate cortex activation induced by ostracism. Soc Cogn Affect Neurosci 5: 385–391. Available: http://www.pubmedcentral.nih.gov/articlerender.fcgi?artid=2999754&tool=pmcentrez&rendertype=abstract. Accessed 22 May 2013.

28. Somerville LH, Kelley WM, Heatherton TF (2010) Self-esteem modulates medial prefrontal cortical responses to evaluative social feedback. Cereb Cortex 20: 3005–3013. Available: http://cercor.oxfordjournals.org/content/20/12/3005.short. Accessed 20 March 2013.

29. Spitzer RL, Williams JBW, Gibbon M, First MB (1990) Structured Clinical Interview for DSM-III-R, Patient Edition/Non-patient Edition,(SCID-P/SCID-NP),. Washington, D.C.: American Psychiatric Press, Inc.

30. Bernstein DP, Fink L (1998) Childhood Trauma Questionnaire, a retrospective self-report manual. San Antonio, TX: The Psychological Corporation.

31. Wechsler D (1991) WISC-III: Wechsler intelligence scale for children. Available: http://www.getcited.org/pub/103253171. Accessed 31 May 2013.

32. Wechsler D (1997) WAIS-III: Wechsler adult intelligence scale. San Antonio: Psychological Corporation.

33. First MB, Gibbon M (1997) User's guide for the structured clinical interview for DSM-IV axis I disorders: SCID-1 clinician version. American Psychiatric Publisher.

34. Achenbach TM (1991) Manual for the Child Behavior Checklist.

35. Duijsens IJ, Eurelings-Bontekoe EHM, Diekstra RFW (1996) The VKP, a self-report instrument for DSM-III-R and ICD-10 personality disorders: construction and psychometric properties. Pers Individ Dif 20: 171–182. doi: 10.1016/0191-8869(95)00161-1

36. Arntz A, Wessel I (1996) Jeugd trauma vragenlijst [Dutch version of the childhood trauma questionnaire]. The Netherlands.

37. Thombs BD, Bernstein DP, Lobbestael J, Arntz A (2009) A validation study of the Dutch Childhood Trauma Questionnaire-Short Form: factor structure, reliability, and known-groups validity. Child Abuse Negl 33: 518–523. Available: http://www.ncbi.nlm.nih.gov/pubmed/19758699. Accessed 22 May 2013.

38. Van Harmelen A-L, De Jong PJ, Glashouwer KA, Spinhoven P, Penninx BWJH, et al. (2010) Child abuse and negative explicit and automatic self-associations: the cognitive scars of emotional maltreatment. Behav Res Ther 48: 486–494 Available: http://www.ncbi.nlm.nih.gov/pubmed/20303472.

39. Van Harmelen A-L, Elzinga BM, Kievit RA, Spinhoven P (2011) Intrusions of autobiographical memories in individuals reporting childhood emotional maltreatment. Eur J Psychotraumatol 2. Available: http://www.pubmedcentral.nih.gov/articlerender.fcgi?artid=3402144&tool=pmcentrez&renabstract. Accessed 22 May dertype = 2013.

40. Trickett PK, Mennen FE, Kim K, Sang J (2009) Emotional abuse in a sample of multiply maltreated, urban young adolescents: issues of definition and identification. Child Abuse Negl 33: 27–35. Available: http://www.ncbi.nlm.nih.gov/pubmed/19178945. Accessed 20 May 2013.

41. Van Beest I, Williams KD (2006) When inclusion costs and ostracism pays, ostracism still hurts. J Pers Soc Psychol 91: 918–928. Available: http://www.ncbi.nlm.nih.gov/pubmed/17059310. Accessed 22 May 2013.

42. Sebastian C, Viding E, Williams KD, Blakemore S-J (2010) Social brain development and the affective consequences of ostracism in adolescence. Brain Cogn 72: 134–145. Available: http://www.ncbi.nlm.nih.gov/pubmed/19628323. Accessed 22 May 2013.

43. Lieberman MD, Cunningham W a (2009) Type I and Type II error concerns in fMRI research: re-balancing the scale. Soc Cogn Affect Neurosci 4: 423–428. Available: http://www.pubmedcentral.nih.gov/articlerender.fcgi?artid=2799956&tool=pmcentrez&rendertype=abstract. Accessed 27 February 2013.

44. Meyer ML, Masten CL, Ma Y, Wang C, Shi Z, et al. (2012) Empathy for the social suffering of friends and strangers recruits distinct patterns of brain activation. Soc Cogn Affect Neurosci. Available: http://www.ncbi.nlm.nih.gov/pubmed/22355182. Accessed 3 March 2013.

45. Maldjian JA, Laurienti PJ, Kraft RA, Burdette JH (2003) An automated method for neuroanatomic and cytoarchitectonic atlas-based interrogation of fMRI data sets. Neuroimage 19: 1233–1239. Available: http://www.ncbi.nlm.nih.gov/pubmed/12880848. Accessed 22 May 2013.

46. Brett MA.-L, Valabregue R, Poline J-B (2002) Region of interest analysis using an SPM toolbox. 8th International Conferance on Functional Mapping of the Human Brain. Sendai, Japan. p. abstract 497.

47. Sebastian CL, Fontaine NMG, Bird G, Blakemore S-J, Brito S a De, et al. (2012) Neural processing associated with cognitive and affective Theory of Mind in adolescents and adults. Soc Cogn Affect Neurosci 7: 53–63. Available: http://www.pubmedcentral.nih.gov/articlerender.fcgi?artid=3252629&tool=pmcentrez&rendertype=abstract. Accessed 28 February 2013.

48. Van der Meer L, Costafreda S, Aleman A, David AS (2010) Self-reflection and the brain: a theoretical review and meta-analysis of neuroimaging studies with implications for schizophrenia. Neurosci Biobehav Rev 34: 935–946. Available: http://www.ncbi.nlm.nih.gov/pubmed/20015455. Accessed 21 May 2013.

49. Moran JM, Macrae CN, Heatherton TF, Wyland CL, Kelley WM (2006) Neuroanatomical evidence for distinct cognitive and affective components of self. J Cogn Neurosci 18: 1586–1594. Available: http://www.ncbi.nlm.nih.gov/pubmed/16989558. Accessed 21 May 2013.

50. Grimm S, Ernst J, Boesiger P, Schuepbach D, Hell D, et al. (2009) Increased self-focus in major depressive disorder is related to neural abnormalities in subcortical-cortical midline structures. Hum Brain Mapp 30: 2617–2627. Available: http://www.ncbi.nlm.nih.gov/pubmed/19117277. Accessed 21 May 2013.

51. Blair KS, Geraci M, Smith BW, Hollon N, DeVido J, et al. (2012) Reduced dorsal anterior cingulate cortical activity during emotional regulation and top-down attentional control in generalized social phobia, generalized anxiety disorder, and comorbid generalized social phobia/generalized anxiety disorder. Biol Psychiatry 72: 476–482. Available: http://www.ncbi.nlm.nih.gov/pubmed/22592057. Accessed 4 March 2013.

52. Lemogne C, le Bastard G, Mayberg H, Volle E, Bergouignan L, et al. (2009) In search of the depressive self: extended medial prefrontal network during self-referential processing in major depression. Soc Cogn Affect Neurosci 4: 305–312. Available: http://www.pubmedcentral.nih.gov/articlerender.fcgi?artid=2728628&tool=pmcentrez&rendertype=abstract. Accessed 21 May 2013.

53. Lindquist K a, Wager TD, Kober H, Bliss-Moreau E, Barrett LF (2012) The brain basis of emotion: a meta-analytic review. Behav Brain Sci 35: 121–143. Available: http://www.ncbi.nlm.nih.gov/pubmed/22617651. Accessed 2 March 2013.

54. Beck AT (2008) The evolution of the cognitive model of depression and its neurobiological correlates. Am J Psychiatry 165: 969–977. Available: http://www.ncbi.nlm.nih.gov/pubmed/18628348. Accessed 21 March 2013.

55. Czéh B, Müller-Keuker JIH, Rygula R, Abumaria N, Hiemke C, et al. (2007) Chronic social stress inhibits cell proliferation in the adult medial prefrontal cortex: hemispheric asymmetry and reversal by fluoxetine treatment. Neuropsychopharmacology 32: 1490–1503. Available: http://www.ncbi.nlm.nih.gov/pubmed/17164819. Accessed 7 March 2013.

56. Sánchez MM, Ladd CO, Plotsky PM (2001) Early adverse experience as a developmental risk factor for later psychopathology: evidence from rodent and primate models. Dev Psychopathol 13: 419–449 Available: http://www.ncbi.nlm.nih.gov/pubmed/11523842.

57. Sanchez MM, Alagbe O, Felger JC, Zhang J, Graff AE, et al. (2007) Activated p38 MAPK is associated with decreased CSF 5-HIAA and increased maternal rejection during infancy in rhesus monkeys. Mol Psychiatry 12: 895–897. Available: http://www.ncbi.nlm.nih.gov/pubmed/17895923. Accessed 14 May 2013.

58. Arnsten AFT (2009) Stress signalling pathways that impair prefrontal cortex structure and function. Nat Rev Neurosci 10: 410–422. Available: http://www.pubmedcentral.nih.gov/articlerender.fcgi?artid=2907136&tool=pmcentrez&rendertype=abstract. Accessed 1 March 2013.

59. Lupien SJ, McEwen BS, Gunnar MR, Heim C (2009) Effects of stress throughout the lifespan on the brain, behaviour and cognition. Nat Rev Neurosci 10: 434–445. Available: http://www.ncbi.nlm.nih.gov/pubmed/19401723. Accessed 28 February 2013.

60. McEwen BS, Eiland L, Hunter RG, Miller MM (2012) Stress and anxiety: structural plasticity and epigenetic regulation as a consequence of stress. Neuropharmacology 62: 3–12. Available: http://www.pubmedcentral.nih.gov/articlerender.fcgi?artid=3196296&tool=pmcentrez&rendertype=abstract. Accessed 5 March 2013.

61. Van Harmelen A-L, Van Tol M-J, Van Der Wee NJA, Veltman DJ, Aleman A, et al. (2010) Reduced medial prefrontal cortex volume in adults reporting childhood emotional maltreatment. Biol Psychiatry 68: 832–838 Available: http://www.ncbi.nlm.nih.gov/pubmed/20692648.

62. Dannlowski U, Stuhrmann A, Beutelmann V, Zwanzger P, Lenzen T, et al. (2012) Limbic scars: long-term consequences of childhood maltreatment revealed by functional and structural magnetic resonance imaging. Biol Psychiatry 71: 286–293. Available: http://www.ncbi.nlm.nih.gov/pubmed/22112927. Accessed 4 March 2013.

63. Tomoda A, Sheu Y-S, Rabi K, Suzuki H, Navalta CP, et al. (2011) Exposure to parental verbal abuse is associated with increased gray matter volume in superior temporal gyrus. Neuroimage 54 Suppl 1: S280–6. Available: http://www.pubmedcentral.nih.gov/articlerender.fcgi?artid=2950228&tool=pmcentrez&rendertype=abstract. Accessed 17 May 2013.

64. Egeland B (2009) Taking stock: childhood emotional maltreatment and developmental psychopathology. Child Abuse Negl 33: 22–26. Available: http://www.ncbi.nlm.nih.gov/pubmed/19167068. Accessed 25 March 2013.

65. Gilbert R, Widom CS, Browne K, Fergusson D, Webb E, et al. (2009) Burden and consequences of child maltreatment in high-income countries. Lancet 373: 68–81. Available: http://www.ncbi.nlm.nih.gov/pubmed/19056114. Accessed 28 February 2013.

66. Ellis BJ, Boyce WT, Belsky J, Bakermans-Kranenburg MJ, van Ijzendoorn MH (2011) Differential susceptibility to the environment: an evolutionary--neurodevelopmental theory. Dev Psychopathol 23: 7–28. Available: http://www.ncbi.nlm.nih.gov/pubmed/21262036. Accessed 21 May 2013.

67. Hardt J, Rutter M (2004) Validity of adult retrospective reports of adverse childhood experiences: review of the evidence. J Child Psychol Psychiatry 45: 260–273 Available: http://www.ncbi.nlm.nih.gov/pubmed/14982240.

68. Tonmyr L, Draca J, Crain J, Macmillan HL (2011) Measurement of emotional/psychological child maltreatment: a review. Child Abuse Negl 35: 767–782. Available: http://www.ncbi.nlm.nih.gov/pubmed/22018520. Accessed 20 May 2013.

69. Etkin A, Prater KE, Hoeft F, Menon V, Schatzberg AF (2010) Failure of anterior cingulate activation and connectivity with the amygdala during implicit regulation of emotional processing in generalized anxiety disorder. Am J Psychiatry 167: 545–554. Available: http://www.ncbi.nlm.nih.gov/pubmed/20123913. Accessed 18 March 2013.

There are several supplemental files that are not available in this version of the article. To view this additional information, please use the citation on the first page of this chapter.

CHAPTER 7

REDUCED CINGULATE GYRUS VOLUME ASSOCIATED WITH ENHANCED CORTISOL AWAKENING RESPONSE IN YOUNG HEALTHY ADULTS REPORTING CHILDHOOD TRAUMA

SHAOJIA LU, WEIJIA GAO, ZHAOGUO WEI, WEIWEI WU, MEI LIAO, YUQIANG DING, ZHIJUN ZHANG, AND LINGJIANG LI

7.1 INTRODUCTION

Early life stress, including childhood trauma, which is very common in our society, has been established as a great risk factor in the subsequent development of multiple psychiatric disorders and unfavorable behavior patterns [1]. Previous studies have investigated the possible pathways from early life stress to psychosis, however, to date, little is known about the mechanisms underlying this association [2]. Recently, biological mechanisms such as dysfunction of the hypothalamic-pituitary-adrenal (HPA)

Reduced Cingulate Gyrus Volume Associated with Enhanced Cortisol Awakening Response in Young Healthy Adults Reporting Childhood Trauma. © Lu S, Gao W, Wei Z, Wu W, Liao M, Ding Y, Zhang Z, and Li L. PLoS ONE, 8,7 (2013), doi:10.1371/journal.pone.0069350. Licensed under Creative Commons Attribution License, http://creativecommons.org/licenses/by/3.0/.

axis [3] and altered volumes of specific brain regions [4] after exposures to early life stress have been reported which may help to elucidate the close relationship between early life stress and onset of psychosis.

Although not always consistent, findings for hyperactivity of the HPA axis as indicated by increased cortisol and adrenocorticotropin-releasing hormone (ACTH) responses to a psychological stress task [5] and to the dexamethasone/corticotropin-releasing factor (Dex/CRF) test [3], hyper-secretion of salivary cortisol over the daytime hours [6], and enhanced cortisol awakening response (CAR) [7] among individuals with early life stress independent of psychosis diagnosis have been observed in an emerging body of investigations. With regard to anatomical magnetic resonance imaging studies, increasing evidence has shown a strong link between decreased hippocampus volume and early life stress [4], [8], [9]. Moreover, childhood emotional maltreatment was reported to be associated with profound reductions of medial prefrontal cortex (mPFC) volume [10] and in another study early deprivation was revealed to be correlated with reduced orbital-frontal cortical volume in postinstitutionalized children [11]. In line with these, the extent of gray matter loss in the cingulate cortex was suggested to be related to a history of early adverse events in a group of depression patients as well, although the sample size of the experimental group was relatively small [12].

It has been well known that glucocorticoids have neurotoxic effects in the central nervous system in some circumstances [13]. In addition, it is quite interesting that the altered brain regions reported in subjects with early life stress contain high concentrations of glucocorticoid receptors and act as well- documented roles in regulating HPA activity [14]. Hence, stress-related dysfunction of the HPA axis and atrophy of the regulator regions may precipitate a vicious circle which will result in greater exposure to glucocorticoids and more severe damage to these brain regions. This relationship has been revealed in preclinical studies [15], but it is less explored in clinical samples. Therefore, the aim of the present study was to investigate the changes and associations of the HPA activity and gray matter volumes in young healthy adults with self-reported childhood trauma exposures. To assess the HPA axis activity, the CAR, a reliable biological marker of reflecting the dynamic activity of the HPA axis [16], was administered to evaluate morning cortisol release to awaken. Based on previous findings, we hypothesized that childhood trauma could induce hyperactivity of the HPA axis and decreased gray matter volume in brain regions such as hippocampus, PFC, or cingulate gyrus; and there could be associations between increased cortisol concentrations and atrophy of those brain regions.

7.2 METHODS

7.2.1 PARTICIPANTS

The study group comprised 48 subjects (male/female, 18/30), ages 18–33 years, including 24 subjects with childhood trauma experiences (CT group) and 24 age- and gender- matched subjects without childhood trauma exposures (non-CT group). For assignment to the CT group, individuals must have had experienced chronic moderate-severe trauma exposures (abuse or/and neglect) before the age of 16. All participants were recruited from a survey that we had carried out to investigate the occurrence of childhood trauma in local communities and universities. Subjects responded with no direct reference to childhood trauma as a key variable in the study. All subjects were thoroughly interviewed by two professional psychologists and were free from any current or lifetime history of psychiatric disorders according to *Diagnostic and Statistical Manual of Mental Disorders,* IV Edition (DSM- IV) criteria, as screened with the Structured Clinical Interview for DSM-IV interview (SCID). The general exclusions were as follows: (1) left handedness, (2) standard scores >50 on Zung's self-rating depression scale (SDS) [17] or >40 on Zung's self-rating anxiety scale (SAS) [18], (3) significant medical illness, (4) presence of major sensorimotor handicaps, (5) history of seizures, head trauma, or unconsciousness, (6) intake of any psychotropic medication or hormone, (7) alcohol or substance abuse, (8) women with pregnancy or in lactation or menstrual period, and (9) contraindications to MRI scan, including metallic implants, retractors or braces, and claustrophobia.

7.2.2 ETHICS STATEMENT

This study was approved by the ethic committee of the Second Xiangya Hospital of Central South University. A complete description of the study was provided to every subject, after that written informed consent was obtained from each participant.

7.2.3 ASSESSMENT OF CHILDHOOD TRAUMA

The existence or absence of childhood trauma was determined by the childhood trauma questionnaire (CTQ) in all subjects. The CTQ is a 28-item retrospective self-report questionnaire designed to assess five types of negative childhood experiences by five sub-scales: emotional abuse, emotional neglect, sexual abuse, physical abuse, and physical neglect, respectively. Subjects who score higher than the threshold of a sub-scale are treated as existence of corresponding childhood trauma experience. The cutoffs of each sub-scale for moderate-severe exposure are: 1) emotional abuse ≥ 13, 2) emotional neglect ≥ 15, 3) sexual abuse ≥ 8, 4) physical abuse ≥ 10, and 5) physical neglect ≥ 10. This reliable questionnaire has great internal consistency and criterion-related validity in clinical and community samples; the good internal consistency coefficients for the five subscales are demonstrated in many studies [19], [20], [21], [22].

7.2.4 SALIVARY CORTISOL

Salivary samples were collected on the magnetic resonance imaging scan day (weekdays) using Salivette® collection devices (Sarstedt, Nümbrecht, Germany) at altogether four time-points throughout the morning: immediately upon awakening, 30, 45, and 60 min following awakening. Samples were centrifuged at 3000 rpm (rounds per minute) for 10 min and recovered saliva samples were stored at $-70°C$ until analysis. Participants are instructed not to brush their teeth and eat before completing sampling, or engage in heavy exercise during the collecting hour. Additionally, they were asked to refrain from smoking and from drinks except water [23].

Moreover, all subjects had to report the time they went to bed as well as the time they woke up [24].

The DRG® Salivary Cortisol ELISA Kit, SLV –2390 (Marburg, Germany) was used to measure salivary cortisol concentrations. The range of the assay is between 0.537–80 ng/ml. The intra- and inter-assay variability coefficients are 1.5–4.5% and 5.8–7.5%, respectively. As measures of the CAR, the area under the curve relative to ground (AUCg) and the area under the curve with respect to increase (AUCi) were calculated using the formulas described by Pruessner et al., (2003). The AUCg is a measure of the total cortisol secretion throughout the first hour following awakening, whereas the AUCi values reflect the dynamic of the CAR, which is more related to the sensitivity of the HPA system and focuses on the cortisol changes over time after awakening [25].

7.2.5 MRI ACQUISITION

Imaging data were acquired using a Philips 3.0-T scanner (Philips, Best, The Netherlands) in the Magnetic Resonance Center belonging to the Second Xiangya Hospital of Central South University. Subjects were asked to lie on the scanner and keep eyes closed. A standard birdcage head coil was used, and the restraining foam pads were placed on two sides of the head to minimize head motion while cotton plug was used with the purpose of diminishing the noise. For each subject, a high-resolution T1-weighted sequence using a three-dimensional magnetization prepared rapid acquisition gradient echo sequence was used. Images of the whole brain were acquired in an axial orientation with the following parameters: slice thickness = 1 mm, gap = 0 mm, repetition time = 7.6 ms, echo time = 3.7 ms, inversion time = 795 ms, field of view = 256*256 mm^2, flip angle = 8°, matrix size = 256*256, resolution = 1.0*1.0*1.0, slices = 180, scan time = 2'58".

7.2.6 VOXEL-BASED MORPHOMETRY (VBM) ANALYSIS

All T1-weighted high-resolution anatomical data were preprocessed by using the previous method [26], [27]. The analyses were performed using

the Statistical Parametric Mapping 8 (SPM8) software (Wellcome Department of Imaging Neuroscience, University College London, UK; http://
www.fil.ion.ucl.ac.uk/spm) in a Matlab environment. The VBM8 Toolbox
(http://dbm.neuro.uni-jena.de/vbm.html) was used for preprocessing the
structural images in SPM8 with default parameters. The data was bias-
corrected, tissue classified, and normalized to Montreal Neurological Institute space using linear (12-parameter affine) and non-linear transforma-
tions within a unified model [28]. Then data analyses were performed on
gray matter segment which was multiplied by the non-linear components
derived from the normalization matrix in order to preserve actual gray
matter value locally (modulated gray matter volume). Finally, the modulated gray matter volume was smoothed with a Gaussian kernel of 8 mm
full width at half maximum.

7.2.7 STATISTICAL ANALYSIS

Data analyses were carried out using Statistical Package for the Social
Sciences version 16.0 (SPSS Inc., Chicago, IL, USA). Independent two-
sample t tests and Chi-square tests (χ^2) were respectively used to tests
for the continuous variables and categorical variables between the two
groups. Salivary cortisol samples after awakening were analyzed using
repeated measures ANOVA with time as the within-subjects factor and
group as the between-subjects factor. The potential confounders, such as
age, gender, time of awakening, hours of sleep, body mass index (BMI),
and smoking, were included as covariates in the measures ANOVA to investigate possible effects on cortisol concentrations. Values are given as
mean ± standard deviation. The level of two-tailed statistical significance
was set at $p < 0.05$ for all tests.

For gray matter volume, two-sample t-test on a voxel-by-voxel basis
was performed to determine the difference between the two groups. The
statistical threshold was set at $p < 0.05$, corrected for multiple comparisons
with false discovery rate (FDR) correction.

To evaluate any correlations between CTQ scores or CAR data and gray
matter structural changes in individuals with childhood trauma, whole brain
multiple regression analyses integrated in SPM basic models were performed

at p<0.05 (FDR corrected). Moreover, Spearman rank correlation analysis was used to evaluate the relationship between CTQ scores and CAR data.

7.3 RESULTS

7.3.1 SAMPLE CHARACTERISTICS

As indicated in Table 1, the two groups of subjects did not differ with respect to age (t = −0.075, p = 0.940), gender (χ^2 = 0.000, p = 1.000), educational level (t = −1.407, p = 0.166), BMI (t = 1.839, p = 0.072), SAS score (t = 1.430, p = 0.160), SDS score (t = 1.014, p = 0.316), and smoking status (Fish's exact test = 2.063, p = 0.609). As we would expect, the two experimental groups differed on levels of CTQ and its sub-scales except sexual abuse (t = 3.234~11.38, p<0.01). The most common aspect of childhood trauma experience in the present sample was emotional neglect (17, 70.8%); a proportion of 62.5% (15) of traumatic subjects experienced at least two forms of childhood trauma exposures.

7.3.2 CORTISOL AWAKENING RESPONSE

There was no significant difference between two groups for time of awakening (CT group 6:46±19 min vs non-CT group 6:40±22 min, p < 0.05) or hours of sleep (CT group 7.04±0.69 h vs non-CT group 6.78±0.47 h, p < 0.05). A repeated-measures ANOVA of the morning salivary cortisol concentrations revealed a significant main effect of group (F = 5.111, p = 0.029), a main effect of time (F = 47.60, p = 0.000), and a group * time effect (F = 3.667, p = 0.014). There were no main or interactive effects of age, gender, BMI, smoking, time of awakening or hours of sleep on cortisol levels when introduced as covariates in the ANOVA. Independent two-sample t tests of the individual time points showed that salivary cortisol concentrations were significantly higher in traumatic subjects at 30 and 45 min following awakening (see figure 1A). Meanwhile, a statistically significant increase was observed for AUCg of salivary cortisol secretion in subjects who had self-reported childhood trauma experiences (see figure

1B). With respect to AUCi, the CT group also showed a higher level than the non-CT group with a statistically significant difference (see figure 1C).

7.3.3 VOXEL-BASED ANALYSIS OF MORPHOMETRY

As compared with subjects without childhood trauma, individuals with adverse experiences in childhood showed significant volume reduction in the right middle cingulate gyrus (see Figure 2). However, no region with significantly increased volume in traumatic subjects was observed.

7.3.4 CORRELATIONS

In subjects with childhood trauma, the whole brain linear regression analysis conducted with SPM8 yielded a strong negative association of the right middle cingulate gyrus volume with CAR AUCg (see Table 2), however, there were no brain areas revealing significant correlations with CTQ scores at the defined threshold. Finally, a positive association was observed between CTQ total score and CAR AUCg ($r_s = 0.674$, p = 0.000) in individuals with childhood trauma experiences.

7.4 DISCUSSION

The present study investigated the HPA activity as measured by CAR and brain structural (gray matter volume) changes in young healthy adults with and without childhood trauma experiences. The current results revealed a significantly enhanced CAR and decreased gray matter volume in the right middle cingulate gyrus (BA 24) and furthermore, a significant association between morning salivary cortisol levels after awaking and the right middle cingulate gyrus volume in subjects with childhood trauma. These outcomes together with the previous preclinical findings suggested that stress-related brain structural volume reductions might be the consequences of prolonged exposure to increased levels of glucocorticoids resulting from chronic early life stress [13].

TABLE 1: Demographic and clinical characteristics of all subjects (n = 48).

	CT group, n = 24			non-CT group, n = 24		
	Mean	SD	Range	Mean	SD	Range
Age (Years)	21.5	3.98	18–33	21.5	3.69	18–33
Gender (Male/Female)	9/15			9/15		
Educational level (Years)	14.0	1.30	12–17	14.7	1.92	12–18
BMI (kg/m²)	21.6	2.05	17.7–25.0	20.6	1.62	18.3–23.3
SDS score	36.2	6.06	25–46	34.5	5.30	27–48
SAS score	24.0	4.51	26–40	32.0	4.78	25–40
Smoking, n (%)						
0	21 (87.5)			22 (91.7)		
≤ 10	1 (4.17)			2 (8.3)		
11–20	2 (8.33)			0 (0)		
CTQ score						
Emotional abuse**	9.21	2.36	6–15	6.21	1.22	5–9
Physical abuse**	7.83	2.93	5–14	5.71	1.33	5–9
Sexual abuse	5.46	0.83	5–7	5.38	0.58	5–6
Emotional neglect**	15.2	3.28	7–20	7.38	2.65	5–13
Physical neglect**	10.2	2.72	5–17	5.63	0.93	5–8
Total**	47.9	6.08	39–58	30.2	4.63	25–40
CT exposures, n (%)						
Emotional abuse	2 (8.33)					
Physical abuse	8 (33.3)					
Sexual abuse	0 (0)					
Emotional neglect	17 (70.8)					
Physical neglect	14 (58.3)					
Multiply exposures	15 (62.5)					
Single exposure	9 (37.5)					

BMI, body mass index; CT, childhood trauma; CTQ, childhood trauma questionnaire; SAS, self-rating anxiety scale; SD, standard deviation; SDS, self-rating depression scale.
$**p < 0.01$

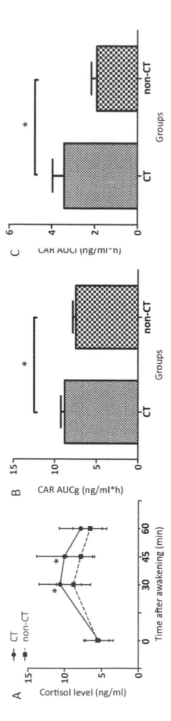

FIGURE 1. Changes in cortisol awakening response (CAR) across the experimental groups. (A) Independent two sample t tests revealed that significant differences of salivary cortisol levels were found between two groups at 30 (t = 2.389, p = 0.021) and 45 min (t = 2.565, p = 0.014) after awakening. Significant increases of cortisol levels were observed in subjects with childhood trauma experiences at those two time points. (B) The CAR area-under-the-curve to ground (AUCg) was significantly differed between two groups (t = 2.335, p = 0.024). Consistent with cortisol levels at 30 and 45 min, subjects who had self-reported childhood trauma showed higher levels of CAR AUCg. (C) With respect to the CAR area-under-the-curve increase (AUCi), significant difference was found as well (t = 2.532, p = 0.016). *Comparison with non-CT group. p <0.05. CT, childhood trauma.

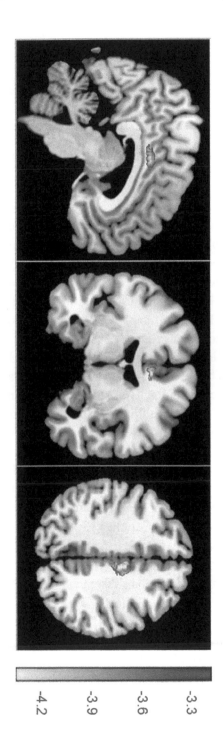

FIGURE 2: Region of different gray matter volume between individuals with and without childhood trauma. Decreased gray matter volume was detected in the right middle cingulate gyrus (Brodman area 24, x = 6, y = −6, z = 39, cluster size = 303, z score = 4.25, puncorrected<0.001, pFDR corrected = 0.047,) in subjects with childhood trauma. The right middle cingulate gyrus is shown in blue.

TABLE 2: Region of decreased gray matter volume in subjects with childhood trauma correlated to CAR AUCg.

Anatomical region	BA	Side	Cluster size	x	y	z	z-score	$p_{uncorrected}$	$p_{FDR\ corrected}$
Middle cingulate gyrus	24	R	419	5	−6	43	4.24	< 0.001	0.047

AUCg, area-under-the-curve to ground; BA, Brodman area; CAR, cortisol awakening response; FDR, false discovery rate.

The enhanced CAR in subjects reporting childhood trauma experiences was a main finding of the present study which indicated that childhood trauma seemed to be a good predictor of hyperactivity of the HPA axis even in the absence of current psychosis diagnosis. This indication was supported by prior studies evaluating the effect of childhood parent loss on adult HPA axis function [29] and investigating the relationship between childhood abuse and HPA axis function using a DEX/CRH test in female borderline personality disorder patients [30], both demonstrating the association between early life stress and hyperactivity of the HPA axis. Our neuroendocrine findings were also generally in accordance with a previous report in young healthy adults with low parental care experiences exhibiting an increased CAR and increased afternoon/evening cortisol outputs [7]. However, hyperactivity of the HPA axis in subjects with early life stress was not always replicated, both diminished cortisol responses in healthy adults reporting significant childhood maltreatment [31] and decreased CAR in healthy college students after early loss experiences [32] were observed in relevant studies. Several potential explanations have been posited to account for this inconsistency, such as sex differences, differences in the timing and severity of childhood trauma, genetic factors, and different concomitant and subsequent psychosocial conditions in later life [32], [33].

In the present study, the right middle cingulate gyrus gray matter volume reduction was detected in subjects with childhood trauma. The cingulate gyrus, a part of the limbic system, plays a critical role in two main neuroanatomic circuits that are believed to be involved in mood regulation

[34]. The extensive connections of the cingulate gyrus with the frontal lobe, the temporal lobe and the striatal structures enable it to be an important area for the integration of emotions [35]. Our data was, at least in part, comparable with findings of reduced gray matter volumes in the cingulate gyrus in both healthy adolescents [36] and adults [4] with high CTQ scores. Interestingly, smaller gray matter volume in the cingulate gyrus was also present in patients with current or remitted depression [37], [38] and post traumatic stress disorder [39] in previous anatomical MRI studies which had focused on this region. In this regard, reduced cingulate gyrus volume before illness onset which contributed a limbic scar induced by long-term childhood trauma experiences in young healthy adults might represent the psychobiological vulnerability for developing psychosis.

It has been demonstrated that hypercortisolism induced by adrenalectomy with high-dose dexamethasone supplementation is associated with a significant reduction in the volume of the cingulate gyrus in rats [40]. In this context, the observed relationship between reduced cingulate gyrus and increased salivary cortisol secretion after awaking in our study extended that preclinical finding. As reported previously, the cingulate gyrus contains high densities of glucocorticoid receptors in most layers [14] which may make it one of the most vulnerable parts to the neurotoxicity of increased glucocorticoids. In addition, the cingulate gyrus also plays a crucial role in the regulation of the HPA axis [41], and thus glucocorticoids induced damage to this area may diminish its ability to exert negative feedback control resulting in more glucocorticoid secretions. This cycle may help to explain the origin of the present association in subjects with childhood trauma, however, our study was cross-sectional designed which precluded causal inferences. Further research will be required to clearly clarify this causal relationship.

Unexpected from our initial expectations, this study did not found any group differences of gray matter volume in hippocampus and PFC that have predominantly been linked to childhood trauma experiences. This may result from heterogeneity of research samples or types of childhood trauma experiences. Previous studies that have detected decreased gray matter volumes in hippocampus and PFC associated with childhood trauma experiences have measured this relationship in patients with depression [9] or in a mixed study sample [10], including healthy adults, patients with depression and/or anxiety disorder. Therefore, volume reductions in hippocampus and PFC may probably sever as results of co-appearance of childhood trauma and psychosis but not childhood trauma alone. In addition, in our sample, the subjects are young (average age = 21.5) which may restrain the time-dependent glucocorticoids induced atrophy of key brain regions. Finally, most studies reporting neuroanatomical correlates of childhood trauma with decreased hippocampus and PFC volumes in adults have focused mainly on the impact of sexual and/or physical abuse [42], [43], [44], but the most frequent aspects of childhood trauma experiences in our samples are physical and emotional neglect, which may also contribute to the differences mentioned above.

7.5 LIMITATIONS

Several limitations must be taken into account when interpreting our findings. First, although it is suitable for CAR and VBM analyses, the sample size in each group is relatively small which restricts us to compare the different types of childhood trauma on HPA axis activity and brain structural changes. Second, childhood trauma experiences are evaluated by a retrospective self-reported questionnaire; hence, the results could be influenced by information biases. Finally, we cannot control for the subjects' compliance with the given sampling time schedule since a prior study has indicated that the participants' compliance with instructions is generally low [45], which may influence the CAR outcomes. However, our CAR results are characterized by a sharp increase in cortisol levels following awaken, peaking at 30 min later which well stands in line with other findings [46].

In conclusion, the present research outcomes suggest that childhood trauma is associated with hyperactivity of the HPA axis and decreased gray matter volume in the right middle cingulate gyrus, which may represent the vulnerability for developing psychosis after childhood trauma experiences. In addition, this study demonstrates that gray matter loss in the cingulate gyrus is related to increased cortisol levels.

REFERENCES

1. MacMillan HL, Fleming JE, Streiner DL, Lin E, Boyle MH, et al. (2001) Childhood abuse and lifetime psychopathology in a community sample. Am J Psychiatry 158: 1878–1883. doi: 10.1176/appi.ajp.158.11.1878

2. Schafer I, Fisher HL (2011) Childhood trauma and psychosis – what is the evidence? Dialogues Clin Neurosci 13: 360–365.

3. Heim C, Mletzko T, Purselle D, Musselman DL, Nemeroff CB (2008) The dexamethasone/corticotropin-releasing factor test in men with major depression: role of childhood trauma. Biol Psychiatry 63: 398–405. doi: 10.1016/j.biopsych.2007.07.002

4. Dannlowski U, Stuhrmann A, Beutelmann V, Zwanzger P, Lenzen T, et al. (2012) Limbic scars: long-term consequences of childhood maltreatment revealed by functional and structural magnetic resonance imaging. Biol Psychiatry 71: 286–293. doi: 10.1016/j.biopsych.2011.10.021

5. Heim C, Newport DJ, Heit S, Graham YP, Wilcox M, et al. (2000) Pituitary-adrenal and autonomic responses to stress in women after sexual and physical abuse in childhood. JAMA 284: 592–597. doi: 10.1001/jama.284.5.592

6. Gunnar MR, Morison SJ, Chisholm K, Schuder M (2001) Salivary cortisol levels in children adopted from romanian orphanages. Dev Psychopathol 13: 611–628. doi: 10.1017/s0954579401003113x

7. Engert V, Efanov SI, Dedovic K, Dagher A, Pruessner JC (2011) Increased cortisol awakening response and afternoon/evening cortisol output in healthy young adults with low early life parental care. Psychopharmacology (Berl) 214: 261–268. doi: 10.1007/s00213-010-1918-4

8. Buss C, Lord C, Wadiwalla M, Hellhammer DH, Lupien SJ, et al. (2007) Maternal care modulates the relationship between prenatal risk and hippocampal volume in women but not in men. J Neurosci 27: 2592–2595. doi: 10.1523/jneurosci.3252-06.2007

9. Vythilingam M, Heim C, Newport J, Miller AH, Anderson E, et al. (2002) Childhood trauma associated with smaller hippocampal volume in women with major depression. Am J Psychiatry 159: 2072–2080. doi: 10.1176/appi.ajp.159.12.2072

10. van Harmelen AL, van Tol MJ, van der Wee NJ, Veltman DJ, Aleman A, et al. (2010) Reduced medial prefrontal cortex volume in adults reporting childhood emotional maltreatment. Biol Psychiatry 68: 832–838. doi: 10.1016/j.biopsych.2010.06.011

11. Pollak SD, Nelson CA, Schlaak MF, Roeber BJ, Wewerka SS, et al. (2010) Neurodevelopmental effects of early deprivation in postinstitutionalized children. Child Dev 81: 224–236. doi: 10.1111/j.1467-8624.2009.01391.x

12. Treadway MT, Grant MM, Ding Z, Hollon SD, Gore JC, et al. (2009) Early adverse events, HPA activity and rostral anterior cingulate volume in MDD. PLoS One 4: e4887. doi: 10.1371/journal.pone.0004887

13. Radley JJ, Sisti HM, Hao J, Rocher AB, McCall T, et al. (2004) Chronic behavioral stress induces apical dendritic reorganization in pyramidal neurons of the medial prefrontal cortex. Neuroscience 125: 1–6. doi: 10.1016/j.neuroscience.2004.01.006

14. Cintra A, Bhatnagar M, Chadi G, Tinner B, Lindberg J, et al.. (1994) Glial and neuronal glucocorticoid receptor immunoreactive cell populations in developing, adult, and aging brain. Ann N Y Acad Sci 746: 42–61; discussion 61–43.

15. Sapolsky RM (2000) Glucocorticoids and hippocampal atrophy in neuropsychiatric disorders. Arch Gen Psychiatry 57: 925–935. doi: 10.1001/archpsyc.57.10.925

16. Mangold D, Marino E, Javors M (2011) The cortisol awakening response predicts subclinical depressive symptomatology in Mexican American adults. J Psychiatr Res 45: 902–909. doi: 10.1016/j.jpsychires.2011.01.001

17. Zung WW, Richards CB, Short MJ (1965) Self-rating depression scale in an outpatient clinic. Further validation of the SDS. Arch Gen Psychiatry 13: 508–515. doi: 10.1001/archpsyc.1965.01730060026004

18. Zung WW (1971) A rating instrument for anxiety disorders. Psychosomatics 12: 371–379. doi: 10.1016/s0033-3182(71)71479-0

19. Bernstein DP, Fink L, Handelsman L, Foote J, Lovejoy M, et al. (1994) Initial reliability and validity of a new retrospective measure of child abuse and neglect. Am J Psychiatry 151: 1132–1136.

20. Scher CD, Stein MB, Asmundson GJ, McCreary DR, Forde DR (2001) The childhood trauma questionnaire in a community sample: psychometric properties and normative data. J Trauma Stress 14: 843–857. doi: 10.1023/a:1013058625719

21. Bernstein DP, Stein JA, Newcomb MD, Walker E, Pogge D, et al. (2003) Development and validation of a brief screening version of the Childhood Trauma Questionnaire. Child Abuse Negl 27: 169–190. doi: 10.1016/s0145-2134(02)00541-0

22. Lu S, Peng H, Wang L, Vasish S, Zhang Y, et al.. (2013) Elevated specific peripheral cytokines found in major depressive disorder patients with childhood trauma exposure: A cytokine antibody array analysis. Compr Psychiatry: in press. doi: 10.1016/j.comppsych.2013.03.026.

23. Bouma EM, Riese H, Ormel J, Verhulst FC, Oldehinkel AJ (2009) Adolescents' cortisol responses to awakening and social stress; effects of gender, menstrual phase and oral contraceptives. The TRAILS study. Psychoneuroendocrinology 34: 884–893. doi: 10.1016/j.psyneuen.2009.01.003

24. Aubry JM, Jermann F, Gex-Fabry M, Bockhorn L, Van der Linden M, et al. (2010) The cortisol awakening response in patients remitted from depression. J Psychiatr Res 44: 1199–1204. doi: 10.1016/j.jpsychires.2010.04.015

25. Pruessner JC, Kirschbaum C, Meinlschmid G, Hellhammer DH (2003) Two formulas for computation of the area under the curve represent measures of total hormone concentration versus time-dependent change. Psychoneuroendocrinology 28: 916–931. doi: 10.1016/s0306-4530(02)00108-7

26. Bitter T, Bruderle J, Gudziol H, Burmeister HP, Gaser C, et al. (2010) Gray and white matter reduction in hyposmic subjects – A voxel-based morphometry study. Brain Res 1347: 42–47. doi: 10.1016/j.brainres.2010.06.003

27. Zhang LJ, Qi R, Zhong J, Xu Q, Zheng G, et al. (2012) The effect of hepatic encephalopathy, hepatic failure, and portosystemic shunt on brain volume of cirrhotic patients: a voxel-based morphometry study. PLoS One 7: e42824. doi: 10.1371/journal.pone.0042824

28. Ashburner J, Friston KJ (2005) Unified segmentation. Neuroimage 26: 839–851. doi: 10.1016/j.neuroimage.2005.02.018

29. Tyrka AR, Wier L, Price LH, Ross N, Anderson GM, et al. (2008) Childhood parental loss and adult hypothalamic-pituitary-adrenal function. Biol Psychiatry 63: 1147–1154. doi: 10.1016/j.biopsych.2008.01.011

30. Rinne T, de Kloet ER, Wouters L, Goekoop JG, DeRijk RH, et al. (2002) Hyperresponsiveness of hypothalamic-pituitary-adrenal axis to combined dexamethasone/corticotropin-releasing hormone challenge in female borderline personality disorder subjects with a history of sustained childhood abuse. Biol Psychiatry 52: 1102–1112. doi: 10.1016/s0006-3223(02)01395-1

31. Carpenter LL, Carvalho JP, Tyrka AR, Wier LM, Mello AF, et al. (2007) Decreased adrenocorticotropic hormone and cortisol responses to stress in healthy adults reporting significant childhood maltreatment. Biol Psychiatry 62: 1080–1087. doi: 10.1016/j.biopsych.2007.05.002

32. Meinlschmidt G, Heim C (2005) Decreased cortisol awakening response after early loss experience. Psychoneuroendocrinology 30: 568–576. doi: 10.1016/j.psyneuen.2005.01.006

33. Heim C, Newport DJ, Mletzko T, Miller AH, Nemeroff CB (2008) The link between childhood trauma and depression: insights from HPA axis studies in humans. Psychoneuroendocrinology 33: 693–710. doi: 10.1016/j.psyneuen.2008.03.008

34. Mayberg HS (1997) Limbic-cortical dysregulation: a proposed model of depression. J Neuropsychiatry Clin Neurosci 9: 471–481.

35. Wang F, Jackowski M, Kalmar JH, Chepenik LG, Tie K, et al. (2008) Abnormal anterior cingulum integrity in bipolar disorder determined through diffusion tensor imaging. Br J Psychiatry 193: 126–129. doi: 10.1192/bjp.bp.107.048793

36. Edmiston EE, Wang F, Mazure CM, Guiney J, Sinha R, et al. (2011) Corticostriatal-limbic gray matter morphology in adolescents with self-reported exposure to childhood maltreatment. Arch Pediatr Adolesc Med 165: 1069–1077. doi: 10.1001/archpediatrics.2011.565

37. Caetano SC, Kaur S, Brambilla P, Nicoletti M, Hatch JP, et al. (2006) Smaller cingulate volumes in unipolar depressed patients. Biol Psychiatry 59: 702–706. doi: 10.1016/j.biopsych.2005.10.011

38. Vasic N, Walter H, Hose A, Wolf RC (2008) Gray matter reduction associated with psychopathology and cognitive dysfunction in unipolar depression: a voxel-based morphometry study. J Affect Disord 109: 107–116. doi: 10.1016/j.jad.2007.11.011

39. Kuhn S, Gallinat J (2013) Gray matter correlates of posttraumatic stress disorder: a quantitative meta-analysis. Biol Psychiatry 73: 70–74. doi: 10.1016/j.biopsych.2012.06.029

40. Cerqueira JJ, Catania C, Sotiropoulos I, Schubert M, Kalisch R, et al. (2005) Corticosteroid status influences the volume of the rat cingulate cortex - a magnetic resonance imaging study. J Psychiatr Res 39: 451–460. doi: 10.1016/j.jpsychires.2005.01.003

41. Diorio D, Viau V, Meaney MJ (1993) The role of the medial prefrontal cortex (cingulate gyrus) in the regulation of hypothalamic-pituitary-adrenal responses to stress. J Neurosci 13: 3839–3847.

42. Stein MB, Koverola C, Hanna C, Torchia MG, McClarty B (1997) Hippocampal volume in women victimized by childhood sexual abuse. Psychol Med 27: 951–959. doi: 10.1017/s0033291797005242

43. Andersen SL, Tomada A, Vincow ES, Valente E, Polcari A, et al. (2008) Preliminary evidence for sensitive periods in the effect of childhood sexual abuse on regional brain development. J Neuropsychiatry Clin Neurosci 20: 292–301. doi: 10.1176/appi.neuropsych.20.3.292

44. Tomoda A, Suzuki H, Rabi K, Sheu YS, Polcari A, et al. (2009) Reduced prefrontal cortical gray matter volume in young adults exposed to harsh corporal punishment. Neuroimage 47 Suppl 2T66–71. doi: 10.1016/j.neuroimage.2009.03.005

45. Gordijn MS, van Litsenburg RR, Gemke RJ, Bierings MB, Hoogerbrugge PM, et al. (2012) Hypothalamic-pituitary-adrenal axis function in survivors of childhood acute lymphoblastic leukemia and healthy controls. Psychoneuroendocrinology 37: 1448–1456. doi: 10.1016/j.psyneuen.2012.01.014

46. Bhagwagar Z, Hafizi S, Cowen PJ (2003) Increase in concentration of waking salivary cortisol in recovered patients with depression. Am J Psychiatry 160: 1890–1891. doi: 10.1176/appi.ajp.160.10.1890

CHAPTER 8

CHILDHOOD MALTREATMENT IS ASSOCIATED WITH LARGER LEFT THALAMIC GRAY MATTER VOLUME IN ADOLESCENTS WITH GENERALIZED ANXIETY DISORDER

MEI LIAO, FAN YANG, YAN ZHANG, ZHONG HE, MING SONG, TIANZI JIANG, ZEXUAN LI, SHAOJIA LU, WEIWEI WU, LINYAN SU, AND LINGJIANG LIL

8.1 INTRODUCTION

Generalized anxiety disorder (GAD) is a common anxiety disorder that usually begins in adolescence, and it affects about 5.7% people in the general population [1]. GAD often co-occurs with major depressive disorder [2] and causes significant distress or impairment in life [3]. However, GAD is less studied compared to other anxiety disorders [4], despite its high prevalence and clinical importance.

The core feature of GAD is pathological anxiety, which is believed to arise from abnormalities in cortical/subcortical interactions based on fear conditioning framework [5], [6], [7]. Sensory fibers from multiple sensory

Childhood Maltreatment Is Associated with Larger Left Thalamic Gray Matter Volume in Adolescents with Generalized Anxiety Disorder. © Liao M, Yang F, Zhang Y, He Z, Song M, Jiang T, Li Z, Lu S, Wu W, Su L, and Li L.PLoS ONE, 8,8 (2013), doi:10.1371/journal.pone.0071898. Licensed under a Creative Commons Attribution License, http://creativecommons.org/licenses/by/3.0/.

modalities arrive at the amygdala passing through the thalamus [6]. The amygdala integrates different information and induces autonomic and behavioral fear response [8]. The thalamus plays an important role in filtering sensory information and emotional regulation [9]. The insular cortex seems to be associated with modulating subjective feeling states and interoceptive awareness [10]. The prefrontal cortex is involved in emotional regulation by down-regulating the activity of the amygdala and related limbic structures [11]. The medial prefrontal cortex and hippocampus are involved in the process of learning and remembering threat stimulus [5], [6]. Deficits in any of these brain regions or connections between these brain regions might result in pathological anxiety [5], [6], [7], [11].

Because of the fewer studies on GAD, the model for the neural circuitry of GAD is extrapolated from findings in other anxiety disorders, with limited empirical data available. Although a few structural neuroimaging studies have been performed in adolescents with GAD, the results are inconsistent. De Bellis et al. [12] observed an increased amygdala volume in GAD patients compared to healthy subjects, whereas another study [13] found a reduced left amygdala volume in adolescents with different anxiety disorders compared to healthy subjects, and a more pronounced decreased amygdala volume in GAD patients as opposed to those with other anxiety disorders. In addition, two studies in adult GAD patients showed larger amygdala [14], [15] and dorsomedial prefrontal cortex [15] in GAD patients relative to healthy subjects. Given the current limited and inconsistent structural neuroimaging data from GAD patients, the first purpose of the present study was to explore alterations of gray matter volumes in adolescent GAD patients.

Childhood maltreatment is highly prevalent with estimations of more than 30% of the adult population having experienced at least one form of maltreatment during childhood [16], and it increases the possibility for developing a variety of mental disorders including anxiety disorders [17]. Maltreatment includes physical, emotional and sexual abuse, as well as physical and emotional neglect [18]. Epidemiological studies have shown that 40% individuals having experienced childhood maltreatment, whether retrospectively or prospectively ascertained, develop anxiety disorders [17]. An earlier age at onset of GAD is significantly related to maltreatment in childhood [19]. Hence, exploring the underpinnings of the relationship between childhood maltreatment and adolescent onset GAD would be helpful in identifying the potential risk markers for this disease.

Recently, more and more studies have focused on the neurobiological consequences of childhood maltreatment. In animal studies, early adverse experiences, such as maternal separation or abuse, induce a series of long-term alterations on cognitive and emotional regulation, hypothalamus-pituitary-adrenal axis function, and brain morphology [20]. Alterations of brain structure, including decreased dendritic spine density, delayed maturation of neurons, altered neuronal structure and synapse formation, and reduced neurogenesis, have been found in the hippocampus, amygdala and prefrontal cortex [21], [22]. Significant morphological microglial activation has been observed in the thalamus and hippocampus in the rodent after stress [23]. In human studies, neuroimaging techniques have been widely used to investigate the changes of brain structure. In healthy subjects [24] or participants regardless of diagnosis [25], [26], [27], childhood maltreatment is frequently associated with reduced gray matter volumes in the hippocampus [24], [25], [26] and prefrontal cortex [24], [26], [27]. A meta-analysis [28] exhibited that amygdala volume in subjects with maltreatment-related posttraumatic stress disorder did not differ from that in healthy controls. However, recently two studies [29], [30] have reported an increased amygdala volume, whereas one study [31] have found a decreased amgdala volume in healthy adolescents who had experienced childhood maltreatment. Besides, reduced gray matter volumes in the insular [24] and thalamus [32], as well as increased gray matter volumes in the superior temporal gyrus [33] have been reported in healthy samples with childhood abuse. Most brain regions mentioned above are involved in anxiety circuitry, and these maltreatment-related gray matter volume changes were investigated in healthy subjects or participants regardless of diagnosis. Why do some subjects having experienced childhood maltreatment eventually develop into GAD, but not the other? Is there any childhood maltreatment related brain structure alteration associated with the occurrence of GAD? The second purpose of the present study, therefore, was to investigate the possible alterations of gray matter volume involved in the association between childhood maltreatment and GAD.

8.2 MATERIALS AND METHODS

8.2.1 SUBJECTS

Twenty-six patients with GAD (14/12, with/without childhood maltreatment) and 25 healthy controls (HCs) (12/13, with/without childhood maltreatment), were enrolled in the present study. All subjects were recruited from local high schools in Hunan Province via advertisements and school notices from Oct. 2011 to Jul. 2012. First, 1885 subjects finished the 41-item self-report questionnaire, the Screen for Child Anxiety Related Emotional Disorders (SCARED) [34], [35]. The SCARED is a reliable and valid screening tool for childhood anxiety disorders, with an optimal total cutoff point score of 25 to separate children with anxiety disorders from those without [34], [35]. Then, 508 subjects with positive SCARED scores and 165 in 1377 subjects with negative SCARED scores were diagnosed with DSM-IV criteria and the Schedule for Affective Disorders and Schizophrenia for School Age Children-Present and Lifetime (K-SADS-PL) version [36] by the same clinician. The K-SADS-PL is a semi-structured instrument to ascertain present and lifetime history of psychiatric disorders. In this study the age range of subjects was 16 to 18, so we only interviewed the adolescent. Inclusion criteria for patients in this study were current first-episode, medication-naive, generalized anxiety disorder without co-morbidity disorders. HCs met criteria for no mental disorders or physical diseases and were selected to match GAD patients on age, gender, and childhood maltreatment. Exclusion criteria for all subjects included current major depression disorder, other anxiety disorders, Tourette's syndrome, conduct disorder, suicidal ideation, lifetime mania, psychosis, or pervasive developmental disorders, mental retardation, any neurological abnormalities, history of seizures, head trauma or unconsciousness, and use of psychoactive substances. All subjects enrolled in this study were medication-naïve, right-handed, and volunteered to participate in this study. Written informed consent was obtained from each adolescent and one of his or her legal guardians after the study had been fully explained. This study was approved by the Ethics Committee of the Second Xiangya Hospital of Central South University, China. Psychological counselors who are responsible for the mental health of the adolescent in these high schools were present in the study. We were asked to give a global evaluation and necessary advices to all participants, including participants who had finished the SCARED but declined to MRI scans and participants who did not meet the inclusion criteria of this study. All potential participants who declined to participate were not disadvantaged in any other way by not participating in the study.

TABLE 1: Demographic, Questionnaire data of adolescent GAD patients and healthy controls.

	GAD (26)		HCs (25)		Statistical value	p
	CM (14)	WCM (12)	CM (12)	WCM (13)		
Age (year)	17.0 ± 0.20	16.67 ± 0.22	16.58 ± 0.22	16.85 ± 0.21	$F_{1,47} = 0.308$	0.582
Sex	7F/7M	6F/6M	6F/6M	6F/7M	$c^2 = 0.020$	0.886
CTQ	46.79 ± 1.35	32.08 ± 1.46	45.00 ± 1.46	32.70 ± 1.40	$F_{1,47} = 0.173$	0.680
Emotional abuse	9.14 ± 2.32	5.67 ± 0.99	7.58 ± 2.84	6.92 ± 1.55	$F_{1,47} = 0.069$	0.794
Emotional neglect	13.64 ± 2.79	7.83 ± 2.04	12.42 ± 3.32	8.31 ± 1.65	$F_{1,47} = 0.281$	0.599
Physical abuse	6.29 ± 2.43	5.25 ± 0.87	6.00 ± 1.21	5.54 ± 0.97	$F_{1,47} = 0.000$	0.997
Physical neglect	11.00 ± 1.88	7.25 ± 1.36	10.42 ± 2.28	7.92 ± 1.89	$F_{1,47} = 0.009$	0.927
Sexual abuse	6.71 ± 1.82	5.08 ± 0.29	5.75 ± 1.49	5.15 ± 0.38	$F_{1,47} = 1.710$	0.197
BDI	9.54 ± 4.61	8.67 ± 4.48	–	–	t = 0.479	0.637
PSWQ	52.5 ± 9.08	57.83 ± 8.64	–	–	t = –1.526	0.140

Means and standard deviations (±) are given. GAD, generalized anxiety disorder; HCs, healthy controls; CTQ, childhood trauma questionnaire; BDI, the Beck Depression Inventory; PSWQ, the Penn State Worry Questionnaire; CM, childhood maltreatment; WCM, without childhood maltreatment.

8.2.2 CLINICAL ASSESSMENT

The Childhood Trauma Questionnaire (CTQ) [37], a 28-item retrospective self-report questionnaire with a total sum score between a minimum

of 25 and a maximum of 125, was administered to assess childhood maltreatment in all subjects. The five CTQ subscales respectively assess five
kinds of childhood maltreatment, including physical abuse, sexual abuse,
emotional abuse, physical neglect, and emotional neglect [37]. Childhood
maltreatment (CM) was defined as a "moderate to severe" score on any of
five subscales. Moderate-severe cutoff scores for each subscale are ≥ 13
for emotional abuse; ≥ 10 for physical abuse or neglect; ≥ 15 for emotional
neglect; ≥ 8 for sexual abuse [38]. The rest of the subjects were considered to be subjects without childhood maltreatment (WCM) according to
the CTQ. Additionally, all the patients were assessed with the Penn State
Worry Questionnaire (PSWQ) [39] and the Beck Depression Inventory
(BDI) [40]. The two questionnaires were introduced to assess anxiety and
depression levels in adolescent GAD patients. Table 1 contains the demographic and clinical measures between GAD patients and HCs with nested
with or without childhood maltreatment comparisons.

8.2.3 STRUCTURAL MAGNETIC RESONANCE IMAGING (MRI) ACQUISITION

MRI examinations were conducted at the Second Xiangya Hospital of
Central South University, in China and performed with a Philips 3.0 Tesla
Scanner, equipped with a SENSE-8 channel head coil. For each participant, T1-weighted high-resolution anatomical images were obtained using
a 3-dimensional (3D) rapid acquisition gradient echo sequence, repetition time (TR) = 7.5 milliseconds, echo time (TE) = 3.7 milliseconds, flip
angle = 8°, field of view = 256 mm×256 mm, slice = 180, voxel size = 1
mm×1 mm×1 mm.

8.2.4 IMAGE ANALYSIS

Image analysis was conducted with SPM8 (http://www.fil.ion.ucl.ac.uk/
spm/) and the VBM8 toolbox (VBM8, version 435; http://dbm.neuro.uni-
jena.de/vbm8/). Individual structural images were preprocessed with the
VBM8 toolbox following the default parameter. T1-weighted images were

corrected for bias-field inhomogeneities, spatially normalized to the Montreal Neurological Institute standard template space, and segmented into gray matter, white matter and cerebrospinal fluid, within a unified model [41] including high-dimensional DARTEL normalization. Gray matter segments were modulated by the non-linear components only, which allows comparing the absolute amount of tissue corrected for individual brain sizes. The voxel resolution after normalization was 1.5 mm×1.5 mm×1.5 mm. The check data quality function was adopted to check homogeneity of gray matter images. No inconsistencies were found among these gray matter images. Finally, the segmented, modulated gray matter images were smoothed by a Gaussian kernel of 8 mm FWHM.

8.2.5 STATISTICAL METHODS

Statistical analysis for the demographic and clinical measures was performed by means of a general linear model with a 2 (diagnosis: GAD vs HCs)×2 (childhood maltreatment: CM vs WCM) comparison, chi-square test or t test, as needed, in SPSS16.

Image statistics were conducted with second-level models in SPM8. The smoothed gray matter images were entered into a voxel-by-voxel general linear model with a 2 (diagnosis: GAD vs HCs)×2 (childhood maltreatment: CM vs WCM) comparison, controlling for age and gender, to assess the diagnosis main effect (GAD > or < HCs), the maltreatment main effect (CM > or < WCM), and the diagnosis-by-maltreatment interaction effects. According to the aim of this study, the diagnosis main effect and the diagnosis-by-maltreatment interaction effects were of particular interest. We defined the amygdala, thalamus, insula, hippocampus and prefrontal cortex (especially the medial prefrontal cortex) as our regions of interest (ROIs), given their important roles in anxiety circuitry. The ROIs were defined according to Tzourio-Mazoyer et al. [42] and the ROIs masks were created by means of the WFU PickAtlas [43]. Then a supplementary whole brain analyses was conducted to examine non-hypothesized regions. For ROIs and whole brain analysis, a family wise error (FWE) rate correction for multiple comparisons was used with a threshold of p < .05. For the brain region of significant diagnosis-by-maltreatment interaction effect, the mean contrast values were extracted from each subject and further analyzed with SPSS16. We conducted a general linear model analysis and simple effect analysis to show the diagnosis-by-maltreatment interaction effect.

To supplement the interaction effect, we further conducted a whole brain regression analysis in SPM8 to investigate the relationship between childhood maltreatment and regional gray matter volume by regressing CTQ scores on the gray matter volume images in the separated groups, as well as in the combined group. A family wise error (FWE) rate correction for multiple comparisons was also used with a threshold of $p < .05$.

8.3 RESULTS

8.3.1 DEMOGRAPHIC AND CLINICAL MEASURES

The results are listed in Table 1. There were no significant differences between the groups in age, gender, CTQ scores, and subscales of the CTQ. No significant differences were found in BDI and PSWQ scores between adolescent GAD patients with or without childhood maltreatment.

8.3.2 STRUCTURAL ALTERATIONS IN GRAY MATTER VOLUMES

We found no diagnosis or maltreatment main effects in all ROIs, controlling for age and gender. However, a significant diagnosis-by-maltreatment interaction effect was observed in the left thalamus ($F_{1,45} = 14.96$; $p = 0.031$, FWE corrected; $x = -8$, $y = -10$, $z = 1$, cluster size = 83 voxels), as shown in Figure 1. No other ROIs showed significant interaction effect.

The whole brain analysis revealed a significant diagnosis main effect in the right putamen ($F_{1,45} = 29.51$; $p = 0.044$, FWE corrected; $x = 27$, $y = 11$, $z = 10$, cluster size = 263 voxels), with larger gray matter volume of the right putamen in adolescent GAD patients compared to healthy controls (GAD > HCs; Figure 2). No maltreatment main effect was found in the whole brain

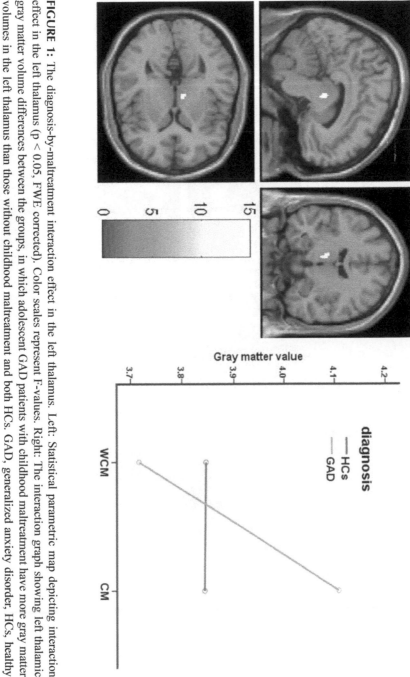

FIGURE 1: The diagnosis-by-maltreatment interaction effect in the left thalamus. Left: Statistical parametric map depicting interaction effect in the left thalamus ($p < 0.05$, FWE corrected). Color scales represent F-values. Right: The interaction graph showing left thalamic gray matter volume differences between the groups, in which adolescent GAD patients with childhood maltreatment have more gray matter volumes in the left thalamus than those without childhood maltreatment and both HCs. GAD, generalized anxiety disorder, HCs, healthy controls, CM, childhood maltreatment, WCM, without childhood maltreatment.

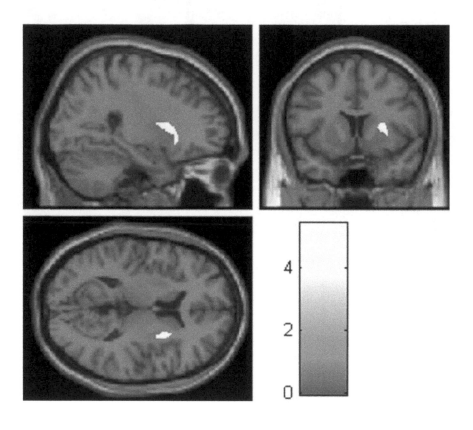

FIGURE 2: Increased right putaminal gray matter volume in adolescent GAD patients compared to healthy controls (p < 0.05, FWE corrected). Color scales represent t-values.

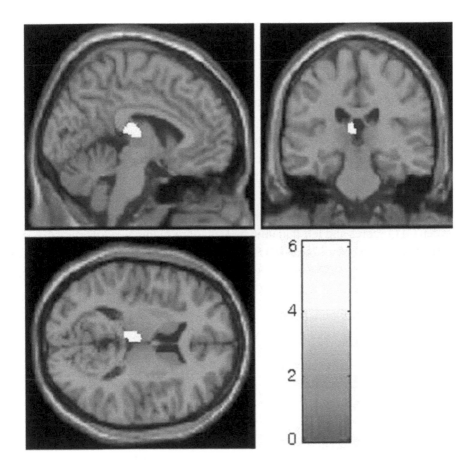

FIGURE 3: Statistical parametric map depicting the positive association between childhood maltreatment (Childhood Trauma Questionnaire [CTQ] scores) and left thalamic gray matter volume (mean contrast values) in GAD patients (p < 0.05, FWE corrected). Color scales represent t-values.

analysis controlling for age and gender. Whole brain analysis found no other brain regions except the left thalamus exhibited significant interaction effect.

Since a significant diagnosis-by-maltreatment interaction effect was observed in the left thalamus, the mean contrast values of this brain region were extracted. The general linear model analysis in SPSS, controlling for age and gender, also showed a significant diagnosis-by-maltreatment interaction effect in the left thalamus, that was adolescent GAD patients with childhood maltreatment had significantly larger gray matter volumes than adolescent GAD patients without childhood maltreatment and both HCs in the left thalamus (diagnosis-by-maltreatment; $F_{1,45}$ = 5.440, p = 0.024) (Figure 1). We compared subjects with childhood maltreatment and those without childhood maltreatment on each diagnosis level. The results exhibited that the maltreatment-related alteration in the left thalamus was only observed in adolescents with GAD (t = −3.514, p = .002), but not in HCs (p > .05).

The regression analysis only yielded a strong positive association between CTQ scores and left thalamic gray matter volume in adolescent GAD patients (x = −5, y = −24, z = 15; t = 6.16, df = 24; $p_{FWE-corrected}$ = 0.036, cluster size = 237), as shown in Figure 3. No brain regions were found to be significantly associated with CTQ scores in HCs or combined group.

8.4 DISCUSSION

We employed high-resolution structural magnetic resonance imaging and voxel-based morphometry approaches to study alterations in gray matter volume, as well as the association between childhood maltreatment and alterations in gray matter volume in adolescent GAD patients in the current study. The results of the present study showed larger gray matter volume in the right putamen in adolescent GAD patients and a diagnosis-by-maltreatment interaction effect in the left thalamus. Further analysis exhibited the significant maltreatment-related alteration in the left thalamus was only found in adolescents with GAD, but not in HCs.

The finding that exhibited no alterations in prior-set ROIs but larger gray matter volumes in an unexpected brain region, the putamen, in GAD subjects seems interesting. To the best of our knowledge, this is the first investigation in GAD reporting putaminal gray matter alterations. The putamen, which belongs to the basal ganglia, has widely spread functional connections with cortical and subcortical areas in the brain [44]. The putamen has been suggested to be related to a number of anxiety disorders and anxiety symptoms, such as GAD [45], social anxiety disorder [45], posttraumatic stress disorder [46], panic disorder [47], obsessive-compulsive disorder [48], lactated-induced anxiety [49] and anxiety symptoms in Parkinson disease [50]. Besides, GAD patients often accompany with somatic symptoms which are associated with sympathetic dysregulation [51]. Previous researches suggested several adrenergic receptors and dopamine receptors exerting an important influence on sympathetic regulation exist in the putamen [52], [53] . Therefore, alterations in the putamen might be associated with somatic anxiety symptoms of GAD.

The amygdala plays an important role in processing emotional valence and generating rapid fear response [5], [6], [8]. Several studies on GAD have reported alterations of gray matter volumes in the amygdala. Two studies in adult GAD patients exhibited larger amygdala [14], [15], whereas the other two studies in adolescent GAD patients yield inconsistent results in the amygdala [12], [13]. Three of these studies investigated GAD patients with co-morbid diagnosis [12], [13], [14], while GAD patients in this study had no co-morbidity disorders. This might partly explain the different findings on the amygada. Although one previous study also examined GAD patients without co-morbid diagnosis, all the subjects were female. Gender differences on amygdala gray matter volume have been reported in many researches [54], [55], [56]. In addition, as we described earlier, childhood maltreatment has been suggested to be associated with alterations of the amygdala volume [29], [30], [31]. All factors mentioned above might complicate the results on alteration of the amygdala.

This is the first study to investigate the possible association between childhood maltreatment and gray matter volumes in GAD patients. Epidemiological evidences have shown that childhood maltreatment would increase the risk of GAD [17], [18]. Our finding suggested a diagnosis-by-maltreatment interaction effect in the left thalamus and revealed a strong positive association between childhood maltreatment and left thalamic gray matter volume only in GAD patients. It partially suggested that the left thalamus might be the childhood maltreatment related brain structure

that would increase the risk of GAD. The diagnosis-by-maltreatment interaction effect in the left thalamus might be the reason why some subjects
with childhood maltreatment develop into GAD but not the others. The
thalamus, a major relay center of the brain with strong reciprocal connections with cortical and subcortical structures, such as the prefrontal
cortex and amygdala, is a critical component of the cortical-(amygdalo)-
thalamic circuits which plays a crucial role, not only in filtering sensory
information, but also in higher cognitive functions and emotional regulation [6], [7], [9], [57]. Changes in the thalamus, which is implicated in
sensory information filtering and alertness [9], [57], [58], might induce
pathological anxiety. Chronic stress increases the state of alertness [59],
which is associated with the thalamus [58]. Consistent with our result, a
Positron Emission Tomography study [60] revealed significantly greater
regional cerebral glucose metabolism in thalamus in adult monkeys who
experienced early life stress compared to controls, and another study [61]
showed that young adults who experienced corporal punishment in childhood exhibited increased cerebral blood volume in the thalamus. Greater
activation or increased volumes in the thalamus might suggest a general
problem with sensory information processing, perhaps indicating hypervigilance, which is thought to be involved in the pathophysiology of GAD.
Structural and functional alterations in the thalamus might reflect a maltreatment-related increase in sensitivity to conditional sensory information
in the environment.

However, one previous study [32] compared 31 physically abused children and 41 non-abused children regardless of mental disorders, and found
reduced bilateral thalamic gray matter volumes. In our study, the more
reported forms of maltreatment were physical and emotional neglect. It is
possible that different forms of childhood maltreatment might be associated with different alteration patterns of the thalamic gray matter volume.
In addition, the study conducted by Hanson et al. [32] investigated the possible linking between physical abuse and neurophysiological alterations in
a general population regardless of mental disorders, whereas we focused
on a possible association between childhood maltreatment and brain deficits in GAD patients. The heterogeneity of the sample might also account
for the inconsistent results.

Brain regions including the hippocampus [24], [25], [26] and prefrontal cortex [24], [26], [27] have been frequently reported to be associated with childhood maltreatment. Preclinical studies have confirmed that early adverse experiences induce alterations of the hypothalamus-pituitary-adrenal axis functions and further result in stress-related changes on the hippocampus [21] and prefrontal cortex [22].We did not find any significant association between childhood maltreatment and the hippocampus and prefrontal cortex at a harsh statistical threshold in this study. However, we found a negative association between childhood maltreatment and left prefrontal gray matter volume in GAD patients and combined group, at a more lenient threshold of $p < .001$, uncorrected. The maltreatment-related alteration in the prefrontal cortex is consistent with previous findings [24], [26], [27]. As concluded in a review, prolonged stress exposure causes architectural changes in prefrontal dendrites [22]. The prefrontal cortex is critically implicated in emotion regulation processes by down-regulating the limbic structures [5], [6], [11]. Volume reduction in the prefrontal cortex could be associated with insufficiencies in emotion regulation and therefore increase the vulnerability for pathological anxiety [22]. The hippocampus did not show any association with childhood maltreatment in any group even at a lenient threshold of $p < .001$, uncorrected. A possible explanation for this phenomenon is delayed effects of early stress on hippocampal development [62]. Reduced hippocampal volume has been consistently reported in adults with histories of childhood maltreatment, but this change has been rarely found in children with childhood maltreatment [28]. Animal studies also suggested that effects of early life stress on hippocampal morphology do not become apparent until adulthood [62].

The findings in the present study showed lateralization, such as larger right putaminal gray matter volume, and a positive association between childhood maltreatment and left thalamic gray matter volume in GAD patients. The reason for such lateralization might be the cases that findings on one side exceed the statistical threshold, while results on the other side did not. In this study, a positive association between childhood maltreatment and right thalamic gray matter volume was found at a more lenient threshold of $p < .005$, uncorrected. This association was not apparent at a more rigorous statistical threshold. However, even at a more lenient threshold of $p < .005$, uncorrected, no difference was found in left putaminal gray matter volume between GAD patients and HCs. The lateralization to the right is consistent with valence lateralization hypothesis, which suggests the right hemisphere is dominant for negative emotions [63]. The lateralization phenomena revealed in this study needs to be further clarified in the future study.

Some limitations of the current study must be acknowledged. First, the sample in this study was relatively small and we only studied first-episode, medication-naive, adolescent GAD patients aged 16 to 18, which might limit the generalizability of our findings. Second, the childhood trauma questionnaire introduced to assess childhood maltreatment is a retrospective self-report questionnaire, which could result in a recall bias. Although one epidemiological study [17] found no differences between prospective and retrospective reports in predicting associations between childhood maltreatment and adult psychopathology, patients with GAD could have a more negative recall bias and a better memory of childhood maltreatment. Third, this is a cross-sectional study, which cannot explain the direct relationships between childhood maltreatment and the occurrence of GAD. Forth, image pre-processing steps such as registration and smoothing in voxel-based morphometry might lead to different results [64].

In conclusion, we reported an increased gray matter volume of the right putamen in subjects with GAD relative to HCs, and a strong positive association between childhood maltreatment and left thalamic gray matter volume only in GAD patients. The increased gray matter volume of the right putamen suggests that pathological change of the putamen may be one of the neural substrates underlying the occurrence of GAD. The thalamus might be involved in the association between childhood maltreatment and the occurrence of GAD. In future studies, the impact of childhood maltreatment should be noted. Since childhood maltreatment is closely associated with GAD and increases the risk of this disorder by modulating brain structures, it seems that neuroimaging studies have been confounded by those multiple effects caused by childhood maltreatment. It should also be noted that there are sensitive periods during which specific brain regions are vulnerable to early adversity [26], and childhood maltreatment-related brain structural alterations might occur at a specific age. Anyway, a longitudinal investigation with a large sample is required to validate the results in our study.

REFERENCES

1. Kessler RC, Berglund P, Demler O, Jin R, Merikangas KR, et al. (2005) Lifetime prevalence and age-of-onset distributions of DSM-IV disorders in the National Comorbidity Survey Replication. Arch Gen Psychiatry 62: 593–602. doi: 10.1001/archpsyc.62.6.593

2. Kessler RC, Gruber M, Hettema JM, Hwang I, Sampson N, et al. (2008) Co-morbid major depression and generalized anxiety disorders in the National Comorbidity Survey follow-up. Psychol Med 38: 365–374. doi: 10.1017/s0033291707002012

3. Hoffman DL, Dukes EM, Wittchen HU (2008) Human and economic burden of generalized anxiety disorder. Depress Anxiety 25: 72–90. doi: 10.1002/da.20257

4. Dugas MJ, Anderson KG, Deschenes SS, Donegan E (2010) Generalized anxiety disorder publications: where do we stand a decade later? J Anxiety Disord 24: 780–784. doi: 10.1016/j.janxdis.2010.05.012

5. Shin LM, Liberzon I (2010) The neurocircuitry of fear, stress, and anxiety disorders. Neuropsychopharmacology 35: 169–191. doi: 10.1038/npp.2009.83

6. Cannistraro PA, Rauch SL (2003) Neural circuitry of anxiety: evidence from structural and functional neuroimaging studies. Psychopharmacol Bull 37: 8–25.

7. Boatman JA, Kim JJ (2006) A thalamo-cortico-amygdala pathway mediates auditory fear conditioning in the intact brain. Eur J Neurosci 24: 894–900. doi: 10.1111/j.1460-9568.2006.04965.x

8. LeDoux JE, Iwata J, Cicchetti P, Reis DJ (1988) Different projections of the central amygdaloid nucleus mediate autonomic and behavioral correlates of conditioned fear. J Neurosci 8: 2517–2529.

9. Herrero MT, Barcia C, Navarro JM (2002) Functional anatomy of thalamus and basal ganglia. Childs Nerv Syst 18: 386–404. doi: 10.1007/s00381-002-0604-1

10. Nagai M, Kishi K, Kato S (2007) Insular cortex and neuropsychiatric disorders: a review of recent literature. Eur Psychiatry 22: 387–394. doi: 10.1016/j.eurpsy.2007.02.006

11. Barbas H, Zikopoulos B, Timbie C (2011) Sensory pathways and emotional context for action in primate prefrontal cortex. Biol Psychiatry 69: 1133–1139. doi: 10.1016/j.biopsych.2010.08.008

12. De Bellis MD, Casey BJ, Dahl RE, Birmaher B, Williamson DE, et al. (2000) A pilot study of amygdala volumes in pediatric generalized anxiety disorder. Biol Psychiatry 48: 51–57. doi: 10.1016/s0006-3223(00)00835-0

13. Milham MP, Nugent AC, Drevets WC, Dickstein DP, Leibenluft E, et al. (2005) Selective reduction in amygdala volume in pediatric anxiety disorders: a voxel-based morphometry investigation. Biol Psychiatry 57: 961–966. doi: 10.1016/j.biopsych.2005.01.038

14. Etkin A, Prater KE, Schatzberg AF, Menon V, Greicius MD (2009) Disrupted amygdalar subregion functional connectivity and evidence of a compensatory network in generalized anxiety disorder. Arch Gen Psychiatry 66: 1361–1372. doi: 10.1001/archgenpsychiatry.2009.104

15. Schienle A, Ebner F, Schafer A (2011) Localized gray matter volume abnormalities in generalized anxiety disorder. Eur Arch Psychiatry Clin Neurosci 261: 303–307. doi: 10.1007/s00406-010-0147-5

16. Scher CD, Forde DR, McQuaid JR, Stein MB (2004) Prevalence and demographic correlates of childhood maltreatment in an adult community sample. Child Abuse Negl 28: 167–180. doi: 10.1016/j.chiabu.2003.09.012

17. Scott KM, McLaughlin KA, Smith DA, Ellis PM (2012) Childhood maltreatment and DSM-IV adult mental disorders: comparison of prospective and retrospective findings. Br J Psychiatry 200: 469–475. doi: 10.1192/bjp.bp.111.103267

18. Gilbert R, Widom CS, Browne K, Fergusson D, Webb E, et al. (2009) Burden and consequences of child maltreatment in high-income countries. Lancet 373: 68–81. doi: 10.1016/s0140-6736(08)61706-7

19. Goncalves DC, Byrne GJ (2012) Sooner or later: age at onset of generalized anxiety disorder in older adults. Depress Anxiety 29: 39–46. doi: 10.1002/da.20881

20. Sanchez MM, Ladd CO, Plotsky PM (2001) Early adverse experience as a developmental risk factor for later psychopathology: evidence from rodent and primate models. Dev Psychopathol 13: 419–449. doi: 10.1017/s0954579401003029

21. Lupien SJ, McEwen BS, Gunnar MR, Heim C (2009) Effects of stress throughout the lifespan on the brain, behaviour and cognition. Nat Rev Neurosci 10: 434–445. doi: 10.1038/nrn2639

22. Arnsten AF (2009) Stress signalling pathways that impair prefrontal cortex structure and function. Nat Rev Neurosci 10: 410–422. doi: 10.1038/nrn2648

23. Sugama S, Fujita M, Hashimoto M, Conti B (2007) Stress induced morphological microglial activation in the rodent brain: involvement of interleukin-18. Neuroscience 146: 1388–1399. doi: 10.1016/j.neuroscience.2007.02.043

24. Dannlowski U, Stuhrmann A, Beutelmann V, Zwanzger P, Lenzen T, et al. (2012) Limbic scars: long-term consequences of childhood maltreatment revealed by functional and structural magnetic resonance imaging. Biol Psychiatry 71: 286–293. doi: 10.1016/j.biopsych.2011.10.021

25. Teicher MH, Anderson CM, Polcari A (2012) Childhood maltreatment is associated with reduced volume in the hippocampal subfields CA3, dentate gyrus, and subiculum. Proc Natl Acad Sci U S A 109: E563–572. doi: 10.1073/pnas.1115396109

26. Andersen SL, Tomada A, Vincow ES, Valente E, Polcari A, et al. (2008) Preliminary evidence for sensitive periods in the effect of childhood sexual abuse on regional brain development. J Neuropsychiatry Clin Neurosci 20: 292–301. doi: 10.1176/appi.neuropsych.20.3.292

27. van Harmelen AL, van Tol MJ, van der Wee NJ, Veltman DJ, Aleman A, et al. (2010) Reduced medial prefrontal cortex volume in adults reporting childhood emotional maltreatment. Biol Psychiatry 68: 832–838. doi: 10.1016/j.biopsych.2010.06.011

28. Woon FL, Hedges DW (2008) Hippocampal and amygdala volumes in children and adults with childhood maltreatment-related posttraumatic stress disorder: a meta-analysis. Hippocampus 18: 729–736. doi: 10.1002/hipo.20437

29. Tottenham N, Hare TA, Quinn BT, McCarry TW, Nurse M, et al. (2010) Prolonged institutional rearing is associated with atypically large amygdala volume and difficulties in emotion regulation. Dev Sci 13: 46–61. doi: 10.1111/j.1467-7687.2009.00852.x

30. Mehta MA, Golembo NI, Nosarti C, Colvert E, Mota A, et al. (2009) Amygdala, hippocampal and corpus callosum size following severe early institutional deprivation: the English and Romanian Adoptees study pilot. J Child Psychol Psychiatry 50: 943–951. doi: 10.1111/j.1469-7610.2009.02084.x

31. Edmiston EE, Wang F, Mazure CM, Guiney J, Sinha R, et al. (2011) Corticostriatal-limbic gray matter morphology in adolescents with self-reported exposure to childhood maltreatment. Arch Pediatr Adolesc Med 165: 1069–1077. doi: 10.1001/archpediatrics.2011.565

32. Hanson JL, Chung MK, Avants BB, Shirtcliff EA, Gee JC, et al. (2010) Early stress is associated with alterations in the orbitofrontal cortex: a tensor-based morphometry investigation of brain structure and behavioral risk. J Neurosci 30: 7466–7472. doi: 10.1523/jneurosci.0859-10.2010

33. Tomoda A, Sheu YS, Rabi K, Suzuki H, Navalta CP, et al. (2011) Exposure to parental verbal abuse is associated with increased gray matter volume in superior temporal gyrus. Neuroimage 54 Suppl 1S280–286. doi: 10.1016/j.neuroimage.2010.05.027

34. Birmaher B, Brent DA, Chiappetta L, Bridge J, Monga S, et al. (1999) Psychometric properties of the Screen for Child Anxiety Related Emotional Disorders (SCARED): a replication study. J Am Acad Child Adolesc Psychiatry 38: 1230–1236. doi: 10.1097/00004583-199910000-00011

35. Su L, Wang K, Fan F, Su Y, Gao X (2008) Reliability and validity of the screen for child anxiety related emotional disorders (SCARED) in Chinese children. J Anxiety Disord 22: 612–621. doi: 10.1016/j.janxdis.2007.05.011

36. Kaufman J, Birmaher B, Brent D, Rao U, Flynn C, et al. (1997) Schedule for Affective Disorders and Schizophrenia for School-Age Children-Present and Lifetime Version (K-SADS-PL): initial reliability and validity data. J Am Acad Child Adolesc Psychiatry 36: 980–988. doi: 10.1097/00004583-199707000-00021

37. Bernstein DP, Stein JA, Newcomb MD, Walker E, Pogge D, et al. (2003) Development and validation of a brief screening version of the Childhood Trauma Questionnaire. Child Abuse Negl 27: 169–190. doi: 10.1016/s0145-2134(02)00541-0

38. Majer M, Nater UM, Lin JM, Capuron L, Reeves WC (2010) Association of childhood trauma with cognitive function in healthy adults: a pilot study. BMC Neurol 10: 61. doi: 10.1186/1471-2377-10-61

39. Meyer TJ, Miller ML, Metzger RL, Borkovec TD (1990) Development and validation of the Penn State Worry Questionnaire. Behav Res Ther 28: 487–495. doi: 10.1016/0005-7967(90)90135-6

40. Beck AT, Beamesderfer A (1974) Assessment of depression: the depression inventory. Mod Probl Pharmacopsychiatry 7: 151–169.

41. Ashburner J, Friston KJ (2005) Unified segmentation. Neuroimage 26: 839–851. doi: 10.1016/j.neuroimage.2005.02.018

42. Tzourio-Mazoyer N, Landeau B, Papathanassiou D, Crivello F, Etard O, et al. (2002) Automated anatomical labeling of activations in SPM using a macroscopic anatomical parcellation of the MNI MRI single-subject brain. Neuroimage 15: 273–289. doi: 10.1006/nimg.2001.0978

43. Maldjian JA, Laurienti PJ, Kraft RA, Burdette JH (2003) An automated method for neuroanatomic and cytoarchitectonic atlas-based interrogation of fMRI data sets. Neuroimage 19: 1233–1239. doi: 10.1016/s1053-8119(03)00169-1

44. Di Martino A, Scheres A, Margulies DS, Kelly AM, Uddin LQ, et al. (2008) Functional connectivity of human striatum: a resting state FMRI study. Cerebral Cortex 18: 2735–2747. doi: 10.1093/cercor/bhn041

45. Guyer AE, Choate VR, Detloff A, Benson B, Nelson EE, et al. (2012) Striatal functional alteration during incentive anticipation in pediatric anxiety disorders. Am J Psychiatry 169: 205–212. doi: 10.1176/appi.ajp.2011.11010006

46. Geuze E, Westenberg HG, Jochims A, de Kloet CS, Bohus M, et al. (2007) Altered pain processing in veterans with posttraumatic stress disorder. Arch Gen Psychiatry 64: 76–85. doi: 10.1001/archpsyc.64.1.76

47. Yoo HK, Kim MJ, Kim SJ, Sung YH, Sim ME, et al. (2005) Putaminal gray matter volume decrease in panic disorder: an optimized voxel-based morphometry study. Eur J Neurosci 22: 2089–2094. doi: 10.1111/j.1460-9568.2005.04394.x

48. Yoo SY, Roh MS, Choi JS, Kang DH, Ha TH, et al. (2008) Voxel-based morphometry study of gray matter abnormalities in obsessive-compulsive disorder. J Korean Med Sci 23: 24–30. doi: 10.3346/jkms.2008.23.1.24

49. Reiman EM, Raichle ME, Robins E, Mintun MA, Fusselman MJ, et al. (1989) Neuroanatomical correlates of a lactate-induced anxiety attack. Arch Gen Psychiatry 46: 493–500. doi: 10.1001/archpsyc.1989.01810060013003

50. Weintraub D, Newberg AB, Cary MS, Siderowf AD, Moberg PJ, et al. (2005) Striatal dopamine transporter imaging correlates with anxiety and depression symptoms in Parkinson's disease. J Nucl Med 46: 227–232.

51. Fisher AJ, Granger DA, Newman MG (2010) Sympathetic arousal moderates self-reported physiological arousal symptoms at baseline and physiological flexibility in response to a stressor in generalized anxiety disorder. Biol Psychol 83: 191–200. doi: 10.1016/j.biopsycho.2009.12.007

52. Isovich E, Mijnster MJ, Flugge G, Fuchs E (2000) Chronic psychosocial stress reduces the density of dopamine transporters. Eur J Neurosci 12: 1071–1078. doi: 10.1046/j.1460-9568.2000.00969.x

53. Flugge G, van Kampen M, Meyer H, Fuchs E (2003) Alpha2A and alpha2C-adrenoceptor regulation in the brain: alpha2A changes persist after chronic stress. Eur J Neurosci 17: 917–928. doi: 10.1046/j.1460-9568.2003.02510.x

54. Cosgrove KP, Mazure CM, Staley JK (2007) Evolving knowledge of sex differences in brain structure, function, and chemistry. Biol Psychiatry 62: 847–855. doi: 10.1016/j.biopsych.2007.03.001

55. Goldstein JM, Seidman LJ, Horton NJ, Makris N, Kennedy DN, et al. (2001) Normal sexual dimorphism of the adult human brain assessed by in vivo magnetic resonance imaging. Cerebral Cortex 11: 490–497. doi: 10.1093/cercor/11.6.490

56. Lenroot RK, Giedd JN (2010) Sex differences in the adolescent brain. Brain Cogn 72: 46–55. doi: 10.1016/j.bandc.2009.10.008

57. Haber SN, Calzavara R (2009) The cortico-basal ganglia integrative network: the role of the thalamus. Brain Res Bull 78: 69–74. doi: 10.1016/j.brainresbull.2008.09.013

58. Posner MI, Petersen SE (1990) The attention system of the human brain. Annu Rev Neurosci 13: 25–42. doi: 10.1146/annurev.ne.13.030190.000325

59. Charmandari E, Tsigos C, Chrousos G (2005) Endocrinology of the stress response. Annu Rev Physiol 67: 259–284. doi: 10.1146/annurev.physiol.67.040403.120816

60. Parr LA, Boudreau M, Hecht E, Winslow JT, Nemeroff CB, et al. (2012) Early life stress affects cerebral glucose metabolism in adult rhesus monkeys (Macaca mulatta). Dev Cogn Neurosci 2: 181–193. doi: 10.1016/j.dcn.2011.09.003

61. Sheu YS, Polcari A, Anderson CM, Teicher MH (2010) Harsh corporal punishment is associated with increased T2 relaxation time in dopamine-rich regions. Neuroimage 53: 412–419. doi: 10.1016/j.neuroimage.2010.06.043

62. Andersen SL, Teicher MH (2004) Delayed effects of early stress on hippocampal development. Neuropsychopharmacology 29: 1988–1993. doi: 10.1038/sj.npp.1300528

63. Wager TD, Phan KL, Liberzon I, Taylor SF (2003) Valence, gender, and lateralization of functional brain anatomy in emotion: a meta-analysis of findings from neuroimaging. Neuroimage 19: 513–531. doi: 10.1016/s1053-8119(03)00078-8

64. Ashburner J, Friston KJ (2000) Voxel-based morphometry–the methods. Neuroimage 11: 805–821. doi: 10.1006/nimg.2000.0582

CHAPTER 9

REDUCED VISUAL CORTEX GRAY MATTER VOLUME AND THICKNESS IN YOUNG ADULTS WHO WITNESSED DOMESTIC VIOLENCE DURING CHILDHOOD

AKEMI TOMODA, ANN POLCARI, CARL M. ANDERSON, AND MARTIN H. TEICHER

9.1 INTRODUCTION

Witnessing domestic violence (WDV) is a highly stressful and potentially traumatic event. Approximately 15.5 million children in the U.S. witness DV annually [1]. Although many parents try to shelter their children from WDV, children in violent homes commonly see, hear, and intervene in episodes of WDV [2]. WDV increases risk for depression [3] and aggression [4], [5] by 2–4 fold, and is a frequent causes of childhood PTSD [6], [7].

Is WDV during childhood associated with enduring effects on brain morphometry? We recently reported a reduction in the integrity of the inferior longitudinal fasciculus (ILF) interconnecting visual cortex to limbic

Reduced Visual Cortex Gray Matter Volume and Thickness in Young Adults Who Witnessed Domestic Violence during Childhood. © *Tomoda A, Polcari A, Anderson CM, and Teicher MH.* PLoS ONE 7(12): e52528. doi:10.1371/journal.pone.0052528. *Licensed under a Creative Commons Attribution License, http://creativecommons.org/licenses/by/3.0/.*

system in a sample of young adults who witnessed interparental violence during childhood [8]. The aim of this study was to investigate whether WDV during childhood was associated with enduring differences in GMV. Voxel based morphometry (VBM) was used to provide an unbiased, even-handed, whole-brain, voxel-by-voxel assessment in a healthy community sample. FreeSurfer, a software program for cortical surface-based reconstruction and analysis [9], [10], was then used in a more focused manner to verify and extend the findings.

9.2 METHODS

9.2.1 ETHICS STATEMENT

This Project has been reviewed and approved by the McLean IRB, Assurance # 00002744. During the review of this Project, the IRB specifically considered (i) the risks and anticipated benefits, if any, to subjects; (ii) the selection of subjects; (iii) the procedures for securing and documenting informed consent; (iv) the safety of subjects; and (v) the privacy of subjects and confidentiality of the data. All participants gave written informed consent prior to participation.

9.2.2 PARTICIPANTS

9.2.2.1 RECRUITMENT AND SCREENING.

Participants were recruited from the community using methods previously detailed [8], [11], [12], [13]. Our goal was to recruit unmedicated right-handed subjects 18–25 years of age who visually witnessed episodes of DV but were not exposed to childhood sexual abuse parental loss, neglect, physical maltreatment or other traumatic events. Subjects were also required to be free from neurological disease or insult, including head trauma resulting in loss of consciousness >5 minutes or migraine headaches. Subjects were excluded with a history of premature birth or birth complications; maternal substance abuse during pregnancy; or medical disorders that could affect brain development.

All potentially eligible subjects from online screenings were invited to the laboratory and underwent detailed evaluation, including Structural Clinical Interviews for DSM-IV Axis I and II psychiatric disorders (SCID) [14]. Exposure to interfamilial violence and other forms of maltreatment were assessed using the 100-item semi-structured Traumatic Antecedents Interview [15]. This interview was designed to evaluate reports of child-hood sexual abuse, physical abuse, WDV, physical or emotional neglect, significant separations or losses, parental verbal abuse, or parental discord [16]. Certified mental health clinicians conducted the assessment and evaluation interviews. A panel of three doctoral-level psychiatric clinicians with extensive experience treating trauma-exposed individuals reviewed information on potential subjects and group assignments were made by full consensus. WDV subjects were selected without regard to psychiatric history, except for alcohol or drug abuse, which were exclusion factors. Selecting subjects meeting criteria for a specific disorder could bias results by only including the most severely affected subjects. Conversely, selecting subjects without any psychiatric history could bias results in the opposite direction. The intent was to recruit a balanced sample that would provide a rigorous test of our proposed hypotheses.

Subjects provided written informed consent prior to completing the online screening instrument, and again before interviews and brain imaging. The present sample overlaps with a previous reported sample used to study fiber tract integrity [8], but includes two more subjects with WDV and three additional controls.

9.2.2.2 SUBJECTS.

Twenty-two subjects (6 males, 16 females; mean age, 21.8±2.4 years) with a history of WDV and 30 healthy age-equivalent controls (8 males, 22 females; mean age, 21.6±2.1 years) who met full criteria were imaged. Subjects in the WDV group reported seeing and hearing years of intense verbal aggression between their parents, which culminated in some years in acts of physical violence. Overall, they reported that they witnessed

3.8±3.5 (mean ±S.D.) years of exposure to interparental physical plus verbal aggression (IP-PA) along with 6.0±4.8 years of exposure to inter-parental verbal aggression without physical violence (IP-VA), for a total duration of 9.8±3.2 years. Controls had no histories of exposure to abuse, traumatic events, or harsh corporal punishment, and did not meet criteria for any major Axis I or Axis II psychiatric disorder. All participants were right-handed and unmedicated.

9.2.3 ASSESSMENTS

9.2.3.1 ABUSE AND TRAUMA RATINGS.

Exposure to parental verbal abuse was assessed with the Verbal Aggression Scale [17], which provides a continuous measure of exposure. Self-report ratings of psychiatric symptoms were obtained using Kellner's Symptom Questionnaire [18]. Ratings of dissociation and 'limbic irritability' were obtained using the Dissociative Experience Scale [19] and limbic system checklist-33 [20].

Childhood poverty may be an important risk factor for psychopathology and affect trajectories of brain development. Young adult subjects were often uncertain about parental income while they were growing up. However, they were well aware of the degree of perceived financial sufficiency, or stress they experienced during this time. This was rated with a Likert item ranging from 1 (much less than enough money for our needs) to 5 (much more than enough money for our needs). Perceived financial sufficiency explained a greater share of the variance in symptom ratings than combined family income.

9.2.3.2 MRI ACQUISITION AND ANALYSIS.

Image analysis was performed on high-resolution, T1-weighted MRI datasets, which were acquired on a Trio Scanner (3T; Siemens AG, Siemens Medical Solutions, Erlangen, Germany). An inversion prepared 3D MPRAGE sequence was used with an eight-element phased-array RF reception coil (Siemens AG). GRAPPA acquisition and processing was used to reduce the scan time, with a GRAPPA factor of 2. Scan parameters were: the sagittal plane, TE/TR/TI/flip = 2.74 ms/2.1 s/1.1 s/12 deg; 3D matrix 256×256×128 on 256×256×170 mm field of view; bandwidth 48.6 kHz; scan time 4:56.

VBM using Statistical Parametric Mapping (SPM8; Wellcome Department of Imaging Neuroscience, University College, London) was conducted as previously described [11], [21], [22]. Briefly, coarsely segmented images were registered to a standard template [23], [24] that conforms to the space defined by the ICBM, NIH P-20 project. Volume changes induced by normalization were adjusted via a modulation algorithm. Spatially normalized images were segmented into gray and white matter and smoothed using a 12-mm full-width half-maximum isotropic Gaussian kernel to generate a statistical parametric map. Between-group differences were analyzed using a general linear model. Potential confounding effects of age, sex, parental education, perceived financial sufficiency, parental verbal abuse, full scale IQ and whole segment GMV were modeled, and attributable variances excluded. Statistical threshold was set at P<0.05 with correction for multiple comparisons at cluster level (height threshold of Z>3.09) because of the increased sensitivity of clusters to detect spatially extended signal changes [25], [26]. Inference testing was based on the theory of Gaussian fields [27]. Potential problems relating to non-isotropic smoothness, which can invalidate cluster level comparisons [23], were corrected by adjusting cluster size from the resel per voxel image [25], [28].

VBM is a potentially powerful technique but it hinges on a number of assumptions, particularly the accuracy of image co-registration [29]. Hence, VBM findings were reevaluated using an independent technique that does not rely on co-registration. Cortical surface-based analysis was performed using FreeSurfer (version 5.1; Laboratory for Computational Neuroimaging, Martinos Center for Biomedical Imaging, Boston, MA) [9], [10], [30]. Each subject's reconstructed brain was converted to an average spherical surface representation that optimally aligned sulcal and gyral features for the individual subject [9], [10]. Subdivision of the cortical ribbon into gyral-based subdivisions resulted in the identification of 82 validated cortical parcellation units per hemisphere. By application of the original deformation algorithms in reverse, ROIs were mapped back on to each unfolded surface [10]. Differences between WDV and control groups in thickness were assessed using analysis of covariance (ANCOVA) with parental education, financial sufficiency, age, gender degree of exposure to PVA, and mean cortical thickness as covariates. Parcellation regions selected for analysis were located in and around the areas of greatest difference identified by VBM. They included visual areas V1, V2 and V5/MT, lingual, fusiform, superior, middle and inferior occipital gyri, occipital pole and cuneus. Correction for multiple comparisons were made using the False discovery rate (FDR) method of Benjamini and Hochberg [31].

9.2.3.3 STATISTICAL ANALYSES.

Data analyses were conducted using R [32]. Differences between groups were evaluated using ANCOVA. The primary hypothesis tested was whether there were significant regional brain differences between subjects who witnessed DV and unexposed psychiatrically healthy controls.

9.2.3.4 SUSCEPTIBLE VERSUS RESILIENT SUBJECTS.

Four additional analyses were performed to extend these findings. In the first we divided the WDV group into susceptible subjects who met criteria for either major depression, PTSD, anxiety disorders, eating disorders or personality disorders following exposure to interparental violence (n = 13) and relatively resilient subjects who have not met criteria for any of these disorders (n = 9). This division was used to determine whether observed group differences in volume or thickness were specific to the susceptible subgroup.

9.2.3.5 WITNESSING INTERPARENTAL VERBAL AGGRESSION VERSUS PHYSICAL AGGRESSION.

Second, we assessed whether witnessing interparental physical aggression (IP-PA) was more strongly associated with alterations in regional GMV or thickness then witnessing episodes of interparental verbal aggression (IP-VA). The relative statistical impact of duration of exposure to IP-VA versus IP-PA on measures of thickness or volume (adjusted for age, gender and total GMV or mean cortical thickness) was determined using multiple regression. Additional regressors included parental education, perceived financial sufficiency and exposure to parental verbal abuse. The variance decomposition method of Lindeman et al., [33], [34] was used to more accurately determine the proportion of the variance attributable to each regressor, as these regressors were not entirely independent (R package relaimpo).

9.2.3.6 SENSITIVE PERIOD ANALYSIS.

Third, the presence of a potential 'sensitive period' when childhood exposure to WDV was most strongly associate with adult measures of volume or thickness was assessed using random forest regression with conditional inference trees ('cforest 'in R package party [35]). This is a new computationally intensive analytic procedure based on decision trees. It is a form of "ensemble learning" in which a large number of unpruned decision trees are generated and their results aggregated. Advantages of random forest regression include: (1) very high classification accuracy; (2) a novel method of determining variable importance; (3) no restrictions regarding the distribution and scaling properties of the data; and (4) high tolerance for multicolinearity [36], [37]. We used a variant of Breiman's approach with conditional trees as the base learners to avoid a potential problem with biased estimates that can emerge when variables differ in range or number of categories [35]. For these analyses 1000 trees were generated with four variables randomly selected for evaluation at each node. Conditional forest regression indicates importance by assessing the decrease in accuracy of the forest's fit following permutation (effective elimination) of a given predictor variable. The more the permutation of a variable decreases accuracy the greater the importance of the variable. Cross-correlations in exposure rates between different ages were also controlled [35]. Exposure to interparental violence from ages 3–16 was coded with 0 for no exposure, 1 for witnessing IP-VA and 2 for witnessing IP-PA in a given year. Conditional forest regression determined the relative importance of exposure at each age on measures of volume or thickness adjusted for age, gender, perceived financial sufficiency, parental education, PVA and total GMV or mean cortical thickness.

9.2.3.7 SYMPTOM RATINGS AND BRAIN MEASURES.

Finally, exploratory correlation analyses assessed whether differences in GMV in the area that differed most significantly between WDV subjects and controls could potentially account for a significant portion of the variance in pre-specified symptom ratings. Both cases and controls were included in the analyses, as we were interested in assessing potential functional correlates of these GMV differences in the general population, not just in subjects with WDV. Examining correlations in only one group, as sometimes advocated, restricts the range of the independent variable and can seriously bias results. Range restrictions deflate correlation coefficients if they reduce the standard deviation of the distribution of scores on one or both variables, and inflate r values if they increase the standard deviation, and should be avoided when possible [38]. Further, the size of the full sample provided sufficient power (0.8) to detect medium size effects (r~0.4). Analysis of the individuals groups only provided sufficient power (0.8) to reliably detect large effects. However, within group analyses were examined for variables that showed a significant relationship with GMV in the entire subject pool to determine if the regressive relationship applied to one group more than the other. Partial correlation analysis was used to minimize the influence of age and gender on the strength of the association.

9.3 RESULTS

9.3.1 DEMOGRAPHICS AND IQ

The two groups were well matched in gender, age, and education (Table 1). There was about a two-year difference in extent of parental education, and significant differences in perceived financial sufficiency and exposure to parental verbal abuse. WDV subjects indicated that their family's financial resources were on average nearly adequate, while controls indicated that they were more than adequate. Subjects in these two groups also had similar mean performance scores on memory tests. There were no significant differences in IQ measures between the groups, which were in the range typical for college students.

9.3.2 SYMPTOM RATINGS AND DIAGNOSES

Subjects in the WDV group had increased ratings of anxiety, depression, somatization, anger-hostility, dissociation and 'limbic irritability'. Thirteen subjects in the WDV group meet DSM-IV criteria for one or more disorders. There were 9 subjects who met criteria (past or current) for major depression. Seven subjects met criteria for an anxiety disorder including four subjects with a history of PTSD. Two subjects had a past history of eating disorders, and one subject met criteria for a personality disorder. There were no differences between groups in their degree of drug use. WDV subjects consumed slightly more alcohol.

9.3.3 VOXEL-BASED MORPHOMETRY

The most prominent VBM finding was a 6.1% decrease in GMV in the right lingual gyrus (BA18; Talairach's coordinates x = 20, y = −103, z = 0, Z = 4.62, P = 0.029, FDR corrected peak level) (Fig. 1). The identified region consisted of a 401 voxel cluster in the right lingual gyrus ~ inferior occipital gyrus (BA18; Talairach's coordinates x = 20–36, y = −103– −92, z = −7–0). No other areas of decreased GMV were found with a corrected cluster or peak probability value that approached significance. Two small regions of decreased GMV were identified in the left cuneus (BA18, x = −20, y = −103, z = −3, cluster size = 216) and right lingual gyrus (BA18, x = 6, y = −78, z = 1, cluster size = 159) at an uncorrected peak level (Z = 3.94 and 3.43, respectively). There was also one small region of increased GMV in WDV subjects in the right thalamus (x = 14, y = −23, z = 12, cluster size = 65) that was significant (Z = 3.47) at an uncorrected peak level.

TABLE 1: Demographic characteristics and assessments or ratings of subjects witnessing domestic violence and unexposed controls.

Characteristics	Unexposured [95% CI] N = 30	Witnessed Domestic Violence N =22	Statistics (ANOVA, other)	p-value
Gender (Males/Females)	8M/22F	6M/16F	Fisher	1
Age (years)	21.6 [20.81–22.39]	21.8 [20.77–22.87]	0.122	0.86
Subject education (years)[a]	14.28 [13.83–14.73]	13.57 [12.12–15.03]	1.18	p > 0.2
Parental education (years)	**16.28 [15.30–17.27]**	**14.50 [13.32–15.68]**	**5.47**	**p < 0.03**
Perceived financial sufficiency	**3.63 [3.38–3.88]**	**2.77 [2.47–3.08]**	**20.25**	**p < 0.00005**
Parental Verbal Abuse Score	**12.53 [9.86–15.20]**	**40.45 [31.68–49.23]**	**49.92**	**p < 10^{-8}**
Memory Assessment Score				
Short-term memory[a,b]	111.9 [105.9–117.8]	108.8 [102.1–115.4]	0.44	p > 0.5
Verbal memory[a,b]	115.6 [109.9–121.3]	113.5 [107.9–119.0]	0.28	p > 0.5
Visual memory[a,b]	111.6 [107.5–115.7]	111.0 [106.3–115.7]	0.05	p > 0.8
Global memory[a,b]	115.8 [111.4–120.1]	114.2 [109.6–118.8]	0.25	p > 0.6
Wechsler Adult Intelligence Scale				
Verbal IQ[a,b]	127.0 [122.2–131.7]	121.8 [116.2–127.4]	2.03	p > 0.1
Performance IQ[a,b]	115.9 [111.2–120.4]	114.3 [109.0—119.7]	0.19	p > 0.6
Full Scal IQ[a,b]	123.6 [118.9–128.2]	120.2 [114.7–125.6]	0.90	p > 0.3
Verbal Comprehension Index[a,b]	127.3 [122.8–131.8]	124.5 [119.3–129.8]	0.66	p > 0.4
Perceptual organization[a,b]	117.9 [112.7–123.1]	116.6 [110.5–122.7]	0.11	p > 0.7
Working Memory Index[a,b]	116.1 [110.1–122.1]	109.7 [102/7–116.8]	1.93	p > 0.1
Processing Speed Index[a,b]	112.3 [107.3–117.4]	108.2 [102.3–114.1]	1.15	p > 0.2
Anxiety[a,b]	**4.21 [2.87–5.55]**	**8.55 [6.59–10.50]**	**15.18**	**p < 0.0004**
Depression[a,b]	**3.03 [1.74–4.33]**	**7.41 [5.39–9.43]**	**15.43**	**p < 0.0003**
Somatization[a,b]	**3.61 [2.14–5.07]**	**7.81 [6.09–9.53]**	**13.61**	**p < 0.0007**
Anger-Hostility[a,b]	**3.31 [2.33–4.29]**	**5.27 [3.88–6.67]**	**5.73**	**p < 0.03**
Limbic System Checklist-33[a,b]	**10.57 [6.87–14.27]**	**23.18 [16.60–29.76]**	**13.6**	**p < 0.0006**
Dissociative Experience Scale[a,b]	**4.58 [3.12–6.03]**	**14.32 [10.20–18.45]**	**26.95**	**p < 10^{-5}**
Drug use (days/month)[a,b]	0.36 [0.030–0.69]	0.27 [0.018–0.52]	0.16	p > 0.6
Alcohol use (drinks/month)[a,b]	**4.84 [3.70–5.98]**	**6.76 [5.25–8.28]**	**4.37**	**p < 0.05**

Adjusted for [a]age, [b]gender

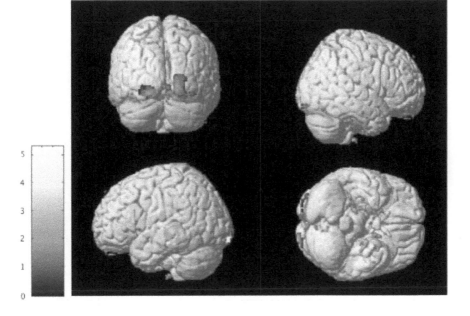

FIGURE 1: Voxel based morphometry results. Significant differences between subjects exposed to high levels of WDV and unexposed controls. Significantly decreased gray-matter densities in WDV subjects were measured in the right and left lingual gyrus (BA18). Color scale: 0–5 represent t-values.

TABLE 2: Mean and 95% confidence intervals for cortical thickness measures in the occipital region for unexposed subjects versus subjects who witnessed domestic violence.

Measures	Unexposed**	WDV Group	Group F	Group q***
right V1	1.76 ;1.71–1.81]	2.68 [1.61–1.74]	2.94	p > 0.2
right V2*	**2.16 [2.13–2.20]**	**2.03 [1.99–2.08]**	**14.65**	**p < 0.007**
right MT	2.58 [2.52–2.65]	2.56 [2.48–2.65]	0.07	p > 0.8
right inferior occipital gyrus and sulcus	2.77 [2.71–2.84]	2.69 [2.61–2.78]	1.53	p > 0.4
right cuneus gyrus	1.87 [1.82–1.93]	1.75 [1.69–1.82]	5.38	p < 0.07
right middle occipital gyrus	2.84 [2.78–2.91]	2.87 [2.79–2.94]	0.13	p > 0.8
right superior occipital gyrus	2.31 [2.24–2.39]	2.40 [2.30–2.49]	1.27	p > 0.4
right lateral fusiform gyrus	2.97 [2.89–3.05]	2.97 [2.97–3.07]	0.005	p > 0.9
right lingual gyrus	**2.14 [2.09–2.19]**	**2.00 [1.94–2.07]**	**8.05**	**p < 0.04**
right occipital pole	2.08 [2.02–2.13]	1.96 [1.89–2.03]	5.31	p < 0.07
left V1	1.68 [1.65–1.72]	1.60 [1.55–1.64]	6.33	p < 0.07
left V2	**2.10 [2.06–2.14]**	**1.97 [1.93–2.02]**	**13.65**	**p < 0.007**
left MT	2.55 [2.50–2.61]	2.55 [2.49–2.61]	0.005	p > 0.9
left inferior occipital gyrus and sulcus	2.53 [2.46–2.61]	2.45 [2.35–2.54]	1.35	p > 0.4
left cuneus gyrus	1.81 [1.76–1.86]	1.79 [1.73–1.85]	0.19	p > 0.8
left middle occipital gyrus	2.77 [2.71–2.83]	2.81 [2.73–2.88]	0.47	p > 0.6
left superior occipital gyrus	2.29 [2.23–2.36]	2.24 [2.16–2.32]	0.69	p > 0.6
left lateral fusiform gyrus	2.95 [2.88–3.01]	2.90 [2.82–2.98]	0.55	p > 0.6
left lingual gyrus	1.99 [1.94–2.05]	1.87 [1.81–1.94]	5.75	p < 0.07
left occipital pole	**2.11 [2.06–2.17]**	**1.96 [1.89–2.03]**	**8.39**	**p < 0.04**

*Regions highlighted in bold differed significant between groups. **Mean values adjusted for age, gender, parental education, perceived financial suffering, exposure to parental verbal abuse and mean cortical thickness. ***P values adjusted for multiple comparisons using false discovery rate (FDR) method of Benjamini and Hochberg.*

9.3.4 CORTICAL THICKNESS MEASURES

FreeSurfer was used to ascertain whether there were significant associations between WDV and thickness of the different portions of the visual cortex. As indicated in Table 2, the right lingual gyrus was about 6.5% thinner in WDV subjects. Further, V2 bilaterally and the left occipital pole were about 6% thinner in subjects witnessing DV.

9.3.5 SUSCEPTIBLE VERSUS RESILIENT RESULTS

Results for the subgroup analysis comparing susceptible versus relatively resilient individuals with WDV are summarized in Table 3. Susceptible and relatively resilient subjects did not differ significantly in their extent of exposure to interparental violence. On average, resilient subjects reported 3.9±2.7 years of exposure to IP-PA and 4.8±3.9 years of exposure to IP-VA versus 4.0±4.0 and 6.2±4.6 years respectively for susceptible subjects (IP-PA: $t = -0.08$, df $= 19.998$, p>0.9; IP-VA: $t = -0.79$, df $= 19.025$, p>0.4; Welch Two Sample t-test). ANCOVA showed a significant main effect of group in the comparison between unexposed, susceptible and relatively resilient subjects in the four regions delineated in Table 2. For these regions (i.e., V2 bilaterally, right lingual gyrus, left occipital pole) it appeared to make no difference whether the subjects were susceptible or resilient, the regions were equally thin. However, this analysis also identified two regions of extrastriate visual cortex (MT/V5 bilaterally) in which there was a small but significant difference between resilient and susceptible individuals. On balance, V5/MT was about 5% thicker in resilient versus susceptible subjects exposed to WDV (both p<0.04).

9.3.6 WITNESSING INTERPARENTAL VERBAL VERSUS PHYSICAL ABUSE RESULTS ON GMV

The strength of the statistical association between duration of exposure to IP-VA versus IP-PA and GMV of the right lingual gyrus is illustrated in Figure 2. Duration of exposure to IP-VA accounted for 19.8% of the variance in GMV of this region. In contrast, duration of exposure to IP-PA accounted for only 3.2% of the variance.

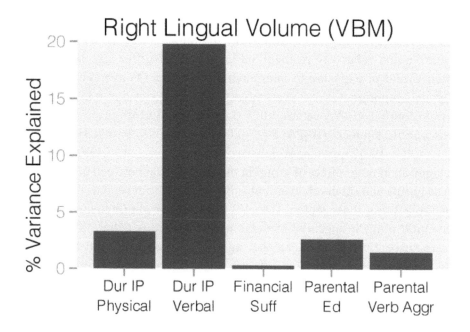

FIGURE 2: Relative importance – linear regression. Linear regression with variance decomposition indicating the percent of variance in adjusted gray matter volume from peak VBM cluster accounted for by regressors. Dur IP Physical – duration of witnessing interparental physical aggression (years), Dur IP Verbal – duration of witnessing interparental verbal aggression without physical violence (years), Financial Suf – perceived financial sufficiency during childhood, Parental Ed – average parental education (years), Parental Verb Aggr – average parental verbal aggression score.

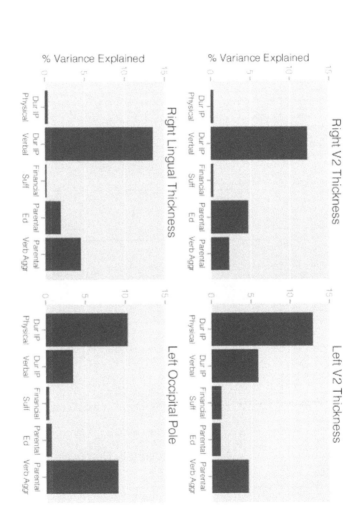

FIGURE 3: Relative importance – linear regression. Linear regression with variance decomposition indicating the percent of variance in adjusted regional cortical thickness measures from FreeSurfer accounted for by regressors. Dur IP Physical – duration of witnessing interparental physical aggression (years), Dur IP Verbal – duration of witnessing interparental verbal aggression without physical violence (years), Financial Suf – perceived financial sufficiency during childhood, Parental Ed – average parental education (years), Parental Verb Aggr – average parental verbal aggression score.

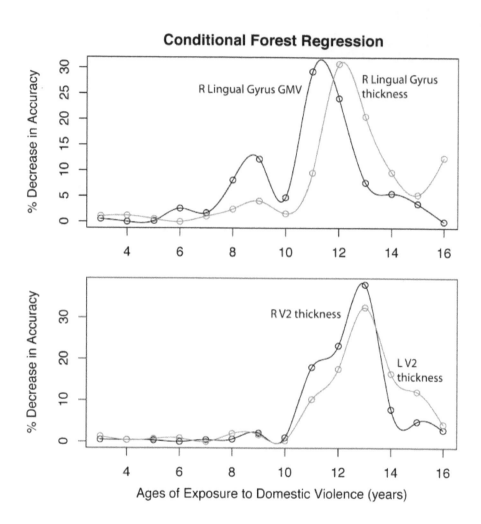

FIGURE 4: Relative importance – conditional forest regression. Relative importance of exposure to interparental aggression during specific years of childhood on adjusted measures of gray matter volume or thickness in right lingual gyrus and V2 bilaterally. Results based on random forest regression with conditional trees indicating the percent decrease in accuracy of the fit following permutation of each exposure age.

9.3.7 WITNESSING INTERPARENTAL VERBAL VERSUS PHYSICAL ABUSE RESULTS ON THICKNESS

Figure 3 illustrates the results of the same analyses applied to the Free-Surfer thickness measures. Duration of exposure to IP-VA accounted for a much greater share of the variance then exposure to IP-PA in thickness measures for right V2 and right lingual gyrus. In contrast, duration of exposure to IP-PA accounted for a greater share of the variance then exposure to IP-VA in thickness measures for left V2 and left occipital pole.

9.3.8 SENSITIVE PERIOD RESULTS

Results of conditional forest regression analyses are illustrated in Figure 4. Overall, exposure age accounted for 17% (p = 0.002) and 13% (p<0.008) of the variance in right lingual gyrus GMV and thickness, respectively. Similarly, exposure age accounted for 14% (p<0.006) and 11% (p<0.02) of the variance in thickness of right and left V2. The importance of exposure at each age was not however uniform. The most important predictors of right lingual gyrus GMV and thickness were exposure at 11–12, and 12–13 years of age, respectively. Similarly, exposures at 12–13 years were key predictors of left and right V2 thickness. Peak sensitivity occurred at age 11 for right lingual gyrus GMV, age 12 for right lingual gyrus thickness and age 13 for V2 thickness bilaterally.

9.3.9 RIGHT LINGUAL GYRUS GMV AND SYMPTOM RATINGS

As indicated in Table 4, there were significant associations between GMV in right lingual gyrus (BA18: Talairach's coordinates x = 6, y = –80, z = –4) and symptoms of dissociation and limbic irritability across the entire subject pool. These statistical associations were most discernible in the LSCL-33 subscales measuring somatization and dissociation. Partial correlations between GMV in right lingual gyrus and LSCL-33 were discernible in the WDV group but not in the control group.

TABLE 4: Partial correlations between gray matter volume in right lingual gyrus and symptom ratings, controlling for age and gender.

	All subjects		WDV Group		Healthy Controls	
Ratings	Correlation	P value	Correlation	P value	Correlation	P value
Dissociation (DES)	**−0.31**	**p < 0.03**	−0.32	p < 0.16	0.32	p < 0.09
Limbic Irritability (LSCL-33)	**−0.36**	**p < 0.009**	**−0.49**	**p < 0.02**	0.17	p > 0.3
LSCL: Somatization	**−0.39**	**p <0.004**	**−0.46**	**p < 0.03**	0.04	p >0.8
LSCL: Hallucination	−0.26	p < 0.06	−0.37	p < 0.10	0.11	p > 0.5
LSCL: Automatisms	−0.25	p <0.08	−0.32	p < 0.15	0.28	p < 0.14
LSCL: Dissociation	**−0.35**	**p = 0.01**	**−0.50**	**p < 0.02**	0.25	p < 0.2
Anxiety (SQ)	−0.26	p = 0.06	−0.30	p < 0.2	0.12	p > 0.5
Depression (SQ)	−0.23	p = 0.10	−0.27	p > 0.2	0.26	p < 0.18
Somatization (SQ)	−0.12	p > 0.4	−0.07	p > 0.7	0.20	p > 0.3
Hostility (SQ)	−0.10	p > 0.5	−0.15	p > 0.5	0.22	p > 0.2

Abbreviations: DES, Dissociative Experience Scale; LSC. L-33, Limbic System Checklist-33; SQ, Symptom Questionnaire; WDV, Witnessing Domestic Violence. Values in bold are stastically significant.

9.4 DISCUSSION

VBM identified a significant association between exposure to WDV and reduced GMV in the right lingual gyrus. Analysis using cortical surface parcellation also provided evidence for statistically significant differences in the thickness of the right lingual gyrus, left occipital pole and V2 bilaterally.

This right lingual gyrus plays a critical role in global aspects of figure recognition [39] and object naming [40]. It is characterized anatomically as a unimodal sensory area [41], though its response to visual stimuli is modulated by other types of sensory information through cortical back projections [41]. It may also be an important substrate for dreaming [42] and colors [43], and is a brain region that consistently shows a reduction in cerebral blood flow after sleep deprivation or disruption [44], [45]. Sleep disruption caused by WDV may diminish activity and blood flow to this region, and consequently alter its developmental trajectory. V2 is a component of early visual cortex that appears to be essential for conscious visual awareness [46]. Similarly, V5/MT plays a critical role in the conscious perception of visual movement [47]. Recent studies suggest that visual awareness depends on feedback from regions of extrastriate visual cortex (e.g., V2, V5/MT) back to V1 [46], [47]. The occipital pole is the most posterior portion of the occipital cortex and is the major component of V1.

These results are similar to findings previously reported in subjects exposed to CSA [48]. Those individuals had reduced GMV in V1 and V2 bilaterally, with most prominent differences appearing in right lingual gyrus and left fusiform gyrus using FreeSurfer [48]. These observations also fit with results of our previous WDV DTI analysis, which showed reduced FA in the left ILF connecting occipital cortex and limbic regions [8]. Reduced occipital GMV was also found by Fennema-Notestine et al., [49] to be associated with a prior history of childhood abuse. Further, diminished activation of right visual association areas was reported in a PET study of women with CSA-related PTSD [50]. Hence, different studies using an array of techniques provide complementary evidence for a potential association between exposure to certain forms of childhood abuse and structure or function of the visual cortex.

Interestingly, we observed a very different pattern of results in a sample of young adults exposed to high levels of parental verbal abuse but not WDV, CSA or PA. VBM revealed an increase in GMV in left superior temporal gyrus (auditory cortex) of verbally-abused subjects [11], and DTI analyses showed reduced FA in the arcuate fasciculus interconnecting Wernicke's and Broca's areas [51]. Together these findings suggest that sensory systems that process and interpret the adverse sensory inputs may be modified by the exposure.

The present findings also revealed a potential sensitive period, between 11–13 years of age, when exposure to WDV exerted maximal effects on GMV or thickness. This fits with our prior observation of a sensitive period between 7–13 years for WDV and myelination of the ILF [8], and with our earlier observation that CSA was associated with a reduction in occipital cortex GMV if it occurred prior to age 12 (in an all female sample) but not after [48]. Research has shown that plasticity of the visual cortex abates following puberty [52], and that human perceptual development remains vulnerable to damage from adverse visual experience until 10 to 13 years of age [53]. The present findings suggest that visual cortex may be particularly vulnerable to WDV during the peripubertal period.

The present findings also expand on our prior observation that visually witnessing IP-VA was associated with greater statistical effects on FA in the ILF then witnessing IP-PA [8]. GMV and thickness of the right lingual gyrus and right V2 in the current study were strongly influenced by duration of exposure to IP-VA but not IP-PA. On the other hand, thickness in left V2 and left occipital pole appeared to be more strongly influenced by duration of exposure to IP-PA then IP-VA. This is an intriguing observation of potential hemispheric differences that, if replicable, suggests a considerable degree of complexity in the association between type of exposure and morphometric measures.

Differences between WDV and controls in these regions were apparent in both relatively resilient and susceptible individuals. We had previously reported that reduced GMV in visual cortex was observable in both susceptible and resilient subjects with CSA [48], and that reduced FA in the ILF in WDV subjects was not mediated by presence or severity of psychiatric symptoms [8]. The presence of discernible (and potentially equivalent) neurobiological abnormalities in psychiatrically-resilient versus susceptible individuals with maltreatment histories appears to be an emerging trend. For example, we recently reported a strong association between severity of exposure to maltreatment and hippocampal subfield volume in dentate gyrus and CA3 that was unrelated to history or severity of depression or PTSD [13]. Similarly, Dannlowski et al. [54] reported reduced hippocampal volume, and increased amygdala reactivity in maltreated subjects without psychopathology. Additional evidence for amygdala hyperreactivity in maltreated individuals without psychopathology has also been reported by McCrory et al. [55] and van Harmelen et al. [56]. This has led to the speculation that these neurobiological correlates of exposure may be more of a risk factor for psychopathology than a consequence.

Interestingly, though reductions in right lingual GMV was observed in both susceptible and resilient subjects, there were significant associations between GMV in this region and self-report ratings of dissociation and limbic irritability. The most strongly associated subscale of the LSCL-33 in the WDV group consisted of items such as "The sensations that events, conversations, or a place was strangely familiar, as if you had experienced or dreamed the situation before" and "The sensation that your mind has left your body, or that you are watching yourself as a detached observer". While resilient subjects exposed to DV had ratings of depression and anxiety that were no greater than controls (and substantially lower than in susceptible individuals), they had dissociative experience scores nearly equal to susceptible subjects and much greater than controls. Hence, although resilient subjects in the WDV group did not experience the most common psychiatric consequences of exposure (depression and anxiety) they did experience heightened levels of dissociation. Alterations in the development of the right lingual gyrus or the ILF [8] (which interconnects lingual gyrus with hippocampus) may play a role in the generation of this phenomenon. This may result from: (1) a problem in the cross-modal influence of other sensory systems on visual perception [41]; (2) impaired integration of visual perceptions and hippocampal contextual memories through the ILF; or (3) intrusion of dream-like visual imagery (associated with right lingual gyrus [42]) into wake time.

The present study is relatively unique in its focus on the potential consequences of exposure to a specific type of abuse. This approach has been useful in revealing similarities and differences between the neurobiological correlates of exposure to childhood sexual abuse [48], [57], parental verbal abuse [11], [51], WDV [8] and harsh corporal punishment [12], [58]. This strength is also a limitation as many abused individuals, particularly those involved in the mental health system, experienced multiple forms of maltreatment. Hamby et al. [59] reported in a nationally representative sample, that 56.6% of youth witnessing interparental violence would, over the course of their lifetime, experience other forms of maltreatment. Hence, our findings are more applicable to the remaining 43%, who experience WDV as their sole form of maltreatment.

Studies of individuals exposed to single types of maltreatment have primarily identified differences in sensory regions or pathways [8], [11], [48], [51]. This observation stands in contrast to findings from other studies, including our own, that combine subjects exposed to one or more types of abuse into a single group. Those studies have predominantly identified alterations in corpus callosum, hippocampus and frontal cortex [13], [54], [60], [61], [62], [63], [64], [65], [66], [67], [68], [69], [70], [71]. It may be the case that chronic exposure to a specific type of adversity primarily affects the development of sensory systems that process or convey the adverse sensory input. In contrast, neurobiological response (or adaptation) to multiple forms of adversity may occur predominantly at limbic or frontal cortical levels and affect interhemispheric communication. This view is compatible with the observation that risk for psychopathology is much greater in individuals exposed to multiple types of maltreatment then single forms [17], [72].

The main limitation of this study is the relatively small sample size. A large, initial sample of 18- to 25-year-olds was surveyed to identify a healthy sample of subjects in the community, as opposed to psychiatric sources, who were exposed only to WDV and to no other forms of trauma or early adversity. Exposure to high levels of WDV but to no other forms of abuse is a relatively common occurrence, reported by about 4% of subjects in this age range [17]. Our findings should generalize to subjects experiencing WDV or WDV plus parental verbal abuse, but no other forms of abuse, as we selected subjects without regard to psychopathology (except substance abuse). It remains to be seen if the same findings emerge in subjects exposed to WDV plus sexual or physical abuse.

Although this study revealed a significant association between a self-reported history of WDV and decreased GMV or thickness in visual cortex, it should be emphasized that the finding is correlational and does not prove that WDV caused the decrease. Prospective longitudinal studies are required to establish a causal relationship. Nevertheless, these findings are consistent with a causal relationship and suggest that exposure to WDV may act as a traumatic stressor to alter the development of the visual cortex. If so, these results underscore efforts to prevent children from exposure to acts of domestic violence and other forms of abuse or neglect.

REFERENCES

1. McDonald R, Jouriles EN, Ramisetty-Mikler S, Caetano R, Green CE (2006) Estimating the number of American children living in partner-violent families. J Fam Psychol 20: 137–142. doi: 10.1037/0893-3200.20.1.137

2. Fantuzzo J, Boruch R, Beriama A, Atkins M, Marcus S (1997) Domestic violence and children: prevalence and risk in five major U.S. cities. J Am Acad Child Adolesc Psychiatry 36: 116–122. doi: 10.1097/00004583-199701000-00025

3. Nicodimos S, Gelaye BS, Williams MA, Berhane Y (2009) Associations between witnessing parental violence and experiencing symptoms of depression among college students. East Afr J Public Health 6: 184–190. doi: 10.4314/eajph.v6i2.51764

4. Abrahams N, Jewkes R (2005) Effects of South African men's having witnessed abuse of their mothers during childhood on their levels of violence in adulthood. Am J Public Health 95: 1811–1816. doi: 10.2105/ajph.2003.035006

5. Casiano H, Mota N, Afifi TO, Enns MW, Sareen J (2009) Childhood maltreatment and threats with weapons. J Nerv Ment Dis 197: 856–861. doi: 10.1097/nmd.0b013e3181be9c55

6. Luthra R, Abramovitz R, Greenberg R, Schoor A, Newcorn J, et al. (2009) Relationship between type of trauma exposure and posttraumatic stress disorder among urban children and adolescents. J Interpers Violence 24: 1919–1927. doi: 10.1177/0886260508325494

7. Silva RR, Alpert M, Munoz DM, Singh S, Matzner F, et al. (2000) Stress and vulnerability to posttraumatic stress disorder in children and adolescents. Am J Psychiatry 157: 1229–1235. doi: 10.1176/appi.ajp.157.8.1229

8. Choi J, Jeong B, Polcari A, Rohan ML, Teicher MH (2012) Reduced fractional anisotropy in the visual limbic pathway of young adults witnessing domestic violence in childhood. Neuroimage 59: 1071–1079. doi: 10.1016/j.neuroimage.2011.09.033

9. Dale AM, Fischl B, Sereno MI (1999) Cortical surface-based analysis. I. Segmentation and surface reconstruction. Neuroimage 9: 179–194. doi: 10.1006/nimg.1998.0395

10. Fischl B, Sereno MI, Dale AM (1999) Cortical surface-based analysis. II: Inflation, flattening, and a surface-based coordinate system. Neuroimage 9: 195–207. doi: 10.1006/nimg.1998.0396

11. Tomoda A, Sheu YS, Rabi K, Suzuki H, Navalta CP, et al. (2011) Exposure to parental verbal abuse is associated with increased gray matter volume in superior temporal gyrus. Neuroimage 54 Suppl 1S280–286. doi: 10.1016/j.neuroimage.2010.05.027

12. Tomoda A, Suzuki H, Rabi K, Sheu YS, Polcari A, et al. (2009) Reduced prefrontal cortical gray matter volume in young adults exposed to harsh corporal punishment. Neuroimage 47 Suppl 2T66–71. doi: 10.1016/j.neuroimage.2009.03.005

13. Teicher MH, Anderson CM, Polcari A (2012) Childhood maltreatment is associated with reduced volume in the hippocampal subfields CA3, dentate gyrus, and subiculum. Proc Natl Acad Sci U S A 109: E563–572. doi: 10.1073/pnas.1115396109

14. First MB, Spitzer RL, Gibbon M, Williams JBW (1997) Structured clinical interview for DSM-IV axis I disorders - clinician version (SCID-CV). Washington, DC: American Psychiatric Press.

15. Herman JL, Perry JC, van der Kolk BA (1989) Traumatic Antecedents Interview. Boston: The Trauma Center.

16. Roy CA, Perry JC (2004) Instruments for the assessment of childhood trauma in adults. J Nerv Ment Dis 192: 343–351. doi: 10.1097/01.nmd.0000126701.23121.fa

17. Teicher MH, Samson JA, Polcari A, McGreenery CE (2006) Sticks, stones, and hurtful words: relative effects of various forms of childhood maltreatment. Am J Psychiatry 163: 993–1000. doi: 10.1176/appi.ajp.163.6.993

18. Kellner R (1987) A symptom questionnaire. Journal of Clinical Psychiatry 48: 268–273.

19. Bernstein DP, Fink L (1998) Childhood Trauma Questionnaire Manual. San Antonio, TX: The Psychological Corporation.

20. Teicher MH, Glod CA, Surrey J, Swett C Jr (1993) Early childhood abuse and limbic system ratings in adult psychiatric outpatients. J Neuropsychiatry Clin Neurosci 5: 301–306.

21. Good CD, Johnsrude IS, Ashburner J, Henson RN, Friston KJ, et al. (2001) A voxel-based morphometric study of ageing in 465 normal adult human brains. Neuroimage 14: 21–36. doi: 10.1006/nimg.2001.0786

22. Good CD, Johnsrude I, Ashburner J, Henson RN, Friston KJ, et al. (2001) Cerebral asymmetry and the effects of sex and handedness on brain structure: a voxel-based morphometric analysis of 465 normal adult human brains. Neuroimage 14: 685–700. doi: 10.1006/nimg.2001.0857

23. Ashburner J, Friston KJ (2000) Voxel-based morphometry–the methods. Neuroimage 11: 805–821. doi: 10.1006/nimg.2000.0582

24. Ashburner J, Friston KJ (2005) Unified segmentation. Neuroimage 26: 839–851. doi: 10.1016/j.neuroimage.2005.02.018

25. Hayasaka S, Phan KL, Liberzon I, Worsley KJ, Nichols TE (2004) Nonstationary cluster-size inference with random field and permutation methods. Neuroimage 22: 676–687. doi: 10.1016/j.neuroimage.2004.01.041

26. Moorhead TW, Job DE, Spencer MD, Whalley HC, Johnstone EC, et al. (2005) Empirical comparison of maximal voxel and non-isotropic adjusted cluster extent results in a voxel-based morphometry study of comorbid learning disability with schizophrenia. Neuroimage 28: 544–552. doi: 10.1016/j.neuroimage.2005.04.045

27. Friston KJ, Holmes A, Poline JB, Price CJ, Frith CD (1996) Detecting activations in PET and fMRI: levels of inference and power. Neuroimage 4: 223–235. doi: 10.1006/nimg.1996.0074

28. Worsley KJ, Andermann M, Koulis T, MacDonald D, Evans AC (1999) Detecting changes in nonisotropic images. Hum Brain Mapp 8: 98–101. doi: 10.1002/(sici)1097-0193(1999)8:2/3<98::aid-hbm5>3.0.co;2-f

29. Bookstein FL (2001) "Voxel-based morphometry" should not be used with imperfectly registered images. Neuroimage 14: 1454–1462. doi: 10.1006/nimg.2001.0770

30. Fischl B, Liu A, Dale AM (2001) Automated manifold surgery: constructing geometrically accurate and topologically correct models of the human cerebral cortex. IEEE Trans Med Imaging 20: 70–80. doi: 10.1109/42.906426

31. Hochberg Y, Benjamini Y (1990) More powerful procedures for multiple significance testing. Stat Med 9: 811–818. doi: 10.1002/sim.4780090710

32. R Development Core Team (2010) R: A Language and Environment for Statistical Computing. Vienna, Austria: R Foundation for Statistical Computing.

33. Lindeman RH, Merenda PF, Gold RZ (1980) Introduction to Bivariate and Multivariate Analysis. Glenview, IL: Scott, Foresman.

34. Grömping U (2007) Estimators of relative importance in linear regression based on variance decomposition. The American Statistician 61: 139–147. doi: 10.1198/000313007x188252

35. Strobl C, Boulesteix AL, Zeileis A, Hothorn T (2007) Bias in random forest variable importance measures: illustrations, sources and a solution. BMC Bioinformatics 8: 25. doi: 10.1186/1471-2105-8-25

36. Breiman L (2001) Random Forests. Machine Learning 45: 5–32. doi: 10.1023/a:1010933404324

37. Cutler DR, Edwards TC, Beard KH, Cutler A, Hess KT, et al. (2007) Random forests for classification in ecology. Ecology 88: 2783–2792. doi: 10.1890/07-0539.1

38. Russo R (2003) Statistics for the Behavioral Sciences: An Introduction: Psychology Press. 192–193 p.

39. Fink GR, Halligan PW, Marshall JC, Frith CD, Frackowiak RS, et al. (1996) Where in the brain does visual attention select the forest and the trees? Nature 382: 626–628. doi: 10.1038/382626a0

40. Kiyosawa M, Inoue C, Kawasaki T, Tokoro T, Ishii K, et al. (1996) Functional neuroanatomy of visual object naming: a PET study. Graefes Arch Clin Exp Ophthalmol 234: 110–115. doi: 10.1007/bf00695250

41. Macaluso E, Frith CD, Driver J (2000) Modulation of human visual cortex by crossmodal spatial attention. Science 289: 1206–1208. doi: 10.1126/science.289.5482.1206

42. Bischof M, Bassetti CL (2004) Total dream loss: a distinct neuropsychological dysfunction after bilateral PCA stroke. Ann Neurol 56: 583–586. doi: 10.1002/ana.20246

43. Howard RJ, ffytche DH, Barnes J, McKeefry D, Ha Y, et al. (1998) The functional anatomy of imagining and perceiving colour. Neuroreport 9: 1019–1023. doi: 10.1097/00001756-199804200-00012

44. Joo EY, Tae WS, Han SJ, Cho JW, Hong SB (2007) Reduced cerebral blood flow during wakefulness in obstructive sleep apnea-hypopnea syndrome. Sleep 30: 1515–1520.

45. Joo EY, Seo DW, Tae WS, Hong SB (2008) Effect of modafinil on cerebral blood flow in narcolepsy patients. Sleep 31: 868–873.

46. Salminen-Vaparanta N, Koivisto M, Noreika V, Vanni S, Revonsuo A (2012) Neuronavigated transcranial magnetic stimulation suggests that area V2 is necessary for visual awareness. Neuropsychologia 50: 1621–1627. doi: 10.1016/j.neuropsychologia.2012.03.015

47. Silvanto J, Lavie N, Walsh V (2005) Double dissociation of V1 and V5/MT activity in visual awareness. Cereb Cortex 15: 1736–1741. doi: 10.1093/cercor/bhi050

48. Tomoda A, Navalta CP, Polcari A, Sadato N, Teicher MH (2009) Childhood sexual abuse is associated with reduced gray matter volume in visual cortex of young women. Biol Psychiatry 66: 642–648. doi: 10.1016/j.biopsych.2009.04.021

49. Fennema-Notestine C, Stein MB, Kennedy CM, Archibald SL, Jernigan TL (2002) Brain morphometry in female victims of intimate partner violence with and without posttraumatic stress disorder. Biol Psychiatry 52: 1089–1101. doi: 10.1016/s0006-3223(02)01413-0

50. Bremner JD, Vermetten E, Vythilingam M, Afzal N, Schmahl C, et al. (2004) Neural correlates of the classic color and emotional stroop in women with abuse-related posttraumatic stress disorder. Biol Psychiatry 55: 612–620. doi: 10.1016/j.biopsych.2003.10.001

51. Choi J, Jeong B, Rohan ML, Polcari AM, Teicher MH (2009) Preliminary evidence for white matter tract abnormalities in young adults exposed to parental verbal abuse. Biol Psychiatry 65: 227–234. doi: 10.1016/j.biopsych.2008.06.022

52. Hubel DH, Wiesel TN (1998) Early exploration of the visual cortex. Neuron 20: 401–412. doi: 10.1016/s0896-6273(00)80984-8

53. Lewis TL, Maurer D (2005) Multiple sensitive periods in human visual development: evidence from visually deprived children. Dev Psychobiol 46: 163–183. doi: 10.1002/dev.20055

54. Dannlowski U, Stuhrmann A, Beutelmann V, Zwanzger P, Lenzen T, et al. (2012) Limbic scars: long-term consequences of childhood maltreatment revealed by functional and structural magnetic resonance imaging. Biol Psychiatry 71: 286–293. doi: 10.1016/j.biopsych.2011.10.021

55. McCrory EJ, De Brito SA, Sebastian CL, Mechelli A, Bird G, et al. (2011) Heightened neural reactivity to threat in child victims of family violence. Current Biology 21: R947–R948. doi: 10.1016/j.cub.2011.10.015

56. van Harmelen AL, van Tol MJ, Demenescu LR, van der Wee NJ, Veltman DJ, et al.. (2012) Enhanced amygdala reactivity to emotional faces in adults reporting childhood emotional maltreatment. Soc Cogn Affect Neurosci.

57. Andersen SL, Tomoda A, Vincow ES, Valente E, Polcari A, et al. (2008) Preliminary evidence for sensitive periods in the effect of childhood sexual abuse on regional brain development. J Neuropsychiatry Clin Neurosci 20: 292–301. doi: 10.1176/appi.neuropsych.20.3.292

58. Sheu YS, Polcari A, Anderson CM, Teicher MH (2010) Harsh corporal punishment is associated with increased T2 relaxation time in dopamine-rich regions. Neuroimage 53: 412–419. doi: 10.1016/j.neuroimage.2010.06.043

59. Hamby S, Finkelhor D, Turner H, Ormrod R (2010) The overlap of witnessing partner violence with child maltreatment and other victimizations in a nationally representative survey of youth. Child Abuse Negl 34: 734–741. doi: 10.1016/j.chiabu.2010.03.001

60. Stein MB, Koverola C, Hanna C, Torchia MG, McClarty B (1997) Hippocampal volume in women victimized by childhood sexual abuse. Psychol Med 27: 951–959. doi: 10.1017/s0033291797005242

61. Bremner JD, Randall P, Vermetten E, Staib L, Bronen RA, et al. (1997) Magnetic resonance imaging-based measurement of hippocampal volume in posttraumatic stress disorder related to childhood physical and sexual abuse–a preliminary report. Biol Psychiatry 41: 23–32. doi: 10.1016/s0006-3223(96)00162-x

62. Vermetten E, Schmahl C, Lindner S, Loewenstein RJ, Bremner JD (2006) Hippocampal and amygdalar volumes in dissociative identity disorder. Am J Psychiatry 163: 630–636. doi: 10.1176/appi.ajp.163.4.630

63. Vythilingam M, Heim C, Newport J, Miller AH, Anderson E, et al. (2002) Childhood trauma associated with smaller hippocampal volume in women with major depression. Am J Psychiatry 159: 2072–2080. doi: 10.1176/appi.ajp.159.12.2072

64. De Bellis MD, Keshavan MS, Clark DB, et al. (1999) A.E. Bennett Research Award. Developmental traumatology. Part II: Brain development. Biol Psychiatry 45: 1271–1284. doi: 10.1016/s0006-3223(99)00045-1

65. De Bellis MD, Keshavan MS, Frustaci K, Shifflett H, Iyengar S, et al. (2002) Superior temporal gyrus volumes in maltreated children and adolescents with PTSD. Biol Psychiatry 51: 544–552. doi: 10.1016/s0006-3223(01)01374-9

66. Teicher MH, Dumont NL, Ito Y, Vaituzis C, Giedd JN, et al. (2004) Childhood neglect is associated with reduced corpus callosum area. Biol Psychiatry 56: 80–85. doi: 10.1016/j.biopsych.2004.03.016

67. Jackowski AP, Douglas-Palumberi H, Jackowski M, Win L, Schultz RT, et al. (2008) Corpus callosum in maltreated children with posttraumatic stress disorder: a diffusion tensor imaging study. Psychiatry Res 162: 256–261. doi: 10.1016/j. pscychresns.2007.08.006

68. Carrion VG, Weems CF, Reiss AL (2007) Stress predicts brain changes in children: a pilot longitudinal study on youth stress, posttraumatic stress disorder, and the hippocampus. Pediatrics 119: 509–516. doi: 10.1542/peds.2006-2028

69. Richert KA, Carrion VG, Karchemskiy A, Reiss AL (2006) Regional differences of the prefrontal cortex in pediatric PTSD: an MRI study. Depress Anxiety 23: 17–25. doi: 10.1002/da.20131

70. Teicher MH, Samson JA, Sheu YS, Polcari A, McGreenery CE (2010) Hurtful words: association of exposure to peer verbal abuse with elevated psychiatric symptom scores and corpus callosum abnormalities. Am J Psychiatry 167: 1464–1471. doi: 10.1176/appi.ajp.2010.10010030

71. Edmiston EE, Wang F, Mazure CM, Guiney J, Sinha R, et al. (2011) Corticostriatal-Limbic Gray Matter Morphology in Adolescents With Self-reported Exposure to Childhood Maltreatment. Arch Pediatr Adolesc Med 165: 1069–1077. doi: 10.1001/ archpediatrics.2011.565

72. Anda RF, Felitti VJ, Bremner JD, Walker JD, Whitfield C, et al. (2006) The enduring effects of abuse and related adverse experiences in childhood: A convergence of evidence from neurobiology and epidemiology. Eur Arch Psychiatry Clin Neurosci 256: 174–186. doi: 10.1007/s00406-005-0624-4

Table 3 is not available in this version of the article. To view this additional information, please use the citation on the first page of this chapter.

PART III

EARLY TRAUMA
AND PHYSICAL HEALTH

CHAPTER 10

ASSOCIATION BETWEEN ADVERSE CHILDHOOD EXPERIENCES AND DIAGNOSIS OF CANCER

MONIQUE J. BROWN, LEROY R. THACKER, AND STEVEN A. COHEN

10.1 INTRODUCTION

Adverse childhood experiences (ACEs) represent a child's exposure to negative events, including emotional, physical and sexual abuse, domestic violence, absence of a parent because of divorce or separation, and a family/household member's mental illness, incarceration, and/or substance abuse [1], [2]. Research has shown that majority of adults have experienced at least one ACE, [3] with prevalence estimates ranging from 65% to 87% [4], [5]. The Centers for Disease Control and Prevention (CDC) has found that 59.4% of Behavioral Risk Factor Surveillance System (BRFSS) respondents in 2009 reported having at least one ACE, approximately 25% of adults reported experiencing ≥3 ACEs and 8.7% reported ≥5 ACEs [2]. These high prevalence estimates highlight the importance for additional

Association between Adverse Childhood Experiences and Diagnosis of Cancer. © Brown MJ, Thacker LR, and Cohen SA. PLoS ONE **8**,6 (2013), doi:10.1371/journal.pone.0065524. Licensed under a Creative Commons Attribution License, http://creativecommons.org/licenses/by/3.0/.

efforts, locally, state-wide, and nationally, to help in the reduction and prevention of child maltreatment, and associated family dysfunction. There is also a need for services to treat outcomes, especially stress-related health illnesses, associated with ACEs [2], from childhood through adulthood.

Research has shown that ACEs are linked to multiple adverse health outcomes, and are interrelated rather than occurring independently [6]. ACEs have been linked to substance abuse [7]–[11], depression [7],[8],[12],[13], cardiovascular disease [7],[14], diabetes [7], cancer [7],[15],[16], risky sexual behaviors, [17–19], sexually transmitted infections [7],[8],[19] suicidality [7],[8],[13],[17],[18], and premature mortality in adulthood [17]. Researchers have suggested examining multiple ACEs allowing for a potential assessment of a graded relationship between these exposures and health outcomes [6]. However, exploring ACEs using principal component analysis (PCA), a more robust method, will account for the loading of each adverse experience. Using this strategy will allow for aggregation of ACE components that are likely to cluster together.

Stressful and/or traumatic experiences during childhood may also have neurodevelopmental impacts that persist over the lifespan [7]. The public health impact of ACEs has only been recently evaluated systematically through large epidemiologic efforts [20]. A strong relationship has been shown between stress and/or trauma during childhood, and smoking behavior [21],[22] with a threefold increase in the risk of lung cancer for participants who had ≥6 ACEs [15]. However, the increase in the risk of lung cancer could only be partially explained by the relationship between ACEs and smoking. Therefore, this finding suggests that the association between ACEs and cancer may be attributable to other mechanisms in which stressors and trauma during childhood negatively affect health [15]. In a prospective study, adverse events occurring between ages six and eight, and cumulative adversity from birth to age eight were found to be associated with inflammation at age ten [23].

While the link between ACEs and adverse health outcomes, including lung cancer, has been demonstrated, the association between ACEs and diagnosis of all cancers has not been examined. The main aims of this study were to: 1) Examine the association between ACEs and all cancers; and 2) Examine the association between ACEs, and childhood cancer and adulthood cancer. We hypothesize that we will see a positive association between ACEs, and overall, childhood and adulthood cancer. To our knowledge, this is the first study to examine the association between ACEs and all cancers, childhood and adulthood cancer using a population-based sample.

10.2 METHODS

10.2.1 ETHICS STATEMENT

The Virginia Commonwealth University Institutional Review Board provided a formal written waiver for the current study as publicly available, anonymous, secondary data were used. Verbal consent was provided to interviewers for the Behavioral Risk Factor Surveillance System survey [25].

10.2.2 DATA SOURCE AND SAMPLE

This cross-sectional study used data from the 2010 Behavioral Risk Factor Surveillance System (BRFSS) survey. The BRFSS is the largest ongoing telephone health survey, conducted in all US states, the District of Columbia, Puerto Rico, Guam and the U.S. Virgin Islands [24]. Established by the CDC, the BRFSS provides data on a variety of health issues among the non-institutionalized adult population (age 18 years and older). In addition to core questions that are asked by every state, there are optional modules that cover other health topics or ask more in-depth questions on a topic that was included among the core questions. The 2010 survey included an optional ACE module [25]. Wisconsin was the only state that also included the cancer survivorship module [25]. Respondents in Wisconsin who refused, or were not asked questions relating to ACEs and cancer diagnosis were not eligible (n = 551). The resultant sample size was 4,230.

10.2.3 OPERATIONAL DEFINITION
OF ADVERSE CHILDHOOD EXPERIENCES

The optional ACE module included questions about adverse, stressful and/ or traumatic events experienced as a child (Table 1). We created three component scores of the ACEs using principal component analysis (PCA). The use of adverse events during childhood in research has shown to be dependable as the test-retest reliability in the responses to questions about ACEs has been found to be good to excellent [26].

TABLE 1: Types of Adverse Childhood Experiences, related BRFSS questions and operational definitions.

Questions	Operational Definition	Abbreviation
1. Did you live with anyone who was depressed, mentally ill or suicidal?	No vs. Yes	DEPRS
2. Did you live with anyone who was a problem drinker or alcoholic?	No vs. Yes	DRINK
3. Did you live with anyone who used illegal street drugs or who abused prescription medications?	No vs. Yes	DRUGS
4. Did you live with anyone who served time or was sentenced to serve time in a prison, jail, or other correctional facility?	No vs. Yes	PRISN
5. Were your parents separated or divorced?	No vs. Yes	DIVRC
6. How often did your parents or adults in your home ever slap, hit, kick, punch, or beat each other up?	Never vs. At least once	PUNCH
7. How often did a parent or adult in your home ever hit, beat, kick, or physically hurt you in any way? Do not include spanking.	Never vs. At least once	HURT
8. How often did a parent or adult in your home ever swear at you, insult you, or put you down?	Never vs. At least once	SWEAR
9. How often did anyone at least 5 years older than you or an adult ever touch you sexually?	Never vs. At least once	TOUCH
10. How often did anyone at least 5 years older than you or an adult try to make you touch them sexually?	Never vs. At least once	TTHEM
11. How often did anyone at least 5 years older than you or an adult force you to have sex?	Never vs. At least once	HVSEX

10.2.4 OPERATIONAL DEFINITION OF CANCER

The Cancer Survivorship module, which was also optional, included several questions on cancer diagnosis and treatment. Cancer was selected as previous research has suggested a link between ACEs and different types of cancer ,[15,16]. Cancer diagnosis was defined by: 1) "Have you ever been told by a doctor, nurse, or other health professional that you had cancer?" which elicited a binary (yes/no) response; and 2) "At what age were you told that you had cancer?" Age of cancer diagnosis was then separated into 1–17 years (childhood cancer), [27–29] and 18 years or older (adulthood cancer). Self-reported data on cancer diagnosis has been shown to have high validity with moderate to excellent agreement to medical chart information, and has been shown to be useful for epidemiologic research, [27,28].

10.2.5 POTENTIAL CONFOUNDERS

Potential confounders considered were either associated with ACEs or cancer diagnosis, as reported in the literature. In previous studies, sociodemographic variables were included as covariates [20]. Older age is a risk factor for cancer, [29,30] and age was included as a continuous covariate. Sex and race/ethnicity were also included as covariates in the models as they are strong predictors for certain types of cancers. For example, aside from non-melanoma skin cancer, breast cancer is the most common cancer among women, and prostate cancer is the most common cancer among men, [31–33]. From 1999 to 2008, White women had the highest breast cancer incidence rates while Black women had the highest mortality rates [34]. Men tend to have higher colorectal cancer incidence and mortality rates compared to women with Black men having higher colorectal cancer incidence and mortality rates compared to men of other racial/ethnic groups [35]. Income, educational level, and marital status have also been included in previous studies as a covariate examining the link between ACEs and adverse outcomes, [15,20]. Socioeconomic status may partially explain the association of traumatic events during childhood with chronic illness [36]. Insurance status was also considered as a potential confounder as adverse experiences during childhood is linked to health [37], which is ultimately influenced by access to healthcare and insurance status.

TABLE 2: Distribution of socio-demographic characteristics of BRFSS respondents by Exposure to Adverse Childhood Experiences.

Confounders	ACEs	No ACEs	P-value*
	N = 2,505	N = 1,725	
	WN = 2,463,391	WN = 1,490,356	
	%	%	
Age			
18–34	36.6	20.1	<0.0001
35–49	25.8	24.8	
50+	37.6	55.1	
Gender			
Female	50.5	50.6	0.9689
Male	49.5	49.4	
Race/Ethnicity			
White	87.6	93.5	0.0001
Black	3.5	1.6	
Hispanic	3.0	0.8	
Other[a]	5.9	4.2	
Income (Annual)			
<$15,000	5.4	2.1	<0.0001
$15,000–<$50,000	52.2	44.8	
$50,000+	42.4	53.0	
Education[b]			
<HS Graduate	5.0	4.6	0.1437
HS Graduate	32.9	29.2	
>HS Graduate	62.1	66.2	
Marital Status			
Married	58.8	69.5	<0.0001
Not married	41.2	30.5	
Insurance Status			
Insured	87.1	93.8	<0.0001
Not insured	12.9	6.2	

N = frequency; WIN = weighted frequency. *P-value shows differences between respondents exposed to ACEs and unexposed to ACEs. [a]The group "Other," "Native American/Alaska Native" and "Native Hawaiian/Other Pacific Islander." [b]Education categories: <high school graduate, high school graduate, >high school graduate.

TABLE 3: Adverse Childhood Experiences Pattern derived by Principal Component Analysis.

Component 1		Component 2		Component 3	
ACE	Load	ACE	Load	ACE	Load
TOUCH	0.40	DRUGS	0.31	SWEAR	0.52
TTHEM	0.40	DRNK	0.31	HURT	0.46
HVSEX	0.38	PUNCH	0.29	DIVRC	0.37
HURT	0.30	PRISN	0.28	TOUCH	0.03
PUNCH	0.30	DIVRC	0.25	DRUGS	−0.46
DRUGS	0.29	DEPRS	0.20	PRISN	−0.41
DRNK	0.28	HURT	0.14	PUNCH	−0.06
DEPRS	0.26	SWEAR	0.07	HURT	−0.05
DIVRC	0.25	TOUCH	−0.44	SWEAR	−0.04
PRISN	0.24	TTHEM	−0.42	TTHEM	−0.02
SWEAR	0.07	HVSEX	−0.39	HVSEX	−0.01

10.2.6 ANALYTIC APPROACH

All analyses considered the BRFSS complex multistage sampling strategy [38]. Weighted prevalence estimates were obtained for ACEs and cancer. Proc surveyfreq procedures were used for bivariate analyses to examine the association between age, gender, race/ethnicity, income, educational status, marital status and insurance status, and exposure to ACEs (Table 2). P values were used to determine statistically significant differences between exposure to overall ACEs and no exposure to ACEs.

PCA was used to obtain a score (Component 1) taking into consideration the shared information across the ACEs; and shape components (Components 2 and 3) that were independent of Component 1. We observed the component loading pattern (the correlation coefficients of the original variables with the components) (Table 3). Multivariable logistic regression models were used to provide adjusted odds ratios (OR) and 95% confidence intervals (CI) for the association between the three principal components of ACEs and overall cancer, childhood cancer, and adulthood cancer adjusting for age, gender, race/ethnicity, income, educational status, marital status, and insurance status (Table 4). A variable was considered to be a potential confounder based on the literature review done a priori, as well as analyses shown in Table 2 comparing individuals exposed to ACEs to those not exposed to ACEs.

TABLE 4: Multiple Logistic Regression of Principal Components of Adverse Childhood Experiences with Cancer Diagnosis in Adulthood.

Component	Overall		Childhood		Adulthood	
	cOR[a]	aOR[b]	cOR[a]	aOR[b]	cOR[a]	aOR[b]
	(95% CI)	(95% CI)	(95% CI)	(95% CI)	(95% CI)	(95% CI)
Component 1	1.02	1.15	1.35	0.75	1.02	**1.21**
	(0.88–1.17)	(0.97–1.37)	(0.47–3.91)	(0.48–1.18)	(0.89–1.18)	**(1.03–1.43)**
Component 2	0.90	1.06	1.05	0.88	0.86	1.03
	(0.77–1.06)	(0.86–1.31)	(0.76–1.45)	(0.43–1.81)	(0.74–1.01)	(0.83–1.29)
Component 3	1.10	0.95	1.35	2.52	1.10	0.93
	(0.88–1.39)	(0.76–1.18)	(0.47–3.91)	(0.47–13.4)	(0.87–1.39)	(0.74–1.17)

[a]Crude odds ratio. [b]Adjusted odds ratio controlling for age, gender, race/ethnicity, income, educational status, marital status, and insurance status.

10.3 RESULTS

10.3.1 WEIGHTED PREVALENCE ESTIMATES

The weighted prevalence of ACEs was 62.3%. Approximately 10% of respondents reported having had a diagnosis of cancer, with 9.3% reporting their first diagnosis of cancer in adulthood and 0.2% reporting their first diagnosis of cancer during childhood. Among respondents who reported a diagnosis of cancer, 98.5% reported having their first diagnosis during adulthood (at age 18 or older).

10.3.2 DISTRIBUTION OF CHARACTERISTICS ACROSS EXPOSURE GROUPS

Table 2 shows the distribution of socio-demographic characteristics of BRFSS respondents by exposure and non-exposure to ACEs. There were statistically significant differences in the proportions of respondents who reported ACEs compared to those who did not by age, race/ethnicity, income, education, marital status, and insurance. More than half of the respondents who did not report exposure to ACEs were 50 years or older. Of the respondents who reported ACEs, 87.6% were White compared to 93.5% of the respondents who did not report ACEs. Among those respondents who reported ACEs, a higher percentage of Black, Hispanic or Other respondents reported ACEs compared to respondents in these racial/ethnic groups who did not report ACEs. There was no statistically significant difference by gender. Approximately a third of respondents who reported ACEs were high school graduates and just under a third of respondents had at least some college education.

10.3.3 COMPONENT LOADING PATTERN

Table 3 shows the ACE pattern derived from PCA. All the values for each ACE were positively correlated with Component 1. The sexual abuse variables (TOUCH, TTHEM and HVSEX) had the highest weights for Component 1. Variables that indicated ACEs that were not directed towards the child (DEPRS, DRINK, DRUGS, PRISN, DIVRC, and PUNCH) had the highest positive weights for Component 2; and the two variables that indicated psychological and physical abuse towards the child (HURT and SWEAR, respectively) had the highest positive weights for Component 3.

10.3.4 ASSOCIATION BETWEEN ACES AND FIRST DIAGNOSIS OF CANCER

Table 4 shows the association between ACEs and diagnosis of all cancers, childhood cancer and adulthood cancer. There were no statistically significant results using the unadjusted models. However, after adjusting for age, sex, race/ethnicity, income, education and marital status, Component 1 was significantly associated with being diagnosed with cancer in adulthood (adjusted OR: 1.21; 95% CI: 1.03–1.43). Components 2 and 3 were not significantly associated with diagnosis of cancer.

10.4 DISCUSSION

While many studies have reported on ACEs and specific types of cancer, to our knowledge, no study has reported on the association between ACEs and all cancers, differentiating between cancer in childhood and adulthood. The complex sampling scheme also provides population-based estimates. Among the respondents, approximately one in ten respondents reported ever having been diagnosed with cancer. The prevalence estimate of cancer derived from the current study of 10% is higher than the estimate of 4.2% derived from the National Cancer Institute (NCI) ,[39,40] and census population estimates [41]. Component 1, for which the sexual abuse variables had the highest weights, was significantly associated with cancer in adulthood.

This is the first study to examine adverse childhood experiences using PCA. By using PCA, we were able to use the data to derive three components of ACEs. Brown et al. (2010) found a positive association between ACEs and risk of lung cancer, which coincides with the findings in the current study. However, the current study considered prevalence of all cancers overall, in childhood and adulthood, in a cross-sectional study with a population-based sample while Brown et al. (2010) looked at incident lung cancer using a prospective study design [15].

Evidence suggests that ACEs have been associated with risky sexual behaviors, [17]–[19] sexually transmitted infections, [7],[8],[19] chronic diseases such as cancer, [15],[16] obesity, [42] depression [7],[8],[12],[13], and smoking [42],[43]. Research has also shown that ACEs are associated with risk factors for chronic disease [15] such as chronic obstructive pulmonary disease [44], liver disease [4], and ischemic heart disease

[14]. ACEs can be viewed as an initial exposure that may result in adverse health outcomes during adulthood. However, exposure to adversity during childhood is a likely source of, not only chronic but also acute stressors. [45] These acute stressors may alter fundamental biological functions, therefore, adversely impacting health, even before the onset of adulthood. In utero exposure to adverse conditions can result in adverse health outcomes [46] throughout the life course. However, in the current study, we did not find an association between ACEs and childhood cancer. These findings suggest that ACEs may be related to adulthood cancer via alternate pathways as at present, we do not have biological models linking ACEs to cancer. One possible pathway is through the association of ACEs with risk factors for chronic diseases that may negatively affect general health and contribute to disease progression [15], which can predispose an individual for cancer in adulthood. However, more research is needed to explore this potential mechanism.

This study must be considered with limitations in mind. First, the study was cross-sectional. Based on the survey questions, we were not able to determine age of exposure to adverse childhood experiences. Therefore, it is possible that a diagnosis of cancer in childhood could have occurred before exposure to ACEs. However, since the majority of cancers were adulthood cancers (98.5%), the ambiguity of temporal sequence would have only applied to 1.5% of the cancer cases in the overall category. It is also possible that there could have been underreporting of ACEs due to desirability bias, which would result in non-differential misclassification of the exposure. However, if this were the case, then the estimates produced would have been underestimates of the "true" association.

The study also had several strengths. We are able to obtain population-based estimates of the prevalence of cancer and exposure to ACEs using a comprehensive, nationally representative database. This is the first study to use PCA to assess the association of ACEs with an adverse health outcome, and the first study to examine ACEs and overall, childhood and adulthood cancer. The use of PCA also allowed for aggregation of variables that were likely to cluster together without relying on subjective construction of factors.

10.5 CONCLUSIONS

The results show an association between Component 1, for which the sexual abuse had the highest weights, and adulthood cancer. A reduction in ACEs, perhaps sexual abuse ACEs, could reduce adverse health outcomes, such as cancer. ACEs screening should be implemented during routine healthcare examinations for children and adults so as to help in identifying patients who may be at risk for chronic illnesses. Future research should focus on sexual abuse ACEs and other adverse health outcomes to determine if individuals who experienced these specific types of ACEs would be at risk for other illnesses.

REFERENCES

1. Anda RF, Butchart A, Felitti VJ, Brown DW (2010) Building a framework for global surveillance of the public health implications of adverse childhood experiences. Am J Prev Med 39: 93–98. doi: 10.1016/j.amepre.2010.03.015
2. Centers for Disease Control and Prevention (CDC) (2010) Adverse childhood experiences reported by adults – five states, 2009. MMWR Morb Mortal Wkly Rep 59: 1609–1613.
3. Lanoue M, Graeber DA, Helitzer DL, Faecett J (2012) Negative Affect Predicts Adults' Ratings of the Current, but Not Childhood, Impact of Adverse Childhood Events. Community Ment Health J [Epub ahead of print].
4. Dong M, Anda RF, Dube SR, Giles WH, Felitti VJ (2003) The relationship of exposure to childhood sexual abuse to other forms of abuse, neglect, and household dysfunction during childhood. Child Abuse Negl 27: 625–639. doi: 10.1016/s0145-2134(03)00105-4
5. LaNoue M, Graeber D, de Hernandez BU, Warner TD, Helitzer DL (2012) Direct and indirect effects of childhood adversity on adult depression. Community Ment Health J 48: 187–192. doi: 10.1007/s10597-010-9369-2
6. Dong M, Anda RF, Felitti VJ, Dube SR, Williamson DF, et al. (2004) The interrelatedness of multiple forms of childhood abuse, neglect, and household dysfunction. Child Abuse Negl 28: 771–784. doi: 10.1016/j.chiabu.2004.01.008
7. Felitti VJ, Anda RF, Nordenberg D, Williamson DF, Spitz AM, et al. (1998) Relationship of childhood abuse and household dysfunction to many of the leading causes of death in adults. The Adverse Childhood Experiences (ACE) Study. Am J Prev Med 14: 245–258. doi: 10.1016/s0749-3797(98)00017-8
8. Jewkes RK, Dunkle K, Nduna M, Jama PN, Puren A (2010) Associations between childhood adversity and depression, substance abuse and HIV and HSV2 inci-

dent infections in rural South African youth. Child Abuse Negl 34: 833–841. doi: 10.1016/j.chiabu.2010.05.002

9. Oladeji BD, Makanjuola VA, Gureje O (2010) Family-related adverse childhood experiences as risk factors for psychiatric disorders in Nigeria. Br J Psychiatry 196: 186–191. doi: 10.1192/bjp.bp.109.063677

10. Strine TW, Dube SR, Edwards VJ, Prehn AW, Rasmussen S, et al. (2012) Associations between adverse childhood experiences, psychological distress, and adult alcohol problems. Am J Health Behav 36: 408–423. doi: 10.5993/ajhb.36.3.11

11. Wu NS, Schairer LC, Dellor E, Grella C (2010) Childhood trauma and health outcomes in adults with comorbid substance abuse and mental health disorders. Addict Behav 35: 68–71. doi: 10.1016/j.addbeh.2009.09.003

12. Chapman DP, Whitfield CL, Felitti VJ, Dube SR, Edwards VJ, et al. (2004) Adverse childhood experiences and the risk of depressive disorders in adulthood. J Affect Disord 82: 217–225. doi: 10.1016/j.jad.2003.12.013

13. Pickles A, Aglan A, Collishaw S, Messer J, Rutter M, et al. (2010) Predictors of suicidality across the life span: the Isle of Wight study. Psychol Med 40: 1453–1466. doi: 10.1017/s0033291709991905

14. Dong M, Giles WH, Felitti VJ, Dube SR, Williams JE, et al. (2004) Insights into causal pathways for ischemic heart disease: adverse childhood experiences study. Circulation 110: 1761–1766. doi: 10.1161/01.cir.0000143074.54995.7f

15. Brown DW, Anda RF, Felitti VJ, Edwards VJ, Malarcher AM, et al. (2010) Adverse childhood experiences are associated with the risk of lung cancer: a prospective cohort study. BMC Public Health 10: 20. doi: 10.1186/1471-2458-10-20

16. Van der Meer LB, van Duijn E, Wolterbeek R, Tibben A (2012) Adverse childhood experiences of persons at risk for Huntington's disease or BRCA1/2 hereditary breast/ovarian cancer. Clin Genet 81: 18–23. doi: 10.1111/j.1399-0004.2011.01778.x

17. Brown DW, Anda RF (2009) Adverse childhood experiences: origins of behaviors that sustain the HIV epidemic. AIDS 23: 2231–2233. doi: 10.1097/qad.0b013e3283314769

18. Klein H, Elifson KW, Sterk CE (2007) Childhood neglect and adulthood involvement in HIV-related risk behaviors. Child Abuse Negl 31: 39–53. doi: 10.1016/j.chiabu.2006.08.005

19. Hillis SD, Anda RF, Felitti VJ, Nordenberg D, Marchbanks PA (2000) Adverse childhood experiences and sexually transmitted diseases in men and women: a retrospective study. Pediatrics 106: E11. doi: 10.1542/peds.106.1.e11

20. Vander Weg MW (2011) Adverse childhood experiences and cigarette smoking: the 2009 Arkansas and Louisiana Behavioral Risk Factor Surveillance Systems. Nicotine Tob Res 13: 616–622. doi: 10.1093/ntr/ntr023

21. Anda RF, Croft JB, Felitti VJ, Nordenberg D, Giles WH, et al. (1999) Adverse childhood experiences and smoking during adolescence and adulthood. JAMA 282: 1652–1658. doi: 10.1001/jama.282.17.1652

22. Jun HJ, Rich-Edwards JW, Boynton-Jarrett R, Austin SB, Frazier AL, et al. (2008) Child abuse and smoking among young women: the importance of severity, accumulation, and timing. J Adolesc Health 43: 55–63. doi: 10.1016/j.jadohealth.2007.12.003

23. Slopen N, Kubzansky LD, McLaughlin KA, Koenen KC (2012) Childhood adversity and inflammatory processes in youth: A prospective study. Psychoneuroendocrinology.
24. Centers for Disease Control and Prevention (CDC) Behavioral Risk Factor Surveillance System website. Available: http://www.cdc.gov/brfss/index.htm. Accessed 2012 Aug 7.
25. Centers for Disease Control and Prevention (CDC) Behavioral Risk Factor Surveillance System: Questionnaires Modules by Category, 2010 website. Available: http://apps.nccd.cdc.gov/BRFSSModules/ModByCat.asp?Yr=2010. Accessed 2012 Aug 6.
26. Dube SR, Williamson DF, Thompson T, Felitti VJ, Anda RF (2004) Assessing the reliability of retrospective reports of adverse childhood experiences among adult HMO members attending a primary care clinic. Child Abuse Negl 28: 729–737. doi: 10.1016/j.chiabu.2003.08.009
27. Gupta V, Gu K, Chen Z, Lu W, Shu XO, et al. (2011) Concordance of self-reported and medical chart information on cancer diagnosis and treatment. BMC Med Res Methodol 11: 72. doi: 10.1186/1471-2288-11-72
28. Stavrou E, Vajdic CM, Loxton D, Pearson SA (2011) The validity of self-reported cancer diagnoses and factors associated with accurate reporting in a cohort of older Australian women. Cancer Epidemiol 35: e75–80. doi: 10.1016/j.canep.2011.02.005
29. Centers for Disease Control and Prevention (CDC) Breast Cancer Risk by Age website. Available: http://www.cdc.gov/cancer/breast/statistics/age.htm. Accessed 2012 Aug 7.
30. Centers for Disease Control and Prevention (CDC) Prostate Cancer Risk by Age website. Available: http://www.cdc.gov/cancer/prostate/statistics/age.htm. Accessed 2012 Aug 7.
31. Centers for Disease Control and Prevention (CDC) Breast Cancer Statistics website. Available: http://www.cdc.gov/cancer/breast/statistics/index.htm. Accessed 2012 Aug 7.
32. Centers for Disease Control and Prevention (CDC) Cancer Prevention and Control: Cancer Among Men website. Available: http://www.cdc.gov/cancer/dcpc/data/men.htm. Accessed 2012 Aug 7.
33. Centers for Disease Control and Prevention (CDC) Cancer Prevention and Control: Cancer Among Women website. Available: http://www.cdc.gov/cancer/dcpc/data/women.htm. Accessed 2012 Aug 7.
34. Centers for Disease Control and Prevention (CDC). Breast Cancer Rates by Race/Ethnicity website. Available: http://www.cdc.gov/cancer/breast/statistics/race.htm. Accessed 2012 Aug 7.
35. Centers for Disease Control and Prevention (CDC) Colorectal (Colon) Cancer Rates by Race and Ethnicity website. Available: http://www.cdc.gov/cancer/colorectal/statistics/race.htm. Accessed 2012 Aug 7.
36. Mock SE, Arai SM (2010) Childhood trauma and chronic illness in adulthood: mental health and socioeconomic status as explanatory factors and buffers. Front Psychol 1: 246. doi: 10.3389/fpsyg.2010.00246
37. Cambois E, Jusot F (2011) Contribution of lifelong adverse experiences to social health inequalities: findings from a population survey in France. Eur J Public Health 21: 667–673. doi: 10.1093/eurpub/ckq119
38. Centers for Disease Control and Prevention (CDC) Behavioral Risk Factor Surveillance System: 2010 BRFSS Overview website. Available: http://www.cdc.gov/brfss/technical_infodata/surveydata/2010.htm. Accessed 2012 Aug 14.

39. American Cancer Society. Cancer Prevalence: How Many People Have Cancer website. Available: http://www.cancer.org/Cancer/CancerBasics/cancer-prevalence. Accessed 2012 Aug 14.
40. National Cancer Institute website. Available: http://seer.cancer.gov/csr/1975_2008/. Accessed 2012 Aug 14.
41. U.S. Census Bureau, 2010 Census. Profile of General Population and Housing Characteristics: 2010 website. Available: http://factfinder2.census.gov/faces/tableservices/jsf/pages/productview.xhtml?pid=DEC_10_DP_DPDP1&prodType=table. Accessed 2012 Aug 14.
42. Dube SR, Cook ML, Edwards VJ (2010) Health-related outcomes of adverse childhood experiences in Texas, 2002. Prev Chronic Dis 7: A52.
43. Ford ES, Anda RF, Edwards VJ, Perry GS, Zhao G, et al. (2011) Adverse childhood experiences and smoking status in five states. Prev Med 53: 188–193. doi: 10.1016/j.ypmed.2011.06.015
44. Anda RF, Brown DW, Dube SR, Bremner JD, Felitti VJ, et al. (2008) Adverse childhood experiences and chronic obstructive pulmonary disease in adults. Am J Prev Med 34: 396–403. doi: 10.1016/j.amepre.2008.02.002
45. Kelly-Irving M, Mabile L, Grosclaude P, Lang T, Delpierre C (2012) The embodiment of adverse childhood experiences and cancer development: potential biological mechanisms and pathways across the life course. Int J Public Health.
46. Barker DJ, Osmond C (1986) Infant mortality, childhood nutrition, and ischaemic heart disease in England and Wales. Lancet 1: 1077–1081. doi: 10.1016/s0140-6736(86)91340-1

CHILDHOOD ADVERSITY AS A RISK FOR CANCER: FINDINGS FROM THE 1958 BRITISH BIRTH COHORT STUDY

MICHELLE KELLY-IRVING, BENOIT LEPAGE, DOMINIQUE DEDIEU, REBECCA LACEY, NORIKO CABLE, MELANIE BARTLEY, DAVID BLANE, PASCALE GROSCLAUDE, THIERRY LANG, AND CYRILLE DELPIERRE

11.1 BACKGROUND

Strong evidence for socio-economic differences in cancer survival and mortality has been shown for many cancers and across populations [1]. In 2004, cancer was among the leading four causes of death in high income countries [2,3], and its increase is typically associated with the third stage of epidemiological transition [4]. Cancer prevalence and mortality is likely to become increasingly predominant in high-income countries and by consequence so is their contribution to health inequalities.

The causes of cancer are complex and multifactorial occurring over a lengthy time. Explanations for mechanisms linking adverse events early in life and cancer have been provided from animal models [5,6]. In humans, early life exposure to Adverse Childhood Experiences (ACE), like trauma, abuse or maltreatment in childhood has been linked both 'indirectly', through tobacco and alcohol use or via 'direct' associations [7] to alterations of the brain structure and neurobiological stress-response systems which in turn have consequences for health and emotional well-being [8]. Studies have described associations between retrospectively collected ACE and health outcomes such as liver disease [9], ischaemic heart disease [10], obesity [11,12], perceived health [13] and psychopathology [14] as well as premature mortality [15,16]. Regarding cancer, data are sparser and inconsistent [7,12,17]. One of the reasons for this inconsistency is that the ACE measure used is usually self-reported by adults asked about trauma and adversity they may have experienced during childhood. Such questions are inevitably vulnerable to various forms of recall bias [18] one example being that individuals' memories of the past may be influenced by their later health [19].

In the present study we aim to determine if exposure to ACE, determined using prospectively collected data, is linked with the onset of cancer in adulthood. The causes of cancers are indeed diverse, however our aim here is to understand the development of susceptibility to this disease at its earliest phase, at the common roots of all cancers [20]. To do this we take a lifecourse approach, considering the whole life span. We hypothesise that the physiological embedding of stress induced by adversity in childhood is linked with biological susceptibility as well as with the development of stress-reducing health behaviours favouring the development of cancers in adult life [21,22]. The objective of this study is to examine the relationship between psychosocial adversity in childhood and cancer after controlling for the effects of material disadvantage, health behaviours and education level using a large prospective cohort study.

11.2 METHODS

Data are from the 1958 National Child Development Study (NCDS) which included all live births during one week in 1958 (n = 18558) in Great Britain. Subsequent data collections (sweeps) were carried out on cohort members aged 7, 11, 16, 23, 33, 42, 46 and 50. At the last sweep, carried out in 2008, 9790 individuals participated in the self-reported question-naire and face to face interview, representing 53% of the original sample. The NCDS has been described in detail elsewhere [23].

11.2.1 ETHICS & DATA ACCESS

Written informed consent was obtained from the parents for childhood measurements and ethical approval for the adult data collection was ob-tained from the National Research Ethics Advisory Panel. NCDS data are open access datasets available to non-profit research organisations.

11.2.2 CANCER IN THE COHORT

The outcome variable of interest in this study was self-reported cancer between the age of 33 and 50. We did not use information available on cancer at the age of 23, as we were interested in controlling for health be-haviours at this age. Two main types of variables were used across the data sweeps. At ages 46 and 50 respondents were asked to report any medical conditions among which cancers can be identified. At ages 33 and 42 re-spondents were asked to report if they had 'ever had cancer'. The outcome variable 'cancer' constructed for the purpose of this study was categorised as follows: 'yes' corresponds to individuals who ever reported having had cancer at ages 33, 42, 46 or 50 (n=444); 'no' corresponds to individuals who explicitly reported not having a cancer at age 33 and 42, and who did not have an ICD coded cancer among their reported medical conditions at the ages of 46 and 50 (n=5694); 'missing' corresponds to individuals who did not respond to any questions on cancer, and respondents who occasionally responded 'no' or who were missing. These are individuals who we could not exclude as having had a cancer based on their self-report (n=11943). Based on this conservative definition respondents had to ac-tively respond 'yes' to having had a cancer or report a cancer as a medical condition. Conversely they had to consistently respond 'no' to questions on cancer and not report cancer as a medical condition. All other cases were defined as 'missing'. Further details on how cancer was coded in the cohort are available from the authors.

11.2.3 ADVERSE CHILDHOOD EXPERIENCES (ACE)

There are many ways in which adversity can be conceptualised [24-27]. We have attempted to construct a theoretical framework prior to extracting any data, in order to create a measurement with a robust content validity. We have identified ACE as a set of traumatic and stressful psychosocial conditions that are out of the child's control, that tend to co-occur and often persist over time [12,14,24,27]. We have restricted ACE to intra-familial events or conditions in the child's immediate environment causing chronic stress responses. In our definition ACE is distinguished from events or conditions linked to the socioeconomic and material environment.

Information was extracted from the study via variables collected at age 7, 11, 16 from questions posed to the child's parent or their teacher. Sources of adversity were divided into six categories:

1. Child in care: child has ever been in public/ voluntary care services or foster care at age 7, 11 or 16.
2. Physical neglect: child appears undernourished/ dirty aged 7 or 11, information collected from the response from child's teacher to the Bristol Social Adjustment Guide.

Household dysfunction, as described by Felitti et al. [12], is a dimension of adversity consisting of four categories each contributing to the score:

3. Offenders: The child lived in a household where a family member was in prison or on probation (age 11 y) or is in contact with probation service at 7 or 11 y; the child has ever been to prison or been on probation at 16 y.

4. Parental separation: The child has been separated from their father or mother due to death, divorce, or separation at 7, 11 or 16 y.
5. Mental illness: Household has contact with mental health services at 7 or 11 y; Family member has mental illness at 7 & 11 or 16 y.
6. Alcohol abuse: Family member has alcohol abuse problem at 7 y.

Exposure to adversity was identified by a positive response to any of the above categories. Respondents were excluded if they had missing data for all six categories. Respondents were considered as having no adversities if they answered 'no' all the categories or if they answered 'no' to one or more category and the other categories were missing. ACE was measured by counting the reports of: child in care, physical neglect, offenders, parental separation, mental illness and alcohol abuse. A three category variable was then constructed (0 adversities/1 adversity/2or+adversities).

11.2.4 EARLY LIFE SOCIOECONOMIC AND BIOLOGICAL CONFOUNDERS

To examine the relationship between ACE and cancer, prior confounding variables potentially associated with both ACE and cancer need to be included in the initial multivariate model. Among the variables available at baseline, collected from the cohort member's mothers via a questionnaire at birth, we identified those most likely to be social or biological confounding factors based on the literature. Household and parental characteristics were included: mother's age at birth, overcrowding (people per room), mother's partner's social class (recoded into manual vs non-manual), and if this was unavailable the mother's father's social class was used mother's education level (left school before/after minimum leaving age), and maternal smoking during pregnancy (no smoking, sometimes, often, heavy). The respondent's characteristics and birth variables were also included: sex, gestational age at birth, parity, birth weight, foetal distress, problems during pregnancy and breastfeeding (no, one month or less, more than one month). To control for health problems in childhood, a childhood pathologies variable was constructed using data collected at ages 7, 11 and 16 y. It was based both on mother's report and medical examinations including congenital conditions, moderate/severe disabilities, chronic respiratory or circulatory conditions, sensory impairments and special schooling (childhood pathology: yes/no).

11.2.5 MEDIATORS ACROSS THE LIFECOURSE

To determine whether any observed associations between ACE and cancer were due to adult mediating factors, the following were added to the models: respondent's educational attainment at 23y (A level/ O level/ no qualification), respondent's occupational social class at 23 y (non-manual active/ manual active/ inactive). The 'malaise inventory' was used to identify symptoms of depression at the age of 23. It was based on a set of 24 questions indentifying symptoms, if the respondent reported experiencing more than seven of the symptoms they were considered as having psychological malaise (no malaise/ malaise), characterized by symptoms of depression and/or anxiety. The health behaviour variables included were: alcohol consumption at 23 y (normal drinking (women: between 1–14 units in the previous week, men: between 1–21 units in the previous week)/ abstinence (reported not consuming any alcohol in the previous week)/ heavy drinking (women: >14 units in the previous week, men: >21 units in the previous week [28]), smoking status at 23 y (never smoked/ past smoker/ current smoker), and BMI (kg/m^2) categorised using the WHO cut-offs age 23 y [29]. Adult life-style variables are available at other points along the lifecourse, however in our models, these adult variables at the age of 23 are a proxy for behavioural patterns in early adulthood predating reports of cancer. Controlling for them serves as a first step to understanding possible mechanisms. In the women's model we also created a variable identifying the age at which cohort members had their first pregnancy, a known risk factor for breast cancer. This was determined based on variables at age 33 y, and 50 y. The resulting variable is in three categories (before 33 y/ after 33 y/ no information on pregnancy).

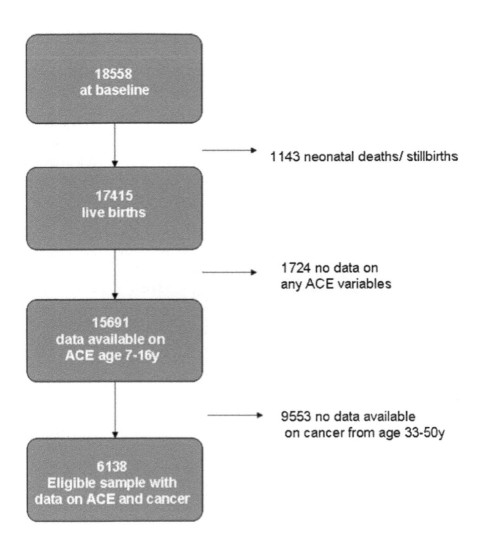

FIGURE 1: Flowchart showing selection of the subsample used for this study.

TABLE 1: Characteristics of the subsample* at birth, during childhood and in adulthood

		Men % (n)	Women % (n)
		46.2 (2836)	53.8 (3302)
Cancer	No	96.7 (2743)	89.4 (2951)
	Yes	3.3 (93)	10.6 (351)
ACE	0	74.4 (2110)	75.2 (2483)
	1	20.0 (566)	19.4 (641)
	2 or more	5.6 (160)	5.4 (178)
Mother's education	Stayed at school after min age	27.3 (773)	26.5 (875)
	Left school at or before min age	69.5 (1970)	70.0 (2310)
	Missing	3.3 (93)	3.5 (117)
Parental social class	Non-manual	30.9 (875)	27.2 (931)
	Manual	65.7 (1862)	68.0 (2246)
	Missing	3.5 (99)	3.8 (125)
Overcrowded household	<1.5 people per room	85.5 (2424)	83.4 (2775)
	≥1.5 people per room	9.1 (257)	11.3 (374)
	Missing	5.5 (155)	5.2 (173)
Parity	Primiparous	36.5 (1034)	37.6 (1242)
	One	32.2 (913)	28.6 (945)
	Two or more	28.6 (812)	30.7 (1013)
	Missing	2.7 (77)	3.1 (102)
Fœtal distress	No	87.4 (2478)	89.1 (2943)
	Yes	9.9 (281)	7.8 (256)
	Missing	2.7 (77)	3.1 (103)
Problems during pregnancy	No	72.7 (2063)	70.7 (2333)
	Yes	24.5 (695)	26.2 (865)
	Missing	2.8 (78)	3.2 (104)
Smoking in pregnancy	No	66.6 (1888)	65.6 (2165)
	Sometimes	5.4 (152)	5.0 (164)
	Moderately	13.8 (390)	14.3 (472)
	Heavily	10.4 (296)	11.0 (364)
	Missing	3.9 (110)	4.1 (137)
Breastfeeding	Yes for more than 1 month	44.6 (1264)	43.4 (1432)
	Yes for up to one month	22.3 (631)	23.1 (763)
	No	27.4 (776)	27.3 (901)

TABLE 1: *Cont.*

		Men % (n)	Women % (n)
		46.2 (2836)	53.8 (3302)
	Missing	5.8 (165)	6.2 (206)
Childhood pathologies	No	65.0 (1843)	69.4 (2291)
	Yes	22.6 (640)	18.5 (612)
	Missing	12.5 (353)	12.1 (399)
Education level 23 y	A levels or higher	25.6 (726)	24.3 (801)
	O levels	39.4 (1117)	45.1 (1489)
	No qualifications	34.2 ((970)	29.1 (961)
	missing	0.8 (23)	1.5 (51)
Social class 23 y	Non-manual active	38.6 (1094)	54.3 (1792)
	Manual active	45.4 (1287)	12.8 (422)
	Inactive	12.9 (365)	31.1 (1027)
	Missing	3.17 (90)	1.9 (61)
Alcohol consumption 23 y	Normal	70.1 (1988)	56.4 (1862)
	Abstinence	4.2 (118)	7.4 (244)
	Heavy	17.6 (498)	8.9 (293)
	missing	8.2 (232)	27.4 (903)
Smoking 23 y	Never	29.3 (832)	33.4 (1102)
	Past	33.9 (960)	27.9 (921)
	Current	36.1 (1023)	37.3 (1230)
	missing	0.7 (21)	1.5 (49)
Psychological malaise 23 y	No	96.1 (2725)	89.5 (2956)
	Yes	3.0 (85)	8.9 (295)
	Missing	0.9 (26)	1.5 (51)
Body mass index 23 y	Normal	75.3 (2136)	76.5 (2526)
(Kg/m^2 WHO cut-offs)	Underweight	2.3 (64)	7.0 (231)
	Overweight	16.8 (477)	11.3 (374)
	Obese	3.9 (110)	2.6 (86)
	missing	1.7 (49)	2.6 (85)
		Mean (s.e.)	Mean (s.e.)
Gestational age in days	(missing n=2986)	279.9 (0.2)	280.4 (0.2)
Birthweight Kg	(missing n=1775)	3.4 (0.0)	3.2 (0.1)
Mother's age at birth	(missing n = 1156)	27.5 (0.1)	27.6 (0.1)

Subsample of individuals who provided information used to create adversity and cancer variables.

TABLE 2: Descriptive statistics on ACE, and the lifecourse covariates by reported cancer for men and women

% (n)		Men		Women	
		Cancer 'yes' % (N=93)	P*	Cancer 'yes' % (n=351)	P*
Cancer		3.3		10.6	
Characteristics in terms of reported cancer					
ACE	0	3.1		9.1	
	1	3.5	0.586	13.0	0.004
	2 or more	5.0	0.189	23.0	<0.0001
Mother's education	Stayed at school after min age	2.6		9.1	
	Left school at or before min age	3.6	0.112	11.2	0.058
Parental social class	Non-manual	2.7		8.6	
	Manual	3.6	0.176	11.5	0.010
Overcrowded household	<1.5 people per room	3.1		10.0	
	≥1.5 people per room	4.9	0.251	15.4	0.002
Parity	Primiparous	3.9		10.0	
	One	2.5	0.151	10.5	0.588
	Two or more	3.3	0.320	11.5	0.261
Fœtal distress	No	3.2		10.5	
	Yes	4.2	0.561	12.0	0.591
Problems during pregnancy	No	3.5		10.2	
	Yes	2.6	0.171	11.9	0.212
Smoking in pregnancy	No	3.1		10.0	
	Sometimes	5.2	0.039	11.6	0.458
	Moderately	4.6	0.045	11.2	0.300
	Heavily	1.8	0.215	13.5	0.072
Breastfeeding	Yes for more than 1 month	3.2		9.7	

TABLE 2: *Cont.*

% (n)		Men		Women	
		Cancer 'yes' % (N=93)	P*	Cancer 'yes' % (n=351)	P*
	Yes for up to one month	3.4	0.884	11.1	0.275
	No	3.4	0.934	11.7	0.161
Childhood patholo-gies	No	3.4		9.9	
	Yes	3.0	0.411	13.3	0.016
Education level 23 y	A levels or higher	2.6		8.3	
	O levels	3.0	0.628	9.2	0.209
	No qualifications	4.0	0.113	14.6	<0.0001
Social class 23 y	Non-manual active	2.9		9.4	
	Manual active	3.4	0.320	11.5	0.222
	Inactive	3.6	0.628	12.2	0.018
Alcohol consump-tion 23 y	Normal	3.1		10.0	
	Abstinence	4.1	0.373	10.9	0.578
	Heavy	3.6	0.820	14.1	0.031
Smoking 23 y	Never	3.0		7.2	
	Past	3.0	0.797	9.2	0.214
	Current	3.7	0.504	14.6	<0.0001
Psychological mal-aise 23 y	No	3.2		10.0	
	Yes	3.4	0.999	16.8	0.001
Body mass index 23 y	Normal	2.7		10.4	
(Kg/m2 WHO cut-offs)	Underweight	4.2	0.703	13.9	0.092
	Overweight	4.5	0.159	10.7	0.662
	Obese	7.3	0.036	6.1	0.285
Age at first preg-nancy	Before 33			9.9	
	After 33			18.4	<0.0001
	No information on pregnancy			8.4	0.225

p-values comparing 'yes' v 'no' cancers for each covariate.

TABLE 3: Multivariate logistic regression models using data obtained from multiple imputation: men (n = 2836)

		Model 1 early life factors		Model 2 + childhood pathologies and ACEs		Model 3 + adult mediating factors	
		OR (95% CI)	p	OR (95% CI)	p	OR (95% CI)	p
Mother's education	Left school after min age						
	Left school at or before min age	1.26 (0.73-2.17)	0.412	1.25 (0.72-2.16)	0.427	1.18 (0.67-2.09)	0.563
Father's social class	Non-manual						
	Manual	1.16 (0.68-1.96)	0.587	1.14 (0.67-1.94)	0.620	1.07 (0.62-1.86)	0.810
Overcrowded household	<1.5 people per room						
	>=1.5 people per room	1.43 (0.76-2.68)	0.265	1.40 (0.75-2.64)	0.295	1.36 (0.72-2.59)	0.344
Gestational age	(days)	1.01 (0.99-1.03)	0.444	1.01 (0.99-1.03)	0.445	1.01 (0.99-1.03)	0.444
Parity	Primiparous						
	One	0.73 (0.42-1.26)	0.263	0.72 (0.42-1.25)	0.245	0.71 (0.41-1.23)	0.215
	Two or more	0.95 (0.52-1.73)	0.855	0.90 (0.48-1.66)	0.726	0.85 (0.45-1.60)	0.613
Birthweight	(kg)	0.77 (0.49-1.21)	0.259	0.78 (0.49-1.23)	0.279	0.75 (0.47-1.20)	0.231
Fœtal distress	No						
	Yes	1.38 (0.72-2.65)	0.336	1.37 (0.71-2.63)	0.348	1.35 (0.70-2.6)	0.376
Problems during pregnancy	No						
	Yes	0.69 (0.41-1.18)	0.175	0.69 (0.41-1.18)	0.180	0.71 (0.42-1.21)	0.210
Smoking during pregnancy	No						

TABLE 3: *Cont.*

		Model 1 early life factors		Model 2+childhood pathologies and ACEs		Model 3+adult mediating factors	
		OR (95% CI)	p	OR (95% CI)	p	OR (95% CI)	p
	Sometimes	1.62 (0.75-3.49)	0.222	1.59 (0.74-3.44)	0.238	1.52 (0.70-3.33)	0.294
	Moderately	1.38 (0.8-2.37)	0.250	1.36 (0.79-2.35)	0.270	1.36 (0.78-2.36)	0.281
	Heavily	0.54 (0.21-1.37)	0.194	0.52 (0.2-1.34)	0.176	0.50 (0.20-1.28)	0.149
Mother's age at birth	(years)	0.99 (0.95-1.04)	0.731	0.99 (0.95-1.04)	0.809	1.00 (0.95-1.04)	0.887
Breastfed	Yes, for more than 1 month						
	Yes, for up to one month	1.03 (0.58-1.81)	0.924	1.04 (0.59-1.82)	0.905	1.04 (0.59-1.84)	0.898
	No	1.00 (0.59-1.68)	0.991	0.99 (0.59-1.68)	0.982	0.96 (0.56-1.64)	0.885
Childhood pathologies	No						
	Yes			0.87 (0.51-1.49)	0.623	0.83 (0.48-1.43)	0.500
ACE	No adversities						
	One adversity			1.11 (0.66-1.87)	0.687	1.03 (0.61-1.76)	0.904
	Two or more adversities			1.50 (0.68-3.29)	0.314	1.39 (0.62-3.11)	0.419
Education level at 23 y	A levels or higher						
	O levels					1.00 (0.51-1.96)	0.995
	No qualifications					1.24 (0.56-2.73)	0.599
Social class 23 y	Non-manual active						

TABLE 3: *Cont.*

		Model 1 early life factors		Model 2 + childhood pathologies and ACEs		Model 3 + adult mediating factors	
		OR (95% CI)	p	OR (95% CI)	p	OR (95% CI)	p
	manual active					0.92 (0.5-1.69)	0.788
	Inactive					1.06 (0.50-2.24)	0.885
Smoking 23 y	Never						
	Past					1.02 (0.55-1.87)	0.962
	Current					1.21 (0.68-2.18)	0.516
Alcohol 23 y	Normal						
	Abstain-ance					1.33 (0.49-3.59)	0.573
	Heavy					1.10 (0.62-1.94)	0.743
Psychological malaise 23 y	No						
	Yes					0.95 (0.23-3.92)	0.938
BMI groups	Normal						
	Under-weight					1.40 (0.33-5.89)	0.647
	Over-weight					1.58 (0.90-2.78)	0.113
	Obese					2.72 (1.17-6.32)	0.020

11.2.6 MISSING DATA, IMPUTATION AND STATISTICAL ANALYSIS

Cohort follow-up has been good over time, 84% participated at the age of 16, with a gradual dwindling in participation throughout adulthood (72% at age

23, and 65% at age 42) when participants were most likely to have moved [23]. Refusal rates have been low, with 7% at age 23 and 13% at age 42 for example [23]. In the latest data sweep (2008) 54% (n=9790) of the original sample participated, and for some variables missing data presents a considerable challenge. The sample used for this study is described in Figure 1.

To control for possible bias due to missing data, we imputed data for covariates with missing data using the multiple imputation program ICE in STATA v11. Twenty imputations were conducted taking the missing at random (MAR) assumption for the covariates only (mother's education, father's social class, overcrowding, birthweight, gestational age, parity, smoking during pregnancy, mother's age at birth, breastfeeding, child pathologies, educational attainment at 23, social class at 23, malaise inventory at 23, drinking at age 23 and smoking at age 23, age at 1st pregnancy for women). Neither the exposure variable of interest (ACE) nor the cancer variable were included in the multivariate imputation model.

Bivariate crosstabulations were carried out on the imputed data using logistic regression to obtain p-values adjusted for the artificially inflated sample size. As the timing of cancer events was unknown, multivariate logistic regression analyses were carried out on the data obtained from multiply imputed data. Three models were run separately by sex, entering the variables chronologically as they would occur over the lifecourse. First early life socioeconomic circumstances and perinatal variables were entered. In model 2, childhood pathologies and ACE were added to the model. Finally, model 3 additionally controlled for education, social class, psychological malaise, health behaviours at 23 y, and for women, age at first pregnancy.

11.3 RESULTS

Descriptive statistics are presented in Table 1 for the subsample (n=6138) who provided information on whether or not they had had a cancer in the last four data sweeps (ages 33, 43, 46 and 50). Among respondents who answered questions on cancer, 3.3% of men and 10.6% of women reported having had cancer by the age of 50. In total 444 self-reported cancers were identified in the cohort reported between the ages of 33 and 50, 80% being reported by women (n=351). Based on the available information on cancer types reported by the cohort members (see Additional file 1), breast and cervical cancers were most often reported (19% and 15% respectively), followed by skin cancers (11%). Nevertheless, most cancers remained 'undefined '(39%) The majority of men (74%) and women (75%) had had no adverse childhood experiences, based on the prospectively collected information at ages 7, 11 and 16 y. The distribution of ACEs was similar for both sexes, with 5.6% of men and 5.4% of women having two or more ACEs.

Table 2 shows bivariate analyses between ACE, the covariates and cancer for men and women. In men, no relationship is apparent between ACE and cancer. The multivariate models for men (Table 3) show no association between early life socioeconomic variables or perinatal variables and cancer. ACE was not significantly associated with cancer, and neither were social variables at age 23 y, or smoking and alcohol age 23 y.

In women (Table 2), a bivariate relationship between ACE and cancer was shown. The proportion of women reporting a cancer before 50 increased from 9.1% for those with no ACEs to 13.0% for those with one ACE (p= 0.004), and 23.0% for those with 2+ACEs (p<0.001). Both manual parental social class and overcrowding in the household at birth were significantly related to reporting a cancer before 50 y, and mother's education was of borderline significance. Women who had experienced childhood pathologies were also more likely to have had a cancer than those without childhood pathologies (13.3% vs 9.9%, p=0.016). Many of the female respondent's adult variables were also related to having had a cancer before 50 y in the bivariate analyses. Low educational attainment, being inactive in terms of occupation, being a heavy alcohol drinker, being a smoker and having symptoms of depression/ anxiety, and having a first pregnancy after the age of 33 y, were also related to having had a cancer before 50.

The multivariate models for women (Table 4) show that. ACE was significantly associated with reporting cancer before the age of 50 after control for potential confounding variables (model 2). The graded association showed that women with one ACE had a 40% increase in the odds of reporting cancer (p=0.016), and those with two or more ACEs had a 2.5 increase in the odds of reporting a cancer versus women with no ACEs (p< 0.001). When information on adult behaviours, social characteristics and psychological malaise was added to the model this association weakened, so that women with two or more ACEs had a 2.1 increase in the odds of reporting a cancer versus those with no ACEs ($p < 0.001$). Women who had their first pregnancy after the age of 33 y had a significantly increased risk of having cancer before 50 y ($p < 0.001$). Furthermore, women who were smokers at age 23 y had a 70% increase in the odds of having cancer before 50 versus non-smokers ($p = 0.001$), and women who were heavy drinkers at age 23 y also had an increased risk of having cancer before 50, however this result was of borderline statistical significance ($p = 0.055$). Having symptoms of depression/anxiety increased the odds of having a cancer before 50 by 35%, however this result was not significant at the 5% level ($p = 0.1$).

11.4 DISCUSSION

The main finding from this study was that psychosocial adversity in childhood was related to cancer incidence before 50 y among women, after adjusting for prior confounding factors and potential mediators, in a large prospective cohort. An accumulation of ACE remained a strong predictor of cancer in women, after taking important potential mediating factors at age 23 y into account, including smoking and drinking. Women who experienced two or more ACE doubled their risk of having a cancer before 50 relative to women who had had no childhood adversities. There was a tendency towards a graded association between childhood adversity and adult cancer across the ACE categories. These results make a significant contribution to demonstrating and understanding links between ACE and cancer because they use an a priori definition of adversity and identify ACE with prospectively collected data. The strength of the relationship between adversity and cancer was of the same magnitude as that observed between age at first pregnancy and cancer, a well-known risk factor for breast cancer [30].

The two main weaknesses of this study are in the self-reported nature of cancer incidence, and the amount of missing data caused by attrition in the cohort study. These weaknesses are partly addressed by the conservative nature of our cancer variable, and by using estimates obtained from multiple imputations to account for the missing data. We conducted sensitivity analyses by running the model using imputed datasets, using a case–control sub-sample, and using the full cohort dataset without imputation, and found that the relationship between ACE and cancer incidence was stable. Comparisons were made between complete-case analyses and those run on estimates obtained by imputation. The models yielded similar results until the inclusion of variables at age 23 (model 3). The differences observed in the results for model 3 indicate selection bias in the complete case sample, where individuals who had experienced ACEs in childhood were more likely to have missing data at age 23 regarding health behaviours. The multiple imputation model therefore enables adjustment for this bias. We are confident, therefore, that our analyses show a 'real' association, the nature of which needs to be established in further more complex modelling of mediating factors.

The nature of the questions collecting information on cancer varied by data sweep, some being retrospective (have you ever had cancer) while other questions asked about suffering from an illness at the time of the survey. This means that there are information gaps, notably between the age of 33 and 42 where the cohort member could have had cancer and not report that they were suffering from it subsequently. However, this means that our cancer variable is most likely conservative. The validity of self-reported cancer is likely to vary based on the cancer type diagnosed, and tends towards an underreporting of cancers, given the high levels of specificity reported in studies comparing self-reports to registry data [31]. Indeed in studies where cancer registry data have been compared to self-reported information individuals tend to underreport rather than overreport cancer history, and variation in inaccurate cancer reporting varied considerably by cancer type [32]. It is also important to consider the biases linked with overreporting cancers. Individuals whose self-reported cancer was unconfirmed by the Finnish cancer registry were more likely to have accumulated psychosocial strain across their life-course [33]. The probability of misreporting cancer, both over-reporting and under-reporting, has been associated with socio-demographic characteristics: gender, age, BMI, size of household, place of birth, smoking, social participation, educational level, type of employment, alcohol consumption and poor well-being [34]. The sensitivity of self-reported cancer is also likely to be higher among respondents with a high level of education [31]. Stavrou et al. have demonstrated a good level of sensitivity and specificity for self-reported cancer diagnoses in a cohort of older women in Australia, comparing self-reports to registry data [35]. Gupta et al. also describe good concordance between self-reported cancer and medical chart information [36]. The multivariate models also show associations between well known risk factors and cancer before 50, such as age at first pregnancy, smoking and drinking, which contributes to enhancing the validity of the cancer variable.

Women were four times more likely to report having had a cancer than men which is understandable given the age of cohort participants at the time of study. Based on 2007–2009 estimates of cancer incidence in the UK, 44% of cancers diagnosed among women aged 25–49 were breast cancers [37]. Rather than indicating a sex, or gender difference, the lack of association between virtually any of the variables and cancer among men is likely be mainly due to the far lower incidence in men up to age 50 (93 cancers). The distribution of cancer types observed in the cohort will continue to evolve over time, and begin to represent a greater proportion of men due to the increased occurrence of prostate and lung cancer among men >50 years. Linkage will hopefully be made between the cohort and the cancer registries which would significantly increased the validity and reliability of the information, and allow analyses by cancer type in the future, and survival analyses, which are currently not possible due to a lack of information on the timing of events. For these analyses we limited the health behavioural variables to those collected at age 23 rather than using variables available afterwards. This enabled to limit any further problems of missing data, but may lead to an underestimate of the contribution of health behaviours in the associations.

One important strength of this study is in the prospective nature of the information on ACEs. Most studies examining the links between early adversity rely on retrospective questions, which are prone to recall bias, an important issue raised by Korpimaki et al. using data from a working age population and the cancer registry [19]. Using a retrospective questionnaire to identify ACE, and limiting their analyses to individuals whose cancer was diagnosed subsequent to answering questions on ACE to address the problem of recall bias, these authors found no association between working-age cancer and reporting ACE.

TABLE 4: Multivariate logistic regression models using data obtained from multiple imputation: women (n=3302)

		Model 1 early life factors		Model 2 + childhood pathologies and ACEs		Model 3 + adult mediating factors	
		OR (95% CI)	p	OR (95% CI)	p	OR (95% CI)	p
Mother's education	Left school after min age						
	Left school at or before min age	1.06 (0.80-1.41)	0.681	1.06 (0.79-1.41)	0.697	1.07 (0.79-1.45)	0.659
Father's social class	Non-manual						
	Manual	1.17 (0.88-1.56)	0.286	1.11 (0.83-1.49)	0.473	1.13 (0.84-1.53)	0.419
Overcrowded household	<1.5 people per room						
	>=1.5 people per room	1.47 (1.07-2.04)	0.019	1.39 (1.0-1.92)	0.051	1.33 (0.95-1.85)	0.093
Gestational age	(days)	1.00 (0.99-1.01)	0.648	1.00 (0.99-1.01)	0.664	1.00 (0.99-1.01)	0.889
Parity	Primiparous						
	One	1.17 (0.88-1.57)	0.283	1.14 (0.85-1.52)	0.391	1.06 (0.78-1.43)	0.715
	Two or more	1.32 (0.95-1.83)	0.095	1.20 (0.86-1.67)	0.279	1.09 (0.77-1.54)	0.642
Birthweight	(kg)	0.90 (0.70-1.16)	0.425	0.91 (0.71-1.18)	0.488	0.93 (0.72-1.21)	0.585
Fœtal distress	No						
	Yes	1.24 (0.81-1.89)	0.316	1.25 (0.82-1.90)	0.306	1.23 (0.81-1.89)	0.334
Problems during pregnancy	No						

TABLE 4: *Cont.*

		Model 1 early life factors		Model 2 + childhood pathologies and ACEs		Model 3 + adult mediating factors	
		OR (95% CI)	p	OR (95% CI)	p	OR (95% CI)	p
	Yes	1.16 (0.91-1.49)	0.240	1.13 (0.88-1.45)	0.332	1.14 (0.88-1.47)	0.324
Smoking during pregnancy	No						
	Sometimes	1.08 (0.65-1.79)	0.771	0.94 (0.56-1.57)	0.801	0.91 (0.54-1.54)	0.739
	Moderately	1.07 (0.78-1.49)	0.664	1.06 (0.76-1.46)	0.742	1.00 (0.72-1.39)	0.997
	Heavily	1.30 (0.93-1.82)	0.126	1.22 (0.87-1.72)	0.251	1.17 (0.82-1.66)	0.384
Mother's age at birth	(years)	0.98 (0.95-1.00)	0.040	0.98 (0.96-1.00)	0.086	0.98 (0.96-1.01)	0.176
Breastfed	Yes, for more than 1 month						
	Yes, for up to one month	1.10 (0.83-1.46)	0.517	1.09 (0.81-1.45)	0.567	1.08 (0.80-1.45)	0.627
	No	1.14 (0.87-1.51)	0.340	1.09 (0.82-1.44)	0.550	1.08 (0.81-1.43)	0.601
Childhood pathologies	No						
	Yes			1.31 (0.99-1.73)	0.061	1.24 (0.93-1.65)	0.149
ACE	No adversities						
	One adversity			1.40 (1.06-1.83)	0.016	1.30 (0.98-1.72)	0.066

TABLE 4: *Cont.*

		Model 1 early life factors		Model 2 + childhood pathologies and ACEs		Model 3 + adult mediating factors	
		OR (95% CI)	p	OR (95% CI)	p	OR (95% CI)	p
	Two or more adversities			2.46 (1.66-3.65)	<0.001	2.14 (1.42-3.21)	<0.001
Age at 1st pregnancy	<=33y						
>34y					2.24 (1.63-3.08)	<0.001	
	No pregnancies, or no information					0.94 (0.65-1.34)	0.716
Education level at 23 y	A levels or higher						
	O levels					0.95 (0.67-1.34)	0.756
	No qualifications					1.30 (0.87-1.93)	0.198
Social class 23 y	Non-manual active						
	manual active					0.93 (0.64-1.37)	0.726
	Inactive					1.04 (0.77-1.41)	0.796
Smoking 23 y	Never						
	Past					1.27 (0.9-1.78)	0.174
	Current					1.72 (1.25-2.36)	0.001
Alcohol 23 y	Normal						
	Abstainance					1.00 (0.66-1.52)	0.983
	Heavy					1.42 (0.99-2.04)	0.055
Psychological malaise 23 y	No						

TABLE 4: *Cont.*

		Model 1 early life factors		Model 2 + childhood pathologies and ACEs		Model 3 + adult mediating factors	
		OR (95% CI)	p	OR (95% CI)	p	OR (95% CI)	p
	Yes					1.35 (0.94-1.93)	0.099
BMI groups	Normal						
	Under-weight					1.26 (0.81-1.95)	0.299
	Overweight					0.93 (0.63-1.37)	0.706
	Obese					0.47 (0.17-1.24)	0.126

There is currently insufficient power to work on cancers at different sites using the cohort. Furthermore, we take a lifecourse approach to this study and are interested in the pre-cancerous phase where early psychosocial adversity may increase susceptibility to developing cancer earlier. The cause of cancers and their prognoses are indeed diverse, however our aim here is to understand the factors contributing to susceptibility at its earliest phase. The ACE literature indicates that there are early life factors which may increase inflammatory response [38], decrease immune system efficiency [39], increase exposure to viral infections [40], and raise the probability of damaging health behaviours along the lifecourse [41,42]. All of these behavioural and biological processes are known to be involved in cancer development.

The association between stress and cancer development and progression has been shown in biological studies [5]. Stress-related immunological changes bring about declines in natural killer cell activity by depressing their ability to respond to tumour or virally infected cells, and causing a reduction in the body's defences linked to the repair of damaged DNA [43]. Exposure to stressors is known to trigger responses via the central nervous system produced by the hypothalamic-pituitary-adrenal axis (HPA). This activity modifies neuroendocrine pathways which, over the long term, alter the critical physiological mechanisms involved in tumourogenesis [5]. When exposed to chronic stress, "the body remains in a constant state of overdrive" with adverse consequences on the regulation of systems implicated in cancer progression [6]. ACE has also been associated with risky health behaviours such as smoking, alcoholism, early sexual activity, and having multiple sexual partners [8,13,38-41], all of which are 'indirect' risk factors for cancer.

Our hypothesis that adversity in childhood may be linked with biological susceptibility involved in the development of cancers in adult life is reinforced by our findings. It could operate via two main mechanisms: A direct biological effect and an indirect effect via health behaviours (or a combination of both). The fact that ACE is still associated with cancer even after adjusting for behavioural and social mediators favours argument for a direct "biological" role of ACE.

In epidemiological studies, evidence of a direct association between exposure to stress and cancer incidence is mixed and inconclusive. This is likely to be due to the different ways in which stress was conceptualised. A Danish cohort study on 8736 men and women found no direct association between cumulative stressful life events collected retrospectively, mostly during adulthood, and cancer incidence, though they did identify a relationship between stress and unhealthy lifestyles [44]. Ollonen (2005) et al. found support for an overall association between stressful life events across the lifecourse and breast cancer risk in their Finnish case–control study [45]. A meta-analysis of studies on the association between stress and breast cancer did not support an association between stressful life events in adulthood and breast cancer risk [46]. Using linkage to the cancer registry, Fang et al. found that bereaved parents were at increased risk for cancers with an infectious aetiology, especially those linked to infection by the Human Papilloma Virus (HPV) after controlling for confounders. The authors hypothesised that the stress induced by losing a child may accelerate the cancer genesis of an established infection [47]. These studies do not consider the timing of exposure to stressful events in their analyses, an important factor considering the differential effects of physiological stress responses on various areas of the brain depending on when exposure occurs along the lifecourse [48].

11.5 CONCLUSION

Our findings establish an association between ACE and cancer in women using prospective data. The suggestion that cancer risk may be influenced by conditions in the first years of life is potentially important in furthering our understanding of factors contributing to cancer development, and consequently in redirecting scientific research and developing appropriate prevention policies. This nevertheless remains a first step in understanding the relationship between ACE and cancer in the 1958 birth cohort. A second step will require an in-depth exploration of both direct and indirect pathways along which biological, social and psychosocial mechanisms are likely to operate.

REFERENCES

1. Woods LM, Rachet B, Coleman MP: Origins of socio-economic inequalities in cancer survival: a review. Ann Oncol 2006, 17(1):5-19.
2. Lopez AD, Mathers CD, Ezzati M, Jamison DT, Murray CJL: Global and regional burden of disease and risk factors, 2001: systematic analysis of population health data. Lancet 2006, 367(9524):1747-1757.
3. WHO: The global burden of disease: 2004 update. Geneva: World Health Organisation; 2008:160.
4. Omran AR: The epidemiologic transition: A theory of the epidemiology of population change (Reprinted from The Milbank Memorial Fund Quarterly, vol 49, pg 509–38, 1971). Milbank Q 2005, 83(4):731-757.
5. Antoni MH, Lutgendorf SK, Cole SW, Dhabhar FS, Sephton SE, McDonald PG, Stefanek M, Sood AK: Opinion - The influence of bio-behavioural factors on tumour biology: pathways and mechanisms. Nat Rev Cancer 2006, 6(3):240-248.
6. Lutgendorf SK, Sood AK, Antoni MH: Host factors and cancer progression: biobehavioral signaling pathways and interventions. J Clin Oncol 2010, 28(26):4094-4099.
7. Brown DW, Anda RF, Felitti VJ, Edwards VJ, Malarcher AM, Croft JB, Giles WH: Adverse childhood experiences are associated with the risk of lung cancer: a prospective cohort study. BMC Public Health 2010., 10
8. Anda RF, Felitti VJ, Bremner JD, Walker JD, Whitfield C, Perry BD, Dube SR, Giles WH: The enduring effects of abuse and related adverse experiences in child-

hood. A convergence of evidence from neurobiology and epidemiology. Eur Arch Psychiatry Clin Neurosci 2006, 256(3):174-186.

9. Dong M, Dube SR, Felitti VJ, Giles WH, Anda RF: Adverse childhood experiences and self-reported liver disease: new insights into the causal pathway. Arch Intern Med 2003, 163(16):1949-1956.

10. Dong M, Giles WH, Felitti VJ, Dube SR, Williams JE, Chapman DP, Anda RF: Insights into causal pathways for ischemic heart disease: adverse childhood experiences study. Circulation 2004, 110(13):1761-1766.

11. Thomas C, Hypponen E, Power C: Obesity and type 2 diabetes risk in midadult life: The role of childhood adversity. Pediatrics 2008, 121(5):E1240-E1249.

12. Felitti VJ, Anda RF, Nordenberg D, Williamson DF, Spitz AM, Edwards V, Koss MP, Marks JS: Relationship of childhood abuse and household dysfunction to many of the leading causes of death in adults. The adverse childhood experiences (ACE) study. Am J Prev Med 1998, 14(4):245-258.

13. Dube SR, Cook ML, Edwards VJ: Health-related outcomes of adverse childhood experiences in Texas, 2002. Prev Chronic Dis 2010, 7(3):A52.

14. Clark C, Caldwell T, Power C, Stansfeld SA: Does the influence of childhood adversity on psychopathology persist across the lifecourse? a 45-year prospective epidemiologic study. Ann Epidemiol 2010, 20(5):385-394.

15. Brown DW, Anda RF, Tiemeier H, Felitti VJ, Edwards VJ, Croft JB, Giles WH: Adverse childhood experiences and the risk of premature mortality. Am J Prev Med 2009, 37(5):389-396.

16. Kelly-Irving M, Lepage B, Dedieu D, Bartley M, Blane D, Grosclaude P, Lang T, Delpierre C: Adverse childhood experiences and premature all-cause mortality. Eur J Epidemiol 2013, 1-14.

17. Fuller-Thomson E, Brennenstuhl S: Making a link between childhood physical abuse and cancer results from a regional representative survey. Cancer 2009, 115(14):3341-3350.

18. Hardt J, Rutter M: Validity of adult retrospective reports of adverse childhood experiences: review of the evidence. Journal of Child Psychology and Psychiatry 2004, 45(2):260-273.

19. Korpimaki SK, Sumanen MPT, Sillanmaki LH, Mattila KJ: Cancer in working-age is not associated with childhood adversities. Acta Oncol 2010, 49(4):436-440.

20. Balkwill F, Mantovani A: Inflammation and cancer: back to Virchow? Lancet 2001, 357(9255):539-545.

21. Delpierre C, Kelly-Irving M: To what extent are biological pathways useful when aiming to reduce social inequalities in cancer? Eur J Public Health 2011, 21(4):398-399.

22. Kelly-Irving M, Mabile L, Grosclaude P, Lang T, Delpierre C: The embodiment of adverse childhood experiences and cancer development: potential biological mechanisms and pathways across the life course. Int J Public Health 2013, 58(1):3-11.

23. Power C, Elliott J: Cohort profile: 1958 british birth cohort (national child development study). Int J Epidemiol 2006, 35(1):34-41.

24. Anda RF, Butchart A, Felitti VJ, Brown DW: Building a framework for global surveillance of the public health implications of adverse childhood experiences. Am J Prev Med 2010, 39(1):93-98.

25. Benjet C, Borges G, Medina-Mora ME, Zambrano J, Cruz C, Mendez E: Descriptive epidemiology of chronic childhood adversity in Mexican adolescents. J Adolesc Health 2009, 45(5):483-489.
26. Dong M, Anda RF, Felitti VJ, Dube SR, Williamson DF, Thompson TJ, Loo CM, Giles WH: The interrelatedness of multiple forms of childhood abuse, neglect, and household dysfunction. Child Abuse Negl 2004, 28(7):771-784.
27. Rosenman S, Rodgers B: Childhood adversity in an Australian population. Soc Psychiatry Psychiatr Epidemiol 2004, 39(9):695-702.
28. House of Commons Science and Technology Committee: Alcohol guidelines. London: The stationary office; 2012:130.
29. Gallo V, Mackenbach JP, Ezzati M, Menvielle G, Kunst AE, Rohrmann S, Kaaks R, Teucher B, Boeing H, Bergmann MM, et al.: Social inequalities and mortality in Europe – results from a large multi-national cohort. PLoS ONE 2012, 7(7):e39013.
30. Kobayashi S, Sugiura H, Ando Y, Shiraki N, Yanagi T, Yamashita H, Toyama T: Reproductive history and breast cancer risk. Breast Cancer 2012, 19(4):302-308.
31. Navarro C, Chirlaque MD, Tormo MJ, Perez-Flores D, Rodriguez-Barranco M, Sanchez-Villegas A, Agudo A, Pera G, Amiano P, Dorronsoro M, et al.: Validity of self reported diagnoses of cancer in a major Spanish prospective cohort study. J Epidemiol Community Health 2006, 60(7):593-599.
32. Desai MM, Bruce ML, Desai RA, Druss BG: Validity of self-reported cancer history: a comparison of health interview data and cancer registry records. Am J Epidemiol 2001, 153(3):299-306.
33. Korpimaki S, Sumanen M, Suominen S, Mattila K: Self-reported rather than registered cancer is associated with psychosocial strain. BMC Fam Pract 2012., 13
34. Manjer J, Merlo J, Berglund GR: Validity of self-reported information on cancer: determinants of under- and over-reporting. Eur J Epidemiol 2004, 19(3):239-247.
35. Stavrou E, Vajdic CM, Loxton D, Pearson SA: The validity of self-reported cancer diagnoses and factors associated with accurate reporting in a cohort of older Australian women. Cancer Epidemiol 2011, 35(6):E75-E80.
36. Gupta V, Gu K, Chen Z, Lu W, Shu XO, Zheng Y: Concordance of self-reported and medical chart information on cancer diagnosis and treatment. BMC Med Res Methodol 2011., 11
37. The 5 Most Commonly Diagnosed Cancers in Females, Average Percentages and Numbers of New Cases, by Age, UK, 2007–2009. http://www.cancerresearchuk.org/cancer-info/cancerstats/incidence/age/#Cancer
38. Tietjen GE, Khubchandani J, Herial NA, Shah K: Adverse childhood experiences are associated with migraine and vascular biomarkers. Headache: The Journal of Head and Face Pain 2012, 52(6):920-929.
39. Dube SR, Fairweather D, Pearson WS, Felitti VJ, Anda RF, Croft JB: Cumulative childhood stress and autoimmune diseases in adults. Psychosom Med 2009, 71(2):243-250.
40. Shirtcliff EA, Coe CL, Pollak SD: Early childhood stress is associated with elevated antibody levels to herpes simplex virus type 1. Proc Natl Acad Sci 2009, 106(8):2963-2967.

41. Anda RF, Croft JB, Felitti VJ, Nordenberg D, Giles WH, Williamson DF, Giovino GA: Adverse childhood experiences and smoking during adolescence and adulthood. JAMA 1999, 282(17):1652-1658.

42. Anda RF, Whitfield CL, Felitti VJ, Chapman D, Edwards VJ, Dube SR, Williamson DF: Adverse childhood experiences, alcoholic parents, and later risk of alcoholism and depression. Psychiatr Serv 2002, 53(8):1001-1009.

43. Kiecolt-Glaser JK, Glaser R: Psychoneuroimmunology and cancer: fact or fiction? Eur J Cancer 1999, 35(11):1603-1607.

44. Bergelt C, Prescott E, Gronbaek M, Koch U, Johansen C: Stressful life events and cancer risk. Br J Cancer 2006, 95(11):1579-1581.

45. Ollonen P, Lehtonen J, Eskelinen M: Stressful and adverse life experiences in patients with breast symptoms; a prospective case–control study in Kuopio. Finland. Anticancer Res 2005, 25(1B):531-536.

46. Duijts SFA, Zeegers MPA, Van der Borne B: The association between stressful life events and breast cancer risk: a meta-analysis. Int J Cancer 2003, 107(6):1023-1029.

47. Fang F, Fall K, Sparen P, Adami HO, Valdimarsdottir HB, Lambe M, Valdimarsdottir U: Risk of infection-related cancers after the loss of a child: a follow-up study in Sweden. Cancer Res 2011, 71(1):116-122.

48. Lupien SJ, McEwen BS, Gunnar MR, Heim C: Effects of stress throughout the lifespan on the brain, behaviour and cognition. Nat Rev Neurosci 2009, 10(6):434-445.

There is one supplemental file that is not available in this version of the article. To view this additional information, please use the citation on the first page of this chapter.

CHAPTER 12

ADVERSE CHILDHOOD EXPERIENCES AND THE CARDIOVASCULAR HEALTH OF CHILDREN: A CROSS-SECTIONAL STUDY

CHELSEA PRETTY, DEBORAH D. O'LEARY, JOHN CAIRNEY, AND TERRANCE J. WADE

12.1 BACKGROUND

Adverse childhood experiences (ACEs) encompass many possible traumatic and distressing experiences that occur in childhood. Such experiences include traumas such as abuse or neglect but may also include experiences of illness, injury, loss or separation, witnessing a serious event, experiencing a natural disaster and significant changes in the home environment. Research has identified an association between ACEs, such as abuse, household dysfunction, and poverty, and an increased likelihood of developing future health risk factors such as smoking, alcohol and drug use, physical inactivity, and obesity, as well as future chronic illnesses including cardiovascular, lung and liver diseases, and cancer which are, in part, related to these identified risk factors [1-3]. Work by Goodwin &

Adverse Childhood Experiences and the Cardiovascular Health of Children: A Cross-Sectional Study. © Pretty C, O'Leary DD, Cairney J, and Wade TJ; licensee BioMed Central Ltd. BMC Pediatrics *13,208 (2013), doi:10.1186/1471-2431-13-208. Licensed under Creative Commons Attribution 2.0 Generic License, http://creativecommons.org/licenses/by/2.0/.*

Stein (2004), support these results showing that adults who had previously experienced childhood physical abuse, sexual abuse or neglect were 3.7 times more likely to develop cardiovascular disease (CVD) compared to others [4]. Stein and colleagues (2010) similarly showed that the accumulation of greater than three ACEs was associated with hypertension among adults [5]. Childhood factors including adverse events, socioeconomic status, illness, and growth patterns have also been linked to physiological differences in adult cardiovascular systems, accounting for 3.2% of variation of intima media thickness of the carotid artery in men and 2.2% variation in women [6]. Although this is a small effect, the fact that it remains significant after such a long latency period underscores its importance to cardiovascular health.

While previous studies have demonstrated a connection between ACEs and adult chronic illness and conditions, the majority of studies have been retrospective. That is, adults have been asked to reflect back on their childhood using an inventory of possible ACEs to cue their memory [1-5,7,8] but see [9,10]. By relying on retrospective data collected several decades after childhood, there may be an over- or under-estimation of exposure to ACEs. Moreover, it does not identify when these negative health consequences may begin.

Much of the literature linking ACEs to adult chronic illnesses and conditions has focused on extreme events such as sexual abuse [10], and other forms of severe abuse and maltreatment [1-5,7-9]. Besides these most extreme ACEs, there is evidence of a cumulative effect, or dose–response relationship among adults between the number of reported ACEs and the prevalence of health risk behaviours and chronic diseases [1-3]. Work by Felitti et al. (1998) supports this idea, noting that adults who reported four or more ACEs had increased risk of ischemic heart disease, cancer, chronic bronchitis or emphysema, history of hepatitis or jaundice, skeletal fractures, and poor self-rated health [1].

There is also growing evidence that ACEs may be related to CVD through the mediating effect of obesity. For example, with respect to obesity-induced hypertension [11,12], ACEs have been linked to both high blood pressure (BP) and obesity among adults [1,2,5,7-10]. While the majority of these studies utilize body mass index (BMI) as the measure of obesity [1,7,8], a study by Thomas et al. (2008) found that certain severe ACEs were associated with adult central adiposity, measured using waist circumference (WC) [9]. This is an important distinction because central adiposity has been shown to be a strong predictor of hypertension and CVD [13]. Should there be an association between childhood obesity, measured using central adiposity, and ACEs, this may suggest greater CVD risk in adulthood as childhood obesity and HBP are linked to adult obesity [14] and HBP [15]. Most importantly, the effect of ACEs on CVD risk factors has not been studied in children. One exception was a study completed by Noll et al., (2007) who prospectively assessed the effect of ACEs on obesity in childhood, adolescence and young adulthood [10]. However, these researchers only found a relationship between exposure to ACEs and obesity status in young adults [10]. Furthermore, the study was only completed on female sexual abuse victims.

The primary objective of this study was to examine children and the relationship between ACEs and early childhood risk factors for adult CVD, specifically BP, BMI and WC. In addition, we examined whether ACEs were associated with resting heart rate (HR), a marker of parasympathetic and sympathetic activity. Elevated resting HR is associated with obesity-related hypertension, which may be due to reduced parasympathetic [11,16-18] and/or heightened sympathetic activity [17,19,20]. As elevated HR is a predictor of both adult hypertension and CVD and is associated with a hyperkinetic circulation seen in hypertension [21], it may provide an early marker for risk of elevated BP. We also assess whether there is a cumulative effect of ACEs exposure on these childhood CVD risk factors as cumulative exposure to ACEs has been previously linked to chronic diseases among adults. In summary, the findings from our investigation surrounding the relationship between ACEs, resting HR, BMI, WC and systolic BP in a community sample of 11–14 year old children are presented.

12.2 METHODS

12.2.1 SAMPLE

The data used in this study came from the Heart Behavioural and Environmental Assessment Team (HBEAT) study. A community sample of adolescents aged 11 to 14 years (grades 6 to 8) and their parents from one school board in Southern Ontario were asked to participate in the study. The estimated population base was approximately 5 800 students across 50 schools. The study was approved by both university and school district research ethics review boards. Informed written consent was obtained from the parent/guardian and verbal assent was obtained from the child in order to participate in the study.

Participation was voluntary with no exclusionary criteria and sampling occurred in two phases. The initial phase occurred in fall 2007 involving 28 of the 50 randomly selected schools based on 2006 community census grouping. The remaining 22 schools were approached in winter 2008. In total, 1 913 children volunteered to undergo school-based assessment which included a number of anthropometric measures (i.e. BMI) and automated BP. As well, these adolescents took a questionnaire package home of which 1 324 (69.2%) parent questionnaires were returned.

12.2.2 FIELD-TESTING PROTOCOL AND MEASURES

12.2.2.1 BLOOD PRESSURE AND HEART RATE

BP and HR were measured using automatic oscillometric BP units which calculate BP based on the first and fifth Korotkoff sounds (BPM-300, VSM MedTech Devices Inc., Coquitlam, British Columbia, Canada). This unit has been validated for use in children [22]. Students were taken from class in small groups of 8 to 10 to a quiet location in the school. They were asked to relax and to remain silent with their feet flat on the floor sitting upright for about 15 minutes with their arms resting on a table. After 15 minutes, with their right arm positioned at the midpoint of the sternum, BP cuffs were placed on the child's left arm with cuff size based on arm size [23]. The automatic BP unit took six independent measures at 1-minute intervals. The first three measurements were done to familiarize the subject with cuff pressurization and were discarded. The last three systolic BP (SBP), diastolic BP and HR measures were averaged. Two manual oscillatory BP measurements via sphygmomanometer were taken in the event of an error reading on the automated machine.

TABLE 1: Sample descriptive statistics (n = 1,234)

Sample characteristics	% or Mean ± SD
Females (%)	55.0
Child age (years, mean ± SD)	11.8 ± 0.9
Parent history of high blood pressure (%)	17.1
Parent education (%)	
Grade 11 or Less	3.2
Grade 12 or Less	5.6
High school diploma (or GED)	6.6
Partial college/training	19.9
College or University degree	40.8
Graduate or Professional degree	23.9
Family income (mean ± SD)	70 828 ± 31 420
Systolic blood pressure (mmHg, mean ± SD)	93.0 ± 8.7
Heart rate (BPM, mean ± SD))	83.4 ± 11.5
Body mass index (kg/m2 , mean ± SD)	20.6 ± 4.2
Waist circumference (cm, mean ± SD))	72.4 ± 11.6
Height (cm, mean ± SD)	154.2 ± 8.9
Godin-Shephard (METs/week, mean ± SD)°	86.6 ± 64.0
Total ACEs	1.9 ± 1.6
Proportion of sample reporting ACEs‡ (%)	
Death of family member (not parent)	41.6
Lost a pet that they really cared for (died, killed, lost)	34.2
Serious illness or injury in the family	20.8
Conflict or serious argument between parents	19.5
Divorce or separation of parents	18.6
At least one night stay in a hospital	12.4
Serious illness or injury	8.1
Separation from parents	6.9
Badly frightened or attacked by an animal	5.0
Saw someone get badly hurt or die suddenly	4.9
Family member or residence was robbed	4.5
Death of a parent	3.1

Note: SD = standard deviation; GED = general educational development; BPM = beats per minute; MET = metabolic equivalent; ACEs = adverse childhood experiences; ‡Top 12 ACEs reported. ACEs not listed in this table include: in a bad car accident, experienced a natural disaster, moving residence/school or immigrating, separation from sibling or other close family member, a stay in a foster home, other, death of an extended family member or friend, bullying or significant verbal abuse, witnessing serious conflict not between parents. °Godin-Shephard measures leisure time physical activity in a one week period.

12.2.2.2 ANTHROPOMETRICS

Anthropometric measurements were taken for each student in a private location following BP testing. Students were asked to remove shoes prior to testing. Height (cm) was measured using a portable stadiometer (STAT 7X, Ellard Instrumentation Ltd., Monroe, WA, USA). Body mass (kg) was measured using a calibrated electronic medical scale (BWB-800S, Tanita Corporation, Tokyo, Japan). BMI was calculated as mass (kg) divided by the height squared (m^2). WC measures (cm) were taken at the narrowest point of the waist, approximately at the location of the belly button [24]. All measures were taken three times and averaged.

12.2.2.3 PARENT QUESTIONNAIRE PROTOCOL AND MEASURES

Parent questionnaires were sent home with students for their parents to complete. The parent questionnaire included an inventory of ACEs , family income, parental education, and family history of hypertension.

12.2.3 ADVERSE CHILDHOOD EXPERIENCES

Child adverse experiences were identified through parent report using an inventory adapted from the Childhood Trust Events Survey (CTES 2.0 – Caregiver Form) (see Additional file 1), a 26-item inventory adapted from the Traumatic Stress Survey (TSS) [25]. Certain CTES events were removed due to limitations set forth by the school board (i.e., sexual and physical abuse and maltreatment). Parents were asked to respond to 15 events that possibly occurred and perceived to continue to cause the child a great amount of worry or unhappiness. Additional space was provided for parents to identify other events that had caused their child significant worry or unhappiness. Where applicable, these were recoded and additional ACEs were created for analysis. ACEs that were included in this study are detailed in Table 1. Reliability and validity have not been established for the CTES [26].

12.2.4 COVARIATES

Child sex and age (years) were recorded. Family education was based on the maximum level of parental education achieved by any parent (less than grade 11, grade 12, high school diploma or GED, partial college/training, college/ university degree, graduate or professional degree). Household income was calculated using the midpoint value of 14 income categories (under $4 999, $5 000 to $9 999, $10 000 to $14 999, $15 000 to $19 999, $20 000 to $24 999, $25 000 to $29 999, $30 000 to $39 999, $40 000 to $49 999, $50 000 to $59 999, $60 000 to $69 999, $70 000 to $79 999, $80 000 to $89 999, $90 000 to $99 999, $100 000 or more[set at $120 000]) and was treated as a continuous variable. Parental history of hypertension was a dichotomous variable based on parental reporting that identified either parent as having received a diagnosis of hypertension (1) compared to neither parent (0). Child physical activity was measured using the Godin-Shephard Leisure -Time Exercise Questionnaire [27]. This questionnaire was completed by students while they waited for their BP to be measured. Students were asked on average how often they participated in strenuous, moderate and mild exercise within a 7-day period and how often in a week they would sweat from exercising in their leisure time. These results were converted into METs (metabolic equivalent), a measure of energy expenditure, for analysis. The Godin-Shephard has been used to estimate physical activity in children within this age range before and has shown reasonable evidence of reliability and validity [28].

12.2.5 STATISTICAL ANALYSIS

Any subject with one or more missing variables was removed from the study, reducing the sample size from 1 324 to 1 234. Means and standard errors (SE) of all physiological variables were calculated for each frequency category of exposure to ACEs (Figure 1). Overall, these graphs identified a rise in the mean value of HR, WC and BMI at and above four ACEs. This demonstrated a similar pattern as reported previously by Felitti, et al. (1998) of four or more ACEs being a threshold level [1]. As such, ACEs were coded dichotomously to compare those with fewer than 4 ACEs to 4 or more ACEs. Separate regression analyses were run to test the unadjusted and adjusted effect of dichotomous ACEs on all outcomes. Variables included in the adjusted regression models were child age, sex, physical activity as measured by the Godin-Shepherd, parent history of hypertension, family education level, and family income level. Height was also included in the model for SBP and waist circumference was included in the model for HR.

12.3 RESULTS

This sample included slightly more females than males with students averaging 11.8 years old (Table 1). The majority of families had a parent with at least partial college/training education and an average family income of $71 000. Average SBP in the sample was 93.0 (±8.7) mmHg and mean HR was 83 (±12) beats per minute. On average, the children had a BMI of 20.6 (±4.2) kg/m^2 with a WC of 72.4 (±11.6) cm. The modal average of ACEs was one, while the mean was approximately two ACEs and 16.0% of the sample experiencing four or more ACEs. Attrition analysis revealed that students who did not have completed parent questionnaires were significantly taller, more likely to be female, more likely to be from a rural or low-income urban school, and had significantly lower HR than the present sample (data not shown).

Table 2 presents the unadjusted linear regression analyses. Having experienced four or more ACEs was found to be significantly associated with higher HR (b=2.3 bpm, 95% CI (0.6-4.1)), BMI (b=1.4 kg/m^2, 95% CI (0.7-2.0)), and WC (b=3.6 cm, 95% CI (1.8-5.3)). There was no significant effect of ACEs on SBP (b=0.6 mmHg, 95% CI (−0.7-1.9)). Table 3 adjusts for covariates including family education, family income, parental history of hypertension, and child age, sex, and physical activity. Height was added to the SBP model based on the guidelines set forth by the National High Blood Pressure Education Program Working Group on High Blood Pressure in Children and Adolescents (2004) [29]. Even after the addition of these covariates, the effect of ACEs on BMI (b =1.1 kg/m^2, 95% CI (0.5-1.8)) and WC (b =3.1 cm, 95% CI (1.3-4.8)) remained significant. Moreover, controlling for these additional covariates as well as for WC (Table 3) the effect of ACEs on HR remained significant (b=1.8 bpm, 95% CI (0.1-3.6)).

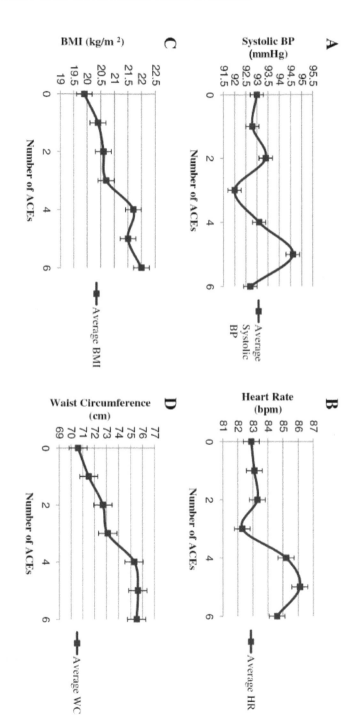

FIGURE 1: Average physiological measures by number of ACEs. A represents the average systolic BP for each ACEs category. B represents the average HR for each ACEs category. C represents the average BMI for each ACEs category. D represents the average WC for each ACEs category. Standard error bars shown. ACEs=Adverse childhood experiences; BP=blood pressure; HR=heart rate; BMI=body mass index; WC=waist circumference.

TABLE 2: Unadjusted effect of adverse childhood experiences (ACE) on cardiovascular risk factors

	SBP b (SE)	HR b (SE)	BMI b (SE)	WC b (SE)
ACEs (4+ vs <4)	0.6(0.7)	2.3** (0.9)	1.4** (0.3)	3.6** (0.9)
R2	0.001	0.055**	0.014**	0.013**

*Note: b values are unstandardized regression coefficients. n = 1234. ACE = adverse childhood experiences; BMI = body mass index; HR = heart rate; SBP = systolic blood pressure; SE = standard error; WC = waist circumference *p < 0.05 and **p < 0.01 (two-tailed).*

TABLE 3: Adjusted effect of adverse childhood experiences (ACE) on cardiovascular risk factors

	SBP b (SE)	HR b (SE)	BMI b (SE)	WC b (SE)
ACEs ($\geq$4 vs <4)	0.34 (0.7)	1.82 (0.9)*	1.13 (0.3)**	3.01 (0.9)**
Child age (years)	−0.24 (0.3)	−1.69 (0.4)**	0.47 (0.1)**	1.55 (0.4)**
Sex	0.84 (0.5)	0.04 (0.7)	0.13 (0.2)	0.56 (0.7)
Height (cm)	0.16 (0.03)**	n/a	n/a	n/a
Waist circumference (cm)	n/a	0.02 (0.03)	n/a	n/a
Godin shephard (METs/ week)	−0.005 (0.004)	−0.01 (0.01)**	−0.001 (0.002)	−0.007 (0.01)
Parent history of HBP	2.03 (0.6)**	0.73 (0.9)	0.88 (0.3)**	3.00 (0.9)**
Family education Level	−0.67 (0.2)**	−0.44 (0.3)	−0.32 (0.10)**	−0.85 (0.3)**
Family income (0,000)	−0.1 (0.08)	−0.2 (0.1)	−0.1 (0.04)**	−0.3 (0.1)
Overall model	R2=0.05**	R2=0.04**	R2=0.05**	R2=0.06**

*Note: b values are unstandardized regression coefficients. n/a = not applicable. ACE = adverse childhood experiences; BMI = body mass index; HBP = hypertension; HR = heart rate; MET = metabolic equivalent; SBP = systolic blood pressure; SE = standard error; WC = waist circumference. n = 1234. *p < 0.05 and **p < 0.01 (two-tailed).*

Finally, additional regression analyses were completed where ACEs were treated as an ordinal variable as opposed to using the threshold. In these models, both BMI and WC remained significant but HR did not (data not shown). This is not surprising upon further investigation of Figure 1 which reveals a more linear relationship between accumulation of ACEs and both WC and BMI. Regression analyses were also completed with clinically significant outcomes (95th percentile of the sample after adjusting for age, sex, and height for SBP and established cut-offs for BMI) for both BMI and SBP (data not shown) [29,30]. Again ACEs were a significant predictor of obesity at the clinical threshold when assessed both as an ordinal variable and using 4 ACEs as a cut-off. Consistent with prior regression analysis, ACEs were not a significant predictor of hypertension status in the sample.

12.4 DISCUSSION

The present study found a threshold effect in which having experienced 4 or greater ACEs is associated with increased resting HR, BMI and WC in this community sample of 11–14 year old adolescents. There does not appear to be a relationship between ACE and SBP in this sample. Furthermore, there appears to be a dose–response relationship between ACE accumulation, BMI and WC where BMI and WC continue to rise with greater numbers of ACEs.

To the authors' knowledge, this is the first study to examine the relationship between HR, SBP and ACEs and the second to look at the association between ACEs and obesity in children [10]. Consistent with the work of Felitti et al. (1998) who evaluated the retrospective health effect of reported ACEs in an adult population, examination of the mean physiological measures with each additional ACE (0 through 6) indicated that four or more ACEs appear to be a threshold exposure for children [1]. While this threshold was not related to higher SBP in this study, having experienced four or more ACEs in childhood was linked to a higher HR, BMI, and WC compared to those children who experienced fewer than four events. Nevertheless, this is in contrast to Noll and colleagues (2007) who did not find an association between obesity and certain ACEs until early adulthood [10]. This discrepancy could be the result of the fact that the present study utilized a general population whereas Noll et al. (2007) focused solely on sexual abuse victims [10]. Furthermore, Noll et al. (2007), utilized wider age ranges, assessing obesity in childhood/early adolescence (6–14 years), middle/late adolescence (15–19 years) and young adulthood (20–27 year) [10]. The findings of the current study suggest that the accumulation of ACEs may accelerate obesity as measured by both BMI and WC, and elevate the sympathetic nervous system as indicated by a higher HR. Moreover, it may be that increased BMI and WC precedes any change in BP, which is supported by previous work linking childhood obesity to the development of hypertension in adulthood [15]. Most importantly, this study highlights the novel finding that ACEs have physiological health consequences that begin much earlier than adulthood. Our findings also coincide with the work of Flaherty, et al. (2009) who found an association between five or more ACEs and some indicators of health problems at age 12, including somatic and other health complaints, and illnesses requiring a doctor's visit [31]. The authors suggested that this age group may be too young to see associations with the negative health behaviours and chronic conditions seen in the ACE Study [1,31]. Our study suggests that we can see the beginnings of these chronic conditions in 11 to 14 year olds. Taken together these studies suggest that ACEs are affecting overall health and prompting physiological changes as early as adolescence.

The significant effect of ACEs on increased HR suggests the existence of a hyperkinetic circulation in these children [20]. Hyperkinetic circulation may be associated with increased sympathetic activity and hypertension in young adults [17,19,20]. There is a current debate in the literature as to whether the elevated HR component of hyperkinetic circulation is driven more by parasympathetic withdrawal [16-19] or by sympathetic over-activation [17,20]. Elevated sympathetic activity is associated with obesity and obesity hypertension. In obesity hypertension, weight gain is seen as the driving force behind elevated BP [11,12]. This weight gain is associated with a rise in sympathetic activity which triggers the renal system to increase sodium retention and therefore, increases blood volume and BP [11,12,32]. Consistent with this, a recent review by Danese & McEwen (2012) focusing on the effect of ACEs on age-related disease [33] proposed that repeated exposure to such events can disrupt the body's allostatic systems which act to maintain stability through changes in one's environment [33]. Prolonged engagement of these systems can lead to structural changes in the amygdala in the brain, prolonged activation of the hypothalamic-pituitary-adrenal axis and sympathetic nervous system, as well as, inflammation [33]. These physiological changes may lead to the development of atherosclerosis and subsequently CVD [33]. When the body engages the hypothalamus-pituitary-adrenal response, it secretes hormones to activate the cardiovascular system in order to cope with stress [34,35]. This causes the sympathetic nervous system to increase its involvement in physiological coping [33,35]. HR is initially raised in a hyperarousal state which can persist, or may cause a child to dissociate from the stress [36]. If the trauma is severe and chronic, the resting state for HR and BP are readjusted, resulting in these children living in a physiological heightened state of arousal including higher heart and respiration rates and muscle vigor [36,37]. This physiological remodeling may explain the exposure threshold of four ACEs since HR was significantly elevated among these children, suggesting alterations in neural regulation of the cardiovascular system.

Although the results of this study suggest an earlier effect of ACEs on the cardiovascular system than previously identified, there are some limitations that need to be addressed. First, the three in-school BP readings occurred successively at one point in time. While three measures were taken, we were unable to account for diurnal changes or longer-term variability. Furthermore, without three independent measures at separate time points, no clinical diagnoses can be made. We can only identify those with elevated BP. Second, this study focused primarily on those aged 11 to 14 (grades 6 to 8). This limited age range may have prohibited seeing an effect of ACEs on BP. If BP is in fact affected by ACEs as suggested by Stein, et al., (2010), it may not be apparent until later in adolescence or early adulthood and may be preceded by increased body weight and sympathetic activity [5]. Further research is necessary to evaluate this hypothesis.

This study included a large and diverse community-level sample. This sample was designed to include different segments of the population, including children in urban and rural areas, and in both low and high income areas. Furthermore, a large inventory of possible ACEs was examined in this study to gain an accurate representation of exposure to ACEs in youth. However, some severe ACEs such as physical, sexual and extreme emotional abuse were not examined in this study. The retrospective studies which have noted the dose–response relationship between ACEs and CVD did include these serious events [1-3,5]. Since it was not possible to account for these events, the true impact of ACEs on these measures may be under-estimated. Moreover, the age of the child when such ACEs occurred was not recorded. Differences in timing between the ACE and the measurement of BP may have accounted for this null result. Furthermore, the average income in this sample was quite high, and therefore may not be entirely representative of the general population. Also, the use of the CTES questionnaire as a measure of ACEs could be seen as a limitation as its validity and reliability have not been established in the literature [26].

12.5 CONCLUSIONS

In conclusion, this study indicates that in a community sample of grade 6 to 8 children, the accumulation of four or more ACEs was significantly associated with higher BMI, WC and resting HR, as well as obesity status, factors shown to be associated with cardiovascular disease among adults. The findings of this study are very important as they highlight the fact that cardiovascular risks identified among adults exposed to ACEs in previous studies actually begin to appear earlier in childhood. This is a novel finding as Noll and colleagues (2007) only found an association between ACEs and obesity in young adult sexual abuse victims and not children [10]. The current findings emphasize that the physiological consequences of ACEs reported in adults are beginning in childhood. Also, the link between elevated resting HR and ACEs which has not been previously shown may indicate alterations in autonomic regulation that have life-long consequences.

Finally, to gain a better understanding of the relationship between ACEs and BP, research should focus on older adolescents, ideally tracking them over time to evaluate whether the effect on BP occurs in later years and whether it is preceded by changes in BMI and WC. Danese & McEwen argue that the adverse effects of ACEs in childhood can be reversible if the child's environment is returned to a stable state [33,38,39]. This suggests that studies should examine how these risks can be mitigated among children and whether this can reduce the potential long-term health consequences shown repeatedly in studies among adults.

REFERENCES

1. Felitti VJ, Anda RF, Nordenberg D, Williamson DF, Spitz AM, Edwards V, Koss MP, Marks JS: Relationship of childhood abuse and household dysfunction to many of the leading causes of death in adults. Am J Prev Med 1998, 14(4):245-258.
2. Danese A, Moffitt TE, Harrington H, Milne BJ, Polanczyk G, Pariante CM, Poulton R, Caspi A: Adverse childhood experiences and adult risk factors for age-related disease: depression, inflammation and clustering risk factors. Arch Pediatr Adolesc Med 2009, 163(12):1135-1143.
3. Dong M, Giles WH, Felitti VJ, Dube SR, Williams JE, Chapman DP, Anda RF: Insights into causal pathways for ischemic heart disease: adverse childhood experiences. Circulation 2004, 110(13):1761-1766.
4. Goodwin RD, Stein MB: Association between childhood trauma and physical disorders among adults in the United States. Psychol Med 2004, 34(3):509-520.
5. Stein DJ, Scott K, Haro Abad JM, Aguilar-Gaxiola S, Alonso J, Angermeyer M, Demytteneare K, de Girolamo G, Iwata N, Posada-Villa J, Kovess V, Lara C, Ormel J, Kessler RC, Von Korff M: Early childhood adversity and later hypertension: data from the world mental health survey. Ann Clin Psychiatry 2010, 22(1):19-28.
6. Lamont D, Parker L, White M, Unwin N, Bennett SM, Cohen M, Richardson D, Dickinson HO, Adamson A, Alberti KG, Craft AW: Risk of cardiovascular disease measured by carotid intima-media thickness at age 49-51: lifecourse study. BMJ 2000, 320(7230):273-278.
7. Gunstad J, Paul RH, Spitznagel MB, Cohen RA, Williams LM, Kohn M, Gordon E: Exposure to early life trauma is associated with adult obesity. Psychiatry Res 2006, 142(1):31-37.
8. Williamson DF, Thompson TJ, Anda RF, Dietz WH, Felitti V: Body weight and obesity in adults and self-reported abuse in childhood. Int J Obes Relat Metab Disord 2002, 26(8):1075-1082.
9. Thomas C, Hyppönen E, Power C: Obesity and type 2 diabetes risk in mid-adult life: the role of childhood adversity. Pediatrics 2008, 121(5):e1240-e1249.
10. Noll JG, Zeller MH, Trickett PK, Putnam FW: Obesity risk for female victims of childhood sexual abuse: a prospective study. Pediatrics 2007, 120(1):e61-e67.
11. Hall JE: Pathophysiology of obesity hypertension. Curr Hypertens Rep 2000, 2(2):139-147.
12. da Silva AA, do Carmo J, Dubinion J, Hall JE: Role of sympathetic nervous system in obesity related hypertension. Curr Hypertens Rep 2009, 11(3):206-211.
13. Björntop P: "Portal" adipose tissue as a generator of risk factors for cardiovascular disease and diabetes. Arteriosclerosis 1990, 10(4):493-496.
14. Craigie AM, Matthews JN, Rugg-Gunn AJ, Lake AA, Mathers JC, Adamson AJ: Raised adolescent body mass index predicts the development of adiposity and a central distribution of body fat in adulthood: a longitudinal study. Obes Facts 2009, 2(3):150-156.
15. Field AE, Cook NR, Gilman MW: Weight status in childhood as a predictor of becoming overweight or hypertensive in early adulthood. Obes Res 2005, 13(1):163-169.

16. Arone LJ, Mackintosh R, Rosenbaum M, Leibel RL, Hirsch J: Autonomic nervous system activity in weight gain and weight loss. Am J Physiol 1995, 269(1):R222-R225.

17. Julius S, Pascual AV, London R: Role of parasympathetic inhibition in the hyperkinetic type of borderline hypertension. Circulation 1971, 44(3):413-418.

18. Hall JE, Crook ED, Jones DW, Wofford MR, Dubbert PM: Mechanisms of obesity-associated cardiovascular and renal diseases. Am J Med Sci 2002, 324(3):127-137.

19. Julius S, Krause L, Schork N, Mejia AD, Jones KA, van de Ven C, Johnson EH, Sekkarie MA, Kjeldsen SE, Petrin J, Schmouder R, Gupta R, Ferraro J, Nazzaro P, Weissfeld J: Hyperkinetic borderline hypertension in Tecumseh. Michigan. J Hypertens 1991, 9(1):77-84.

20. Tjugen TB, Flaa A, Kjeldsen SE: High heart rate as predictor of essential hypertension: The hyperkinetic state, evidence of prediction of hypertension and hemodynamic transition to full hypertension. Prog Cardiovasc Dis 2009, 52(1):20-25.

21. Gillum RF, Makuc DM, Feldman JJ: Pulse rate, coronary heart disease, and death: The NHANES I Epidemiological Follow-up Study. Am Heart J 1991, 121(1):172-177.

22. Mattu GS, Heran BS, Wright JM: Comparison of the automated non-invasive oscillometric blood pressure monitor (BpTRU) with the auscultatory mercury sphygmomanometer in a paediatric population. Blood Press Monit 2004, 9(1):39-45.

23. Pickering TG, Hall JE, Appel LJ, Falkner BE, Graves J, Hill MN, Jones DW, Kurtz T, Sheps SG, Roccella EJ: Recommendations for blood pressure measurement in humans and experimental animals: part 1: blood pressure measurement in humans: a statement for professionals from the subcommittee of professional and public education of the American heart association council on high blood pressure research. Circulation 2005, 111(5):697-716.

24. Wang J, Thornton JC, Bari S, Williamson B, Gallagher D, Heymsfield SB, Horlick M, Kotler D, Laferrère B, Mayer L, Pi-Sunyer FX, Pierson RN Jr: Comparisons of waist circumferences measured at 4 sites. Am J Clin Nutr 2003, 77(2):379-384.

25. Baker DG, Boat BW, Grinvalsky MD, Geracioti TD Jr: Interpersonal and animal-related trauma experiences in female and male military veterans: Implications for program development. Mil Med 1998, 163(1):20-25.

26. Pearl E, Thieken L, Olafson E, Boat B, Connelly L, Barnes J, Putnam F: Effectiveness of community dissemination of parent–child interaction therapy. Psych Trauma: Theory, Res, Pract Pol 2011, 4(2):204-213.

27. Godin GJ, Shepard RJ: A simple method to assess exercise behavior in the community. Can J App Sport Sci 1985, 10(3):141-146.

28. Sallis JF, Condon SA, Goggin KJ, Robby JJ, Kolody B, Alcaraz JE: The development of self-administered physical activity surveys for 4th grade students. Res Q Exerc Sport 1993, 64(1):25-31.

29. National High Blood Pressure Education Program Working Group on High Blood Pressure in Children and Adolescents: The fourth report on the diagnosis, evaluation, and treatment of high blood pressure in children and adolescents. Pediatrics 2004, 114(2):555-576.

30. Cole TJ, Bellizzi MC, Flegal KM, Dietz WH: Establishing a standard definition for child overweight and obesity worldwide: International survey. BMJ 2000, 320(7244):1240-1243.

31. Flaherty EG, Thompson R, Litrownik AJ, Zolotor AJ, Dubowitz H, Runyan DK, English DJ, Everson MD: Adverse childhood exposures and reported child health at age 12. Acad Pediatr 2009, 9(3):150-156.

32. Wofford MR, Anderson DC, Brown CA, Jones DW, Miller ME, Hall JE: Antihypertensive effect of alpha and beta adrenergic blockade in obese and lean hypertensive subjects. Am J Hypertens 2001, 14(7):694-698.

33. Danese A, McEwen BS: Adverse childhood experiences, allostasis, allostatic load, and age-related disease. Physiol Behav 2012, 106(1):29-39.

34. Brotman DJ, Golden SH, Wittstein IS: The cardiovascular toll of stress. Lancet 2007, 370(9592):1089-1100.

35. Tsigos C, Chrousos GP: Hypothalamic-pituitary-adrenal axis, neuroendocrine factors and stress. J Psychosom Res 2002, 53(4):865-871.

36. Perry BD, Pollard R: Homeostasis, Stress, Trauma, and Adaptation: a neurodevelopmental view of Childhood Trauma. Child Adolesc Psychiatr Clin N Am 1998, 7(1):33-51.

37. Perry BD, Pollard RA, Blakley TL, Baker WL, Vigilante D: Childhood trauma, the neurobiology of adaption and use-dependent development of the brain: How states become traits. Inf Mental Hlth J 1995, 16(4):271-291.

38. Fisher PA, Stoolmiller M, Gunnar MR, Burraston BO: Effects of a therapeutic intervention for foster preschoolers on diurnal cortisol activity. Psychoneuroendocinology 2007, 32(8–10):892-905.

39. Brotman LM, Gouley KK, Huang KY, Kamboukos D, Fratto C, Pine DS: Effects of a psychosocial family-based preventive intervention on cortisol response to a social challenge in preschoolers at high risk for antisocial behavior. Arch Gen Psychiatry 2007, 64(10):1172-1179.

CHAPTER 13

IMPACT OF EARLY PSYCHOSOCIAL FACTORS (CHILDHOOD SOCIOECONOMIC FACTORS AND ADVERSITIES) ON FUTURE RISK OF TYPE 2 DIABETES, METABOLIC DISTURBANCES, AND OBESITY: A SYSTEMATIC REVIEW

TERESA TAMAYO, CHRISTIAN HERDER, AND WOLFGANG RATHMANN

13.1 BACKGROUND

In adults, adverse psychosocial factors such as low socioeconomic status (SES), deprivation and traumata have been shown to be associated with type 2 diabetes [1], obesity [2-5], cardiovascular disease [6,7], and unhealthy lifestyle habits [8-11]. A lower social status is related to higher stress levels [12] and poor living conditions [13], which may partly explain these associations. Health disparities can be observed fairly early

Impact of Early Psychosocial Factors (Childhood Socioeconomic Factors and Adversities) on Future Risk of Type 2 Diabetes, Metabolic Disturbances and Obesity: A Systematic Review. © *Tamayo T, Herder C, and Rathman W; licensee BioMed Central Ltd.* BMC Public Health **10**,525 *(2010), doi:10.1186/1471-2458-10-525. Licensed under a Creative Commons Attribution 2.0 Generic License, http://creativecommons.org/licenses/by/2.0/.*

with lower birth weights, an earlier adiposity rebound and higher rates of infant mortality in the low SES groups [14-16].

Data on the association of type 2 diabetes and adverse childhood circumstances is more limited [17]. In two cross-sectional analyses there was evidence that adverse social conditions in childhood are independently associated with an increased risk of metabolic impairments and insulin resistance [18,19]. In a Swedish population-based study, Agardh et al. found low parental education, low family household income and low parental occupational position to be associated with a more than two-fold increased diabetes risk in adulthood, which was attenuated after accounting for adult socioeconomic factors [20].

Thus, the primary aim of this review was to evaluate the risk of psychosocial factors on type 2 diabetes incidence and the role of change in socioeconomic conditions throughout life on the basis of population-based and longitudinal studies. As several factors of childhood environment are closely related and interact with each other [21], we aimed not only to address the relationship of childhood socioeconomic status (CSES) with diabetes incidence, but to investigate a broad range of psychosocial factors. Furthermore, obesity was included as it is considered a key factor for diabetes incidence in youth and as it is also influenced by SES [22]. Since a first comprehensive review [23] on the inverse relationship between SES and obesity in developed countries, these early results have been substantiated in further systematic reviews [2,4]. We included obesity as endpoint in our review and concentrated on longitudinal and population-based studies with particular focus on the analysis of change in weight status and psychosocial circumstances.

13.2 METHODS

13.2.1 SEARCH CRITERIA

We searched Medline in July 2008 and carried out an update of the search in April 2010 applying the search algorithm "(diabetes or insulin resistance or prediabetes or metabolic syndrome) and (SES or social or socioeconomic or psychosocial or income or working status or migration or community or adversities or deprivation or depression or abuse or high risk family or hostility)".

For obesity a search in Medline was carried out in November 2008 (updated in April 2010) using the algorithm "(obesity or overweight) and (socioeconomic or social or deprivation or adversities or childhood socioeconomic or family environment or early life or youth or childhood adversities or deprivation) and (longitudinal or prospective or cohort)".

Both searches were limited to publications after 31 December 1994 in English language. During the 1990s, the diabetes criteria were changed several times, which complicates comparison with earlier publications.

13.2.2 INCLUSION AND EXCLUSION CRITERIA

Studies were included if incident cases of type 2 diabetes were assessed. Cross-sectional studies with data on diabetes prevalence only were excluded. Only population- or community-based studies were included to have a more homogeneous database for this review. Further outcome criteria were insulin resistance, elevated HbA1c values, and the metabolic syndrome because they share risk factors and pathophysiological pathways with type 2 diabetes. Obesity and overweight are considered key risk factors for type 2 diabetes incidence especially in youth, and were furthermore included as outcome parameters, but were analysed separately.

13.2.3 RISK FACTOR SPECIFICATION: CHILDHOOD PSYCHOSOCIAL FACTORS

Studies on incident type 2 diabetes or obesity were included if they contained data on psychosocial factors including a wide range of indicators reflecting the social and psychological conditions under which the participants grew up:

1. basic indicators of CSES: parental education; parental occupation and family income as indicators of economic wealth and stability;

2. further indicators of wealth and deprivation: public housing, housing conditions, house ownership, unemployment;
3. the high-risk family concept (adverse childhood experiences) regarding neglect, abuse and household dysfunction [24];
4. indicators of impaired psychological health in children such as depression or anxiety;
5. indicators of the ability to cope with stressful conditions like coping skills or sense of control [25];
6. indicators of possibly stressful situations: migration, parental stress (e.g. stressful working conditions or parenting stress which are supposed to have implications for children's stress as well);
7. neighbourhood deprivation indexes as indicator for the aggregation of unfavourable circumstances clustered in residential areas. These indexes are based on the social composition of neighbourhoods regarding for example the social status of its habitants, housing and street conditions, mean household sizes per person and other indicators of wealth such as mean household number of cars [26,27].

Studies were excluded if information on obesity and overweight was only available at one time point and no adjustment for previous weight status was done. Studies, which only included type 1 diabetes were also excluded. Self-reported diabetes type may not offer a reliable distinction between the diabetes mellitus types. However, self-reported diabetes type was found in most studies and as type 1 diabetes contributes to a fairly small proportion of the total number of diabetes cases we accepted self-reported diabetes type for the assessment of type 2 diabetes incidence.

13.2.4 DATA EXTRACTION AND QUALITY MANAGEMENT

Eligible studies were assessed by one reviewer (T. T.) and discussed with a second reviewer (W. R.). The final list of variables extracted from the selected studies contained the first author, publication year, country where the study was carried out, cohort size, study duration, age characteristics of participants, measurement of risk factors, definition of diabetes/obesity variables and a brief description of results including effect size. For missing information we contacted some of the authors of the selected studies. Any disagreements regarding numbers, study inclusion, and further analysis were resolved by consensus between the authors (T.T., C.H., W.R.).

13.2.5 ANALYSIS AND QUALITY ASSESSMENT

We restricted the analysis to descriptive measures because of the lack of statistical comparability for most studies. Effect sizes extracted from the publications reflect adjusted results for odds ratio (OR), hazard ratio (HR) and relative risk (RR) regarding confounding (1) for age, sex, BMI, physical activity, smoking and alcohol (for type 2 diabetes incidence) or (2) for birth weight and adult SES (for obesity and overweight). We developed a 5 item quality scale adapted to our purpose with reference to methodological recommendations of the Meta-analysis Of Observational Studies in Epidemiology (MOOSE) group [28,29]. The sum score derived from this scale mainly had to reflect the study design, the examined outcome parameters, the risk factor construction and the control for confounders. Quality was judged taking into account the current state of research on type 2 diabetes and obesity and possible sources of bias. The following criteria were included into the quality score for the (1) diabetes and (2) obesity studies:

- (1&2) Study Duration (D): Highest quality was considered to be obtained from birth cohorts (D = 1) as the future diabetes and obesity risk is influenced strongly by very early health indicators such as birth weight and gestational age. Furthermore, especially in childhood body composition and metabolic functioning vary in different age groups, so that homogeneous age groups as accomplished by birth cohort studies offer another quality advantage [30,31].
- (1&2) Recruitment (R): Our inclusion criteria comprised only studies from which a high grade of representativeness and reproducibility was expected. To further assess the quality of recruitment in the score, population-based or school-based studies with characteristics of a census were rated as of highest quality (R = 1).
- (1&2) Explanatory variables/risk factor specification (E): We expected a source of bias from retrospectively assessed childhood psychosocial factors in the offspring (E = 0). In case parents gave direct information on their oc-

cupational, educational and financial situation we rated the quality of risk factor construction as high (E = 1).

- (1) Outcome parameter diabetes: Highest evidence was expected from blood glucose measurements with internationally valid cut off-levels as defined by the American Diabetes Association or WHO (O = 1) [32]. Self-reported diabetes leads to an underestimation of type 2 diabetes cases due to a high number of undiagnosed cases [33,34]. Therefore, we considered the quality based on self-report as low (O = 0). The plasma-based measure of insulin resistance (HOMA-IR calculated from fasting glucose and insulin levels) is closely related to (pre)diabetes. Thus, evidence from HOMA-IR-based data were considered as high (O = 1). Although HbA1c has been suggested as diagnostic tool for type 2 diabetes (ADA 2010), especially in early diabetes and in prediabetes, HbA1c values lead to misinterpretation which may be also due to a genetic component in the metabolisation of glycated haemoglobin [35-39]. Thus, we rated the quality of HbA1c values as low (O = 0). Metabolic syndrome and type 2 diabetes share some risk factors, but are too different as entities to include the metabolic syndrome as surrogate for type 2 diabetes. We allowed for the term 'metabolic' during Medline search, but rated metabolic syndrome as outcome criteria of low evidence for our study question (O = 0).

- (2) Outcome parameter obesity (O): First, measured weight and height to calculate BMI met our quality demands. Self-reported weight and height on the other hand have been reported to be imprecise in adults and in parents reporting anthropometric data of their children. Standard cut-off values for overweight and obesity served as second indicator for high quality: In children overweight and obesity are defined based on the age and sex-specific 85th and 95th BMI percentiles in growth charts from the National Health and Nutrition Examination Survey (NHANES), the Centers for Disease Control and Prevention (CDC) or the International Obesity Taskforce (IOTF), whereas in adults cut-off values for BMI ≥ 25 kg/m^2 for overweight and ≥ 30 kg/m^2 for obesity were used. The highest score (O = 1) was given only if both criteria were fulfilled.

- (1&2) Confounding (C): Quality of adjustment for confounders was inferred from important behavioural pathways leading to type 2 diabetes. Adjustment of age, sex, BMI, smoking, physical activity and alcohol consumption was required for the highest quality score (C = 1). Adjustment for obesity in the extracted studies was heterogeneous and depended highly on the study question. According to the basic requirements for our review question a high study quality (O = 1) demanded at least control for birth weight or baseline BMI to have possible weight change considered and the adjustment for adult SES to control SES change throughout life.

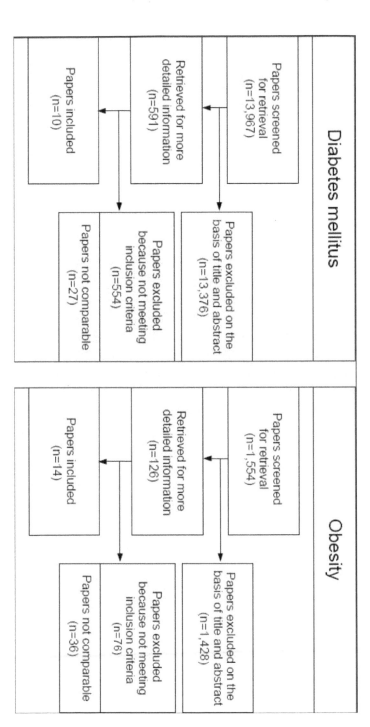

FIGURE 1: Flow diagram of systematic review on type 2 diabetes incidence and on obesity.

13.3 RESULTS

13.3.1 IDENTIFICATION OF RELEVANT STUDIES

The search strategy yielded a total of 19,504 results. After exclusion of 5,537 non-English articles, reviews, animal studies, case reports and articles unrelated to type 2 diabetes, 13,967 abstracts were screened (see also Figure 1). We assumed that social parameters are often treated as confounders and in this case the results are presented in full text articles, but not in abstracts. Therefore, the first screening of abstracts served mainly to identify articles on diabetes incidence (591). Of these, 280 did not meet the study design criteria and were excluded. The further thematic evaluation of the psychosocial relevance of the studies was based on full text articles. Of these 273 were thematically irrelevant mainly because they did not offer information on childhood psychosocial factors. One study retrieved during the update of the search was not added, because it presented race stratified results and there was already non-stratified data available from this study from the first search [40].

For reasons of limited statistical comparability (p.eg. limited presentation of data) 10 studies remained for further descriptive analysis (Table 1). Information on excluded studies with a unique method of analysis such as path models is provided in the discussion, information on excluded studies with solely examined risk factors in the results.

The second search strategy for the outcome obesity retrieved 1,631 results (see also Figure 1). 77 reviews and non-English publications were excluded. 1,428 studies were classified as thematically irrelevant and 76 publications did not meet study design criteria. This led to 50 longitudinal studies of which 36 were not comparable for methodological reasons and risk factor selection. Hence, 14 publications (13 studies) were included for this review and are presented in Table 2.

13.3.2 DESCRIPTION OF INCLUDED STUDIES: DIABETES INCIDENCE

13.3.2.1 STUDY DESIGN

Five publications involving 9,200 cases of incident diabetes were included [41-45]. Furthermore, we identified 147 cases of insulin resistance [46], 953 cases of elevated HbA1c [47,48] and 233 cases of "metabolic malfunctioning" [49] in a total of 199,214 individuals. One study analysed HOMA-IR as continuous variable in 1,167 individuals, but did not present the number of cases with insulin resistance [50]. Hence, overall 10 studies are summarised in Table 1. Four studies were designed as birth cohorts [43,47-49], two were conducted in children or adolescents (age range 3-18 and 14-19 years, respectively) [46,50]. All other studies measured childhood psychosocial factors retrospectively mainly in middle-aged participants.

13.3.2.2 PSYCHOSOCIAL FACTORS

Most of the articles provide data of parents' occupation as basic indicator variable for socioeconomic status (Table 1). Parental education was offered as main SES risk factor in three other studies. However, both SES variables were heterogeneously defined in these studies. Parents' occupation, for instance, was measured in five different classification scales ranging from two to six levels and following three different classification standards [51-53]. Childhood adversity as risk factor was found in one study only [48]. In this study, little parental interest in education was assessed as one aspect of emotional neglect. As this item might be related to formal parental education as well, results are presented in Table 1.

13.3.2.3 OUTCOMES

Self-reported diabetes was the most common outcome measure without restriction to type 2 diabetes. Only one article by Lidfeldt et al. was completely concordant with our review question by defining type 2 diabetes incidence on the basis of fasting plasma glucose levels according to ADA recommendations in an originally diabetes-free cohort [41]. HbA1c levels above 5.8% and 6.0% were used in two studies to define metabolically abnormal cases [47,48]. Insulin resistance according to HOMA-IR levels above 5.8 and 6.0 was the outcome parameter in two more studies [46,50]. One study defined "metabolically normal" cases following the definition of metabolic syndrome from the International Diabetes Federation, but without weight indicators because only participants with BMI ≥ 25 kg/m^2 were included [49].

13.3.3 DESCRIPTION OF INCLUDED STUDIES: OVERWEIGHT AND OBESITY

13.3.3.1 STUDY DESIGN

Thirteen studies met the inclusion criteria for obesity as outcome parameter adding up to a sample size of 70,420 participants with mainly young age groups ranging from 0-19 years. One study included middle-aged persons from 40-60 years and retrospectively assessed paternal occupation [54]. Overall, the study durations ranged from 4-33 years (median: 13 years), five studies were designed as birth cohorts [55-59].

13.3.3.2 PSYCHOLOGICAL FACTORS

The risk factor definition among the obesity studies was similarly heterogeneous as in the diabetes studies. Data on parental occupation was most frequently presented followed by parental education. Family income was additionally included to parental education in four studies [58,60-62]. In one study results on the Home Observation for Measurement of the Environment Short Form Inventory (HOME-SF), a questionnaire measuring emotional and cognitive family environment and parenting abilities, was presented [61,63,64].

13.3.3.3 OUTCOMES

In most studies the outcome regarding obesity and overweight was defined according to standard cut-off values based on BMI values calculated from measured height and weight [54,56,58,60,61,65,66,68,69]. Three studies deduced BMI from self-reported anthropometric measures [55,62,67]. One study analysed dual-energy X-ray absorptiometry (DXA)-assessed fat mass [59]. Obesity incidence was determined only in the National Longitudinal Survey of Youth [61]. One study analysed age at onset of obesity and in a later publication also weight change, but without exclusion of cases with obesity at baseline [62,68]. Analysis of weight change over time was also done in three other studies [58,65,67].

13.3.4 DIABETES INCIDENCE: EFFECT OF CHILDHOOD SOCIOECONOMIC FACTORS

In four studies, the risk of developing type 2 diabetes or metabolic disturbances in the offspring of lower social classes compared to children of higher status was only slightly elevated after adjustment with different effect measures (OR, HR and RR) ranging between 1.08 and 1.7 [41,42,46,48]. In the Alameda County Study, disparities in self-reported diabetes were most striking in overweight women with low childhood socioeconomic position [42]: The odds of developing type 2 diabetes was 3.2 fold higher for them than for their high SEP counterparts (OR: 2.9 (95% CI 1.7; 4.8) vs. 0.9 (95% CI 0.4; 2.2)).

As shown in Table 1, the study with the smallest population size (N = 233) showed the largest protective effect (men OR 0.2 (95% CI 0.05, 0.8); women 0.6 (0.1, 2.7)) of a high social status (father's occupation) [49]. This study was based on middle-aged participants of the Newcastle

Thousand Families Cohort with a BMI exceeding 25 kg/m^2. Furthermore a large protective effect of maternal education (beyond elementary level) was found in a Mexican population (OR 0.6 (95% CI 0.5; 0,8)) [44]. In contrary, low maternal education at high school level or less was a significant factor influencing type 2 diabetes incidence in the Princeton School District Study (ß-coefficient = 4.47 (SE 0.78)).

On the other hand, two studies indicated small, non-significant protective effects of low social class. In a Finnish cohort children at the age of 7 years from blue-collar families had a reduced risk (OR 0.8 (95% CI 0.48; 1.45) of being diagnosed with type 2 diabetes in comparison with their white-collar counterparts [43]. In a British Cohort only women of the lowest classes were less likely to have HbA1c levels exceeding 5.8% (OR 0.8 (95% CI 0.5; 1.4)). In men, no effect of social class was seen in this study (OR 1.1 (95% CI 1.0; 1.8)) [47].

The effect sizes observed in the included studies (except for beta coefficients) are shown in Figure 2. Effects of high social status have been inverted for better visual comparability.

13.3.5 DIABETES INCIDENCE: CHANGE IN SOCIOECONOMIC FACTORS FROM CHILDHOOD TO ADULTHOOD

Furthermore, we were interested in the interaction of childhood and adult psychosocial variables. SES change was explicitly examined in the Nurses Health Study [41]. Participants in this study were relatively homogeneous with respect to their educational level and occupational status as all 100,330 participants were female nurses. Disparities emerged from different childhood socioeconomic positions and from husband's educational level. Improving SES over lifetime (spouse's high educational level and low father's occupational status) with stable high SES as reference resulted in a slightly reduced, albeit non-significant relative risk for type 2 diabetes (RR 0.9 (95% CI 0.7; 1.3), whereas a stable intermediate SES and a declining SES influenced the type 2 diabetes relative risk negatively (RR 1.2 (95% CI 1.06; 1.4) and RR 1.18 (95% CI 1.06; 1.3), respectively). The relative risk in participants with stable low SES was comparable to those participants with stable high SES.

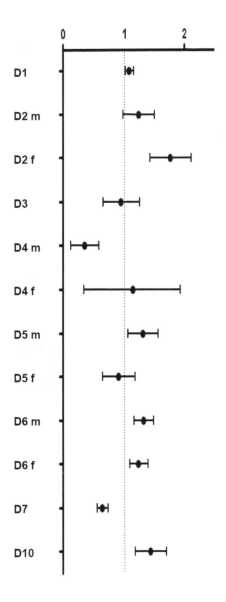

FIGURE 2: Impact of low SES influencing the incidence of type 2 diabetes. Effect sizes given as OR, HR or RR (central point) flanked by lower and upper 95% CI (results of high SES have been inverted, x-axis ends with 8, end points of higher results are not shown). Results of D8 and D9 are given as β-coefficients and are not included in Figure 2. D8 shows no effect of low SES; in D9 a considerably higher risk for type 2 diabetes incidence in the low SES group can be concluded.

Langenberg et al. examined lifetime effects by analysing the change in the influence of the socioeconomic position on HbA1c levels with adjustment for childhood and adult SEP [47]. In both men and women, adult social class had stronger effects than childhood social class on HbA1c values exceeding 5.8%. In women childhood effects on HbA1c even reversed in the fully adjusted model. Langenberg et al. interpret these findings by postulating that low childhood social class continues to influence adult social class as both variables were highly correlated [47].

A high continuity of childhood and adult SEP parameters was also observed in a Finnish cohort, where children from manual classes were more likely to work in manual or lower non-manual occupations in adulthood [46].

13.3.6 DIABETES INCIDENCE: EFFECT OF OTHER PSYCHOSOCIAL FACTORS

Among the retrieved studies there were several unique findings that were not comparable with other studies and other results. However, most of these studies revealed a possible association with future diabetes.

For example psychological factors such as childhood adversities [48] were associated with elevated HbA1c and obesity in one study. After full adjustment, these associations were no longer significant for elevated HbA1c. A further possible psychological risk factors for diabetes among our findings were depression and anxiety [70,71], hostility [72], and sense of coherence [73].

Also, indirect measures of socioeconomic status such as deprived neighbourhoods, and housing conditions [27,28,74,75] are likely to be relevant for future diabetes incidence. Neighbourhood characteristics also have influence on lifestyle habits depending e.g. on the availability of healthy foods and on the number of facilities for physical activity [76].

Overall, more comparable studies are needed to quantify the association of other childhood factors on metabolic impairment.

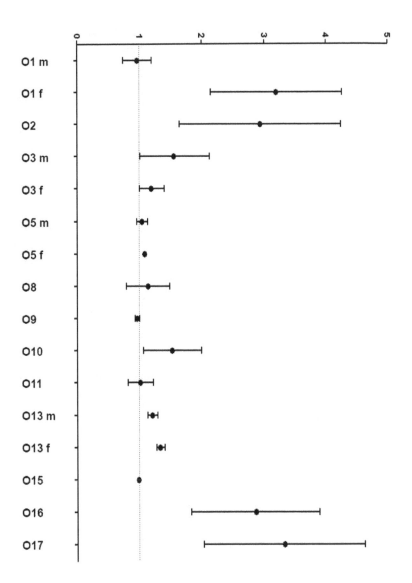

FIGURE 3: Impact of low SES influencing overweight and obesity. Effect sizes given as OR, HR or RR (central point) flanked by lower and upper 95% CI. Results of O6, O7, and O12 are given as β-coefficients and are not included in Figure 3. In O6 a limited effect of low SES is seen; in O7 and O12 a considerably higher risk for type 2 diabetes incidence in the low SES group can be concluded.

13.3.7 OVERWEIGHT AND OBESITY: EFFECT OF CHILDHOOD SOCIOECONOMIC FACTORS

Table 2 gives an overview of the included studies and the observed effect sizes which are furthermore visualised in Figure 3.

Effects were most prominent in two studies regarding income discrepancies: In a national longitudinal survey in the USA the RR in the lowest intracohortal (< 15th percentile) income group for obesity incidence was 2.84 (95% CI 1.39; 5.78) compared to the highest (> 85th percentile) income group during 6 years after the baseline examination [61]. Similar results were observed in a Canadian birth cohort at the follow-up of 4.5 years [58]. At that time point the lowest income group (less than 20,000 CN$/year) had a 2.5 fold increased odds (95% CI 1.3, 4.8) of being overweight at the age of 4.5 years with reference to families with an annual income of 60,000 CN$ or more. These results were obtained after adjustment for gestational age and birth weight. One study observed that belonging to a low income family (less than 26,000 US$/year) was a significant predictor of being obese at an earlier age (β-coefficient: -2.6 (95% CI -3.8; -1.3)) [62]. No effect of family income was seen in another study which included data from the National Longitudinal Surveys Child-Mother files in the USA (OR 1.0 (95% CI 0.99, 1.00)) [57].

Parental education had no direct or a small influence on overweight or obesity outcomes in six studies [57,58,60,61,68,69]. In the National Longitudinal Survey of Youth [61], the Home Observation for Measurement of the Environment-Short Form (HOME-SF) accounted for a modulation of other CSES factors particularly for maternal education. The unadjusted OR for the six-year cumulative incidence of childhood obesity for maternal education lower than high school was 1.47 (95% CI 1.04; 2.1). The fully adjusted OR was lowered to 0.96 (95% CI 0.7; 1.4) when including also HOME-SF as covariable. In contrast, family income seemed to be independent of the HOME-SF in this study [61].

In one study regarding parental occupation gender disparities were prominent [67]. Giskes et al. found a significantly higher risk for baseline overweight and obesity (at 40-60 years) for female participants whose fathers had been working in blue-collar occupations (OR 3.4 (95% CI 1.9;

6.1), father's professional occupation as reference). This effect decreased slightly after adjustment for adult SES (OR 2.8 (95% CI 1.6; 5.2)). These women also gained significantly more weight between baseline and follow-up 13 years later (OR 2.0 (95% CI 1.7; 2.3)). In another study the risk for developing overweight during the 6-year follow-up period was increased for children from blue-collar families (OR 2.4 (95% CI 1.02; 5.4) in comparison with their white-collar counterparts [65]. Furthermore, in two studies the adjusted effects of low status were small [54,55].

Based on results of regression analysis, two studies determined low social status as factor influencing overweight and obesity [56,62], whereas in a third study no such effect was seen (ß-coefficient: -0.05 (SE 0.08)) [66]. In another study, DXA-assessed fat mass was analysed using a slope index of inequality (SII). The results of this method are comparable to regression coefficients and pointed towards a moderate association of maternal education on body fat [59].

However, as these studies only displayed data on regression coefficients, effect sizes cannot be compared to the other studies and are therefore not included in Figure 3.

Taken together, whenever effects were seen in any of the thirteen studies, they pointed towards a deleterious influence of low social status on future risk of overweight and obesity.

13.3.8 OVERWEIGHT AND OBESITY: EFFECT OF CHANGE IN SOCIOECONOMIC FACTORS FROM CHILDHOOD TO ADULTHOOD

The role of change in psychosocial factors in relation to obesity was investigated in the 1958 British birth cohort [55]: Power et al. reported a continuous decrease of the effects of low socioeconomic status in childhood on obesity especially in women with an increase in social position in adulthood. The effect of low social class at the age of seven years in 33-year old women decreased from an unadjusted OR of 1.4 (95% CI 1.3; 1.6) to 1.3 (95% CI 1.1; 1.4) after adjustment for personal education [55].

13.3.9 OVERWEIGHT AND OBESITY: EFFECT OF OTHER PSYCHOSOCIAL FACTORS

We found one study which showed that traumata are associated with obesity [77]. Furthermore, depression and stress [78] are not only relevant for type 2 diabetes, but also for obesity. Also, indirect measures of socioeconomic status such as deprived neighbourhoods, housing conditions [79] and supply for healthy lifestyle habits [80], are likely to be relevant for obesity.

Therefore, psychological factors, and several measures of deprivation in childhood remain an interesting field for further examinations helping to understand the possible pathways leading to obesity and type 2 diabetes.

13.4 DISCUSSION

Based on the included studies and bearing in mind the limited comparability, we can state that psychosocial discrepancies in childhood seem to have an unfavourable impact on future type 2 diabetes incidence. Adjustment for adult SES and BMI attenuated these associations considerably [41,42,47]. This finding raises the question if a favourable life course may be beneficial for the participants' metabolic status. However, detailed life course analysis was rarely carried out among the retrieved studies. Only one study offered indications that improving SES over time seems protective against diabetes and that a decrease in SES is especially harmful [41]. But as only one study provided substantial data on SES change, no valid conclusion can be drawn from these results.

Furthermore, we found the family income [67,58,61] and the father's occupation [[67], in women only; [65]] of relevance for overweight and obesity. Surprisingly, in contrast to our findings on diabetes incidence parental education seemed to have less impact on future obesity risk. However, lower parental education was linked to an earlier age at onset of obesity in one study [62].

13.4.1 LIMITATIONS

Altogether we observe that for both type 2 diabetes and obesity longitudinal, population-based data are scarce. Especially, the life course including childhood indicators of psychosocial status and the role of risk factor clustering is under-investigated. Additionally, analysis of psychosocial factors is carried out heterogeneously and thus exposed to the critique of constructing SES risk factors arbitrarily. Although there are some national consensus statements especially regarding the assessment of the occupational position [51-53], it is difficult to compare these positions because they are attributed with varying amounts of prestige and are influenced by general shifts in composition [81].

13.4.2 IMPLICATIONS FOR RESEARCH

Three steps may help to gain further evidence on the topic: First, basic SES measures in youth and in adulthood should be consequently presented in publications. Second, for a better understanding of the role of psychosocial factors throughout life on the risk of type 2 diabetes, detailed analysis of a broad set of psychosocial factors, their interrelation and their impact on type 2 diabetes are needed. This requires a large-scale systematic analysis applying various psychosocial measures in youth and adulthood that have been found to be of relevance for diabetes and obesity such as depression [82-84], stress [85,86], unemployment [87], lifestyle habits [76,80] deprived neighbourhoods [27,28,74,75,79] and other factors [72,77].

Finally, such a large-scale systematic analysis would also require the application of different analytic approaches. As an example, the interpretation of the results of regression models analysing closely related risk factors throughout lifetime is controversial. Attenuation after adjustment for various risk factors may be attributable to a strong correlation of these risk factors [22] rather than to confounding. An interesting alternative approach may be the analysis of path models. For example Lehman et al. showed that childhood SES had a direct impact on metabolic functioning in the participants by applying path models on data from the Coronary Artery Risk Development in Young Adults study (CARDIA). Furthermore, childhood SES had an impact on early family environment, psychosocial functioning and adult SES. Interestingly, adult SES in this study had no direct influence on metabolic functioning [88]. Additionally, cluster and discriminance analysis can shed further light on the interaction of a broad set of childhood and adult psychosocial indicators accounting also for highly differentiated social milieus with distinct beliefs, behaviours and tastes [89].

13.4.3 IMPLICATIONS FOR PRACTICE

Valid results on the association of childhood socio-economic circumstances and future risk of diabetes and obesity would be important to design targeted and more efficient prevention strategies. Diabetes and obesity prevention may not only profit from educational programmes but also from health politics, from interventions for high-risk families, from coping skills training [90], from empowerment of social networks and from healthy neighbourhoods [91,92].

13.5 CONCLUSION

Taken together, despite the lack of homogeneous data, there is evidence for adverse effects of low psychosocial position in childhood on the risk for type 2 diabetes and obesity in later life. However, more studies and homogeneous standards regarding assessment of exposure and outcome variables and statistical analyses are needed.

REFERENCES

1.   Brown AF, Ettner SL, Piette J, Weinberger M, Gregg E, Shapiro MF, Karter AJ, Safford M, Waitzfelder B, Prata PA, Beckles GL: Socioeconomic Position and Health among Persons with Diabetes Mellitus: A Conceptual Framework and Review of the Literature. Epidemiol Reviews 2004, 26:63-77.

2. Parsons TJ, Power C, Logan S, Summerbell CD: Childhood predictors of adult obesity: a systematic review. Int J Obes Relat Metab Disord 1999, 23(Suppl 8):S1-107.
3. Wang Y, Beydoun MA: The obesity epidemic in the United States-gender, age, socioeconomic, racial/ethnic, and geographic characteristics: a systematic review and meta-regression analysis. Epidemiol Rev 2007, 29:6-28.
4. McLaren L: Socioeconomic status and obesity. Epidemiol Rev 2007, 29:29-48.
5. Shrewsbury V, Wardle J: Socioeconomic status and adiposity in childhood: a systematic review of cross-sectional studies 1990-2005. Obesity (Silver Spring) 2008, 16(2):275-284.
6. Hemingway H, Marmot M: Evidence based cardiology: psychosocial factors in the aetiology and prognosis of coronary heart disease. Systematic review of prospective cohort studies. BMJ 1999, 318(7196):1460-1467.
7. Blane D, Hart CL, Smith GD, Gillis CR, Hole DJ, Hawthorne VM: Association of cardiovascular disease risk factors with socioeconomic position during childhood and during adulthood. BMJ 1996, 313(7070):1434-1438.
8. Kvaavik E, Lien N, Tell GS, Klepp KI: Psychosocial predictors of eating habits among adults in their mid-30s: the Oslo Youth Study follow-up 1991-1999. Int J Behav Nutr Phys Act 2005, 2:9.
9. Osler M, Madsen M, Nybo Andersen AM, Avlund K, McGue M, Jeune B, Christensen K: Do childhood and adult socioeconomic circumstances influence health and physical function in middle-age? Soc Sci Med 2009, 68(8):1425-1431.
10. Mackenbach JP, Stirbu I, Roskam AJ: Socioeconomic inequalities in health in 22 European countries. N Engl J Med 2008, 358(23):2468-2481.
11. Lawlor DA, Davey Smith G, Ebrahim S: Life course influences on insulin resistance: findings from the British Women's Heart and Health Study. Diabetes Care 2003, 26(1):97-103.
12. Meyer IH, Schwartz S, Frost DM: Social patterning of stress and coping: does disadvantaged social statuses confer more stress and fewer coping resources? Soc Sci Med 2008, 67(3):368-379.
13. Connolly V, Unwin N, Sherriff P, Bilous R, Kelly W: Diabetes prevalence and socioeconomic status: a population based study showing increased prevalence of type 2 diabetes mellitus in deprived areas. J Epidemiol Community Health 2000, 54(3):173-177.
14. Lynch JW, Kaplan GA, Cohen RD, Kauhanen J, Wilson TW, Smith NL, Salonen JT: Childhood and adult socio-economic status as predictors of mortality in Finland. Lancet 1994, 343(8896):524-527.
15. Espelt A, Borrell C, Roskam AJ, Rodríguez-Sanz M, Stirbu I, Dalmau-Bueno A, Regidor E, Bopp M, Martikainen P, Leinsalu M, Artnik B, Rychtarikova J, Kalediene R, Dzurova D, Mackenbach J, Kunst AE: Socioeconomic inequalities in diabetes mellitus across Europe at the beginning of the 21st century. Diabetologia 2008, 51(11):1971-1979.
16. Dietz WH: Critical periods in childhood for the development of obesity. Am J Clin Nutr 1994, 59(5):955-959.
17. Kempf K, Rathmann W, Herder C: Impaired glucose regulation and type 2 diabetes in children and adolescents. Diab Metab Res Rev 2008, 24(6):427-437.

18. Lawlor DA, Ebrahim S, Davey Smith G: Socioeconomic position in childhood and adulthood and insulin resistance: cross sectional survey using data from British women's heart and health study. BMJ 2002, 325(7368):805.

19. Lawlor DA, Harro M, Wedderkopp N: Association of socioeconomic position with insulin resistance among children from Denmark, Estonia, and Portugal: cross sectional study. BMJ 2005, 331(7510):183-187.

20. Agardh EE, Ahlbom A, Andersson T, Efendic S, Grill V, Hallqvist J, Ostenson CG: Socio-economic position at three points in life in association with type 2 diabetes and impaired glucose tolerance in middle-aged Swedish men and women. Int J Epidemiol 2007, 36(1):84-92.

21. Ebrahim S, Montaner D, Lawlor DA: Clustering of risk factors and social class in childhood and adulthood in British women's heart and health study: cross sectional analysis. BMJ 2004, 328(7444):861.

22. Goran MI, Ball GD, Cruz ML: Obesity and risk of type 2 diabetes and cardiovascular disease in children and adolescents. J Clin Endocrinol Metab 2003, 88(4):1417-1427.

23. Sobal J, Stunkard AJ: Socioeconomic status and obesity: a review of the literature. Psychol Bull 1989, 105(2):260-275.

24. Felitti VJ, Anda RF, Nordenberg D, Williamson DF, Spitz AM, Edwards V, Koss MP, Marks JS: Relationship of Childhood Abuse and Household Dysfunction to Many of the Leading Causes of Death in Adults: The Adverse Childhood Experiences (ACE) Study. Am J Prev Med 1998, 14(4):245-258.

25. Antonovsky A: Unravelling the Mystery of Health: How People Manage Stress and Stay Well. San Francisco: Jossey-Boss 1987.

26. Carstairs V: Deprivation indices: their interpretation and use in relation to health. J Epidemiol Community Health 1995, 49(Suppl 2):S3-8.

27. Diez Roux AV, Jacobs DR, Kiefe CI: Neighborhood characteristics and components of the insulin resistance syndrome in young adults: the coronary artery risk development in young adults (CARDIA) study. Diabetes Care 2002, 25(11):1976-1982.

28. Stroup DF, Berlin JA, Morton SC, Olkin I, Williamson GD, Rennie D, Moher D, Becker BJ, Sipe TA, Thacker SB: Meta-analysis of observational studies in epidemiology: a proposal for reporting. Meta-analysis Of Observational Studies in Epidemiology (MOOSE) group. JAMA 2000, 283(15):2008-2012.

29. Chida Y, Hamer M: An association of adverse psychosocial factors with diabetes mellitus: a meta-analytic review of longitudinal cohort studies. Diabetologia 2008, 51(12):2168-2178.

30. Karlberg J: A biologically-oriented mathematical model (ICP) for human growth. Acta Paediatr Scand 1989, 350:70-94.

31. Lammi N, Moltchanova E, Blomstedt PA, Tuomilehto J, Eriksson JG, Karvonen M: Childhood BMI trajectories and the risk of developing young adult-onset diabetes. Diabetologia 2009, 52(3):408-414.

32. Expert Committee on the Diagnosis and Classification of Diabetes Mellitus: Report of the Expert Committee on Diagnosis and Classification of Diabetes Mellitus. Diabetes Care 1997, 20(7):1183-1194.

33. Manson JE, Colditz GA, Stampfer MJ: A prospective study of maturity-onset diabetes mellitus and risk of coronary heart disease and stroke in women. Arch Intern Med 1991, 151(6):1141-1147.

34. Field AE, Coakley EH, Must A, Spadano JL, Laird N, Dietz WH, Rimm E, Colditz GA: Impact of overweight on the risk of developing common chronic diseases during a 10-year period. Arch Intern Med 2001, 161(13):1581-1586.

35. Likhari T, Aulakh TS, Singh BM, Gama R: Does HbA1C predict isolated impaired fasting glycaemia in the oral glucose tolerance test in subjects with impaired fasting glycaemia? Ann Clin Biochem 2008, 45(4):418-420.

36. Kilpatrick ES: HbA1c of glucose for diabetes diagnosis? Ann Clin Biochem 2005, 42(3):165-166.

37. Yudkin JS, Forrest RD, Jackson CA, Ryle AJ, Davie S, Gould BJ: Unexplained variability of glycated haemoglobin in non-diabetic subjects not related to glycemia. Diabetologia 1990, 33(4):208-215.

38. Snieder H, Sawtell PA, Ross L, Walker J, Spector TD, Leslie RD: HbA1c levels are genetically determined even in type 1 diabetes: evidence from healthy and diabetic twins. Diabetes 2001, 50(12):2858-2863.

39. Barakat O, Krishnan ST, Dhatariya K: Falsely low HbA1c value due to a rare variant of hemoglobin J-Baltimore. Prim Care Diabetes 2008, 2(3):155-157.

40. Maty SC, James SA, Kaplan GA: Life-course socioeconomic position and incidence of diabetes mellitus among blacks and whites: the Alameda County Study, 1965-1999. Am J Public Health 2010, 100(1):137-45.

41. Lidfeldt J, Li TY, Hu FB, Manson JE, Kawachi I: A prospective study of childhood and adult socioeconomic status and incidence of type 2 diabetes in women. Am J Epidemiol 2007, 165(8):882-889.

42. Maty SC, Lynch JW, Raghunathan TE, Kaplan GA: Childhood socioeconomic position, gender, adult body mass index, and incidence of type 2 diabetes mellitus over 34 years in the Alameda County Study. Am J Public Health 2008, 98(8):1486-1494.

43. Gissler M, Rahkonen O, Järvelin MR, Hemminki E: Social class differences in health until the age of seven years among the finnish 1987 birth cohort. Soc Sci Med 1998, 46(12):1543-1552.

44. Kohler IV, Soldo BJ: Childhood predictors of late-life diabetes: the case of Mexico. Soc Biol 2005, 52(3-4):112-131.

45. Best LE, Hayward MD, Hidajat MM: Life course pathways to adult-onset diabetes. Soc Biol 2005, 52(3-4):94-111.

46. Kivimäki M, Smith GD, Juonala M, Ferrie JE, Keltikangas-Järvinen L, Elovainio M, Pulkki-Råback L, Vahtera J, Leino M, Viikari JS, Raitakari OT: Socioeconomic position in childhood and adult cardiovascular risk factors, vascular structure, and function: cardiovascular risk in young Finns study. Heart 2006, 92(4):474-480.

47. Langenberg C, Kuh D, Wadsworth ME, Brunner E, Hardy R: Social circumstances and education: life course origins of social inequalities in metabolic risk in a prospective national birth cohort. Am J Public Health 2006, 96(12):2216-2221.

48. Thomas C, Hypponen E, Power C: Obesity and type 2 diabetes risk in midadult life: the role of childhood adversity. Pediatrics 2008, 121(5):1240-1249.

49. Hayes L, Pearce MS, Unwin NC: Lifecourse predictors of normal metabolic parameters in overweight and obese adults. Int J Obes (Lond) 2006, 30(6):970-976.

50. Goodman E, Daniels SR, Dolan LM: Socioeconomic disparities in insulin resistance: results from the Princeton School District Study. Psychosom Med 2007, 69(1):61-67.
51. Edwards AM: Classification of occupations. J Am Stat Assoc 1911, 12:618-646.
52. Statistics Finland: Classification of socioeconomic groups 1989. [http://www.stat.fi/meta/luokitukset/ammatti/001-2001/koko_luokitus_en.html] webcite Statistics Finland 1989 Handbooks 17, Helsinki assessed 20 may 2009
53. British Registrar General's Scale, adapted from Stevenson THC: The Vital Statistics of Wealth and Poverty. J R Stat Soc 1928, 91:207-230.
54. Novak M, Ahlgren C, Hammarström A: A life-course approach in explaining social inequity in obesity among young adult men and women. Int J Obes (Lond) 2006, 30(1):191-200.
55. Power C, Manor O, Matthews S: Child to adult socioeconomic conditions and obesity in a national cohort. Int J Obes Relat Metab Disord 2003, 27(9):1081-1086.
56. Laitinen J, Power C, Jarvelin MR: Family social class, maternal body mass index, childhood body mass index, and age at menarche as predictors of adult obesity. Am J Clin Nutr 2001, 74(3):287-294.
57. Salsberry PJ, Reagan PB: Taking the long view: the prenatal environment and early adolescent overweight. Res Nurs Health 2007, 30(3):297-307.
58. Dubois L, Girard M: Early determinants of overweight at 4.5 years in a population-based longitudinal study. Int J Obes (Lond) 2006, 30(4):610-617.
59. Howe LD, Galobardes B, Sattar N, Hingorani AD, Deanfield J, Ness AR, Davey-Smith G, Lawlor DA: Are there socioeconomic inequalities in cardiovascular risk factors in childhood, and are they mediated by adiposity? Findings from a prospective cohort study. Int J Obes (Lond) 2010, in press.
60. Koupil I, Toivanen P: Social and early-life determinants of overweight and obesity in 18-year-old Swedish men. Int J Obes (Lond) 2008, 32(1):73-81.
61. Strauss RS, Knight J: Influence of the home environment on the development of obesity in children. Pediatrics 1999, 103(6):e85.
62. Gordon-Larsen P, Adair LS, Suchindran CM: Maternal obesity is associated with younger age at obesity onset in U.S. adolescent offspring followed into adulthood. Obesity (Silver Spring) 2007, 15(11):2790-2796.
63. Bradley RH: Children's home environments, health, behavior and intervention efforts: a review using the HOME inventory as a marker measure. Genet Soc Gen Psychol Monogr 1993, 119(4):439-490.
64. Bradley RH: The HOME inventory: review and reflections. Adv Child Dev Behav 1994, 25:242-288.
65. Kristensen PL, Wedderkopp N, Møller NC, Andersen LB, Bai CN, Froberg K: Tracking and prevalence of cardiovascular disease risk factors across socio-economic classes: a longitudinal substudy of the European Youth Heart Study. BMC Public Health 2006, 6:20.
66. Elgar FJ, Roberts C, Moore L, Tudor-Smith C: Sedentary behaviour, physical activity and weight problems in adolescents in Wales. Public Health 2005, 119(6):518-524.
67. Giskes K, van Lenthe FJ, Turrell G, Kamphuis CB, Brug J, Mackenbach JP: Socioeconomic position at different stages of the life course and its influence on body

weight and weight gain in adulthood: a longitudinal study with 13-year follow-up. Obesity (Silver Spring) 2008, 16(6):1377-1381.

68. Lee H, Harris KM, Gordon-Larsen P: Life Course Perspectives on the Links Between Poverty and Obesity During the Transition to Young Adulthood. Popul Res Policy Rev 2009, 28(4):505-532.

69. Balistreri KS, Van Hook J: Socioeconomic status and body mass index among Hispanic children of immigrants and children of natives. Am J Public Health 2009, 99(12):2238-2246.

70. Engum A: The role of depression and anxiety in onset of diabetes in a large population-based study. J Psychosom Res 2007, 62(1):31-38.

71. Eaton WW, Armenian H, Gallo J, Pratt L, Ford DE: Depression and risk for onset of type II diabetes. A prospective population-based study. Diabetes Care 1996, 19(10):1097-1102.

72. Räikkönen K, Matthews KA, Salomon K: Hostility predicts metabolic syndrome risk factors in children and adolescents. Health Psychology 2003, 22(3):279-286.

73. Kouvonen AM, Väänänen A, Woods SA, Heponiemi T, Koskinen A, Toppinen-Tanner S: Sense of coherence and diabetes: a prospective occupational cohort study. BMC Public Health 2008, 8:46.

74. Cox M, Boyle PJ, Davey P, Morris A: Does health-selective migration following diagnosis strengthen the relationship between Type 2 diabetes and deprivation? Soc Sci Med 2007, 65(1):32-42.

75. Schootman M, Andresen EM, Wolinsky FD, Malmstrom TK, Miller JP, Yan Y, Miller DK: The effect of adverse housing and neighborhood conditions on the development of diabetes mellitus among middle-aged African Americans. Am J Epidemiol 2007, 166(4):379-387.

76. Wilsgaard T, Jacobsen BK: Lifestyle factors and incident metabolic syndrome. The Tromsø Study 1979-2001. Diabetes Res Clin Prac 2007, 78(2):217-224.

77. Perkonigg A, Owashi T, Stein MB, Kirschbaum C, Wittchen HU: Posttraumatic stress disorder and obesity: evidence for a risk association. Am J Prev Med 2009, 36(1):1-8.

78. Coogan PF, Cozier YC, Krishnan S, Wise LA, Adams-Campbell LL, Rosenberg L, Palmer JR: Neighborhood Socioeconomic Status in Relation to 10-Year Weight Gain in the Black Women's Health Study. Obesity (Silver Spring) 2010, in press.

79. Koch FS, Sepa A, Ludvigsson J: Psychological stress and obesity. J Pediatr 2008, 153(6):839-44.

80. Robinson WR, Stevens J, Kaufman JS, Gordon-Larsen P: The Role of Adolescent Behaviors in the Female-Male Disparity in Obesity Incidence in US Black and White Young Adults. Obesity (Silver Spring) 2009, in press.

81. Liberatos P, Link BG, Kelsey JL: The measurement of social class in epidemiology. Epidemiol Rev 1988, 10:87-121.

82. Liem ET, Sauer PJ, Oldehinkel AJ, Stolk RP: Association between depressive symptoms in childhood and adolescence and overweight in later life: review of the recent literature. Arch Pediatr Adolesc Med 2008, 162(10):981-988.

83. Knol MJ, Twisk JW, Beekman AT, Heine RJ, Snoek FJ, Pouwer F: Depression as a risk factor for the onset of type 2 diabetes mellitus. A meta-analysis. Diabetologia 2006, 49(5):837-845.

84. Brown LC, Majumdar SR, Newman SC, Johnson JA: History of depression increases risk of type 2 diabetes in younger adults. Diabetes Care 2005, 28(5):1063-1067.

85. Turner RJ, Wheaton B, Lloyd DA: The epidemiology of social stress. Am Sociol Rev 1995, 60:104-125.

86. Pouwer F, Kupper N, Adriaanse MC: Does emotional stress cause type 2 diabetes mellitus? A review from the European Depression in Diabetes (EDID) Research Consortium. Discov Med 2010, 9(45):112-118.

87. Laitinen J, Power C, Ek E, Sovio U, Järvelin MR: Unemployment and obesity among young adults in a northern Finland 1966 birth cohort. Int J Obes Relat Metab Disord 2002, 26(10):1329-1338.

88. Lehman BJ, Taylor SE, Kiefe CI, Seeman TE: Relation of childhood socioeconomic status and family environment to adult metabolic functioning in the CARDIA study. Psychosom Med 2005, 67(6):846-854.

89. Bourdieu P: La distinction. Critique sociale du jugement. Éditions de Minuit. Paris 1979.

90. Taylor SE, Stanton AL: Coping resources, coping processes, and mental health. Annu Rev Clin Psychol 2007, 3:377-401.

91. Sallis JF, Saelens BE, Frank LD, Conway TL, Slymen DJ, Cain KL, Chapman JE, Kerr J: Neighborhood built environment and income: examining multiple health outcomes. Soc Sci Med 2009, 68(7):1285-1293.

92. Veugelers P, Sithole F, Zhang S, Muhajarine N: Neighborhood characteristics in relation to diet, physical activity and overweight of Canadian children. Int J Pediatr Obes 2008, 3(3):152-159.

93. Sergeant JC, Firth D: Relative index of inequality: definition, estimation, and inference. Biostatistics 2006, 7:213-224.

There are several tables that are not available in this version of the article. To view this additional information, please use the citation on the first page of this chapter.

CHAPTER 14

EARLY LIFE ADVERSITY AS A RISK FACTOR FOR FIBROMYALGIA IN LATER LIFE

LUCIE A. LOW AND PETRA SCHWEINHARDT

14.1 INTRODUCTION

The causes and underlying pathologies dictating and affecting the development of fibromyalgia (FM) are not yet clear, but as a syndrome affecting between 2 and 4% of the population, with a higher incidence in women [1, 2], it is a hot topic in pain research at this time. As discussed in detail elsewhere in this paper, FM constitutes a chronic pain syndrome, concomitant with a myriad of symptoms, including muscular stiffness and tenderness at specific locations, chronic fatigue, cognitive and mood disturbances, and insomnia. Documented pathophysiologies related to FM include aberrations in neuroendocrine systems, dysfunction in stress regulation and neurotransmitter function, and alterations in brain structure and connectivity. In addition, FM is linked to psychosocial and environmental triggering factors. This paper will explain and discuss some of the potential risk factors of early life adversity (ELA) that display markedly similar outcomes that constitute some of the later symptoms of FM. Pain researchers have clearly shown that noxious events during early life can cause a number of long-lasting changes in pain processing systems in the older organism, which could contribute to the increased pain sensitivity noted in FM patients. In addition, factors such as premature birth and related exposure to stressors, maternal deprivation, and physical or substance abuse in the perinatal period can influence developing neurobiological and psychological states in a number of ways, often causing changes in adulthood similar to the disturbances seen in FM sufferers. Early life pain, hospitalisation, deprivation, emotional trauma, and abuse are discussed in this paper, with speculation about their potential impact upon fibromyalgia.

14.2 RISK FACTOR: PAINFUL EXPERIENCES DURING INFANT DEVELOPMENT

It is well established that the experience of pain during infancy causes long-lasting alterations in pain processing that extend well into childhood and adulthood [3–5]. This adds weight to the need for effective understanding and management of pain in neonates, in order to minimise later consequences of early pain experience (see [6, 7]). Infants born with pre-existing illnesses or under difficult circumstances, born preterm or needing early surgery, may all need hospitalisation for treatment. In addition, the incidence of premature birth and the necessary associated critical care has increased over the last 20 years [8], and the technology to aid survival has meant that infants as young as 24 weeks postmenstrual age (PMA) can survive and develop on the neonatal intensive care unit (NICU). In these clinical settings, multiple painful procedures may be performed daily for routine monitoring, in addition to any necessary surgeries infants require—for example, Simons et al. [9] found that during the first 14 days of hospitalisation, neonates were subject to an average of 14 painful procedures per day. A more recent study [10] confirmed these findings, showing that neonates were exposed to a daily average of 16 painful and/or stressful procedures and that up to 80% of children were not given specific analgesia for these procedures. This high number of procedures, perhaps

repeated over a number of weeks or months, affects the developing nociceptive circuitry of the infant in ways that cause long-lasting changes in
pain processing (see [3]) and could explain some of the abnormalities in
pain processing displayed by FM patients.

14.2.1 THE EFFECT OF PAINFUL STIMULATION ON THE HUMAN NEONATE

Studies of ex-premature children provide compelling evidence for the
long-term effects of human early pain experiences. Walker et al. [11] recently presented data on sensory sensitivity in a cohort of 307 extremely
preterm infants born at less than 26 weeks postmenstrual age (PMA) in
1995 and followed throughout their lives so far (the UK EPICure cohort).
Quantitative sensory testing (QST) established sensory thresholds in these
children at age 11 and showed that these extremely preterm children had
significantly decreased sensitivity to non-noxious mechanical and thermal
stimuli compared to age- and sex-matched term born controls. A similar
result has been seen in 9–12-year-old children who had previously experienced neonatal cardiac surgery—subjects were significantly less sensitive to nonnoxious mechanical and thermal stimuli at both the previously
operated site and noninjured areas [12]. The foundations for this baseline hyposensitivity may be laid whilst children are being cared for on the
NICU—following children with NICU experience at 4, 8, and 18 months
of (corrected) age, dampened pain responses to immunisation and blunted
nociceptive sensitivity to everyday bumps are seen, compared to full-term
controls with no NICU experience [13, 14]. Hermann et al. [15] also found
elevated heat pain thresholds (i.e., decreased sensitivity) in children who
had been hospitalised for a prolonged period as infants and had undergone
repeated painful procedures as part of their treatment.

Importantly and more relevant to FM, when pain-exposed neonates are
reexposed to noxious stimuli in later life, hypersensitivity to the stimulus
is observed. For example, when ex-NICU infants were tested on their perceptual sensitisation to heat pain, where a constant temperature is given
for 30 seconds and the change in perception gauged at the end, neonatally
hospitalised children showed increased sensitisation compared to non-

hospitalised controls, who habituated to the thermal stimulus [15]. Interestingly, heat pain thresholds of all groups of children were increased in the presence of the children's mothers [16], highlighting the importance of social support on the pain experience (which has also been shown to be effective in alleviating the impact of FM [17]). More studies confirm the hypersensitivity seen in ex-premature infants after noxious stimulation in the older child. For example, behavioural sensitivity to noxious mechanical stimuli at the heel persists for at least the first year of life after repeated NICU heel lance experience [18]. Deep somatic and visceral noxious stimulation resulting from early invasive surgery leads to sensitisation of pain responses to later surgery, particularly in regions of the body served by the same spinal nerves as those affected by the initial surgery [19]. These effects are not limited to surgical pain. Children who suffered from burn injuries in infancy (6–18 months of age) showed lower mechanical pain thresholds and greater perceptual sensitisation to both heat and mechanical pain stimuli at sites not originally affected by the burn at ages 9–16 [20]. An interesting study by Buskila and colleagues [21] may be particularly relevant to adult fibromyalgia patients: this study showed that ex-NICU neonates had, as 12–18 year olds, significantly more "tender points" and lower tenderness thresholds than matched full-term children. Seeing as FM diagnosis has been partly based on soreness at a certain number of "tender points", it could be informative to follow these adolescents over time and assess later FM prevalence.

Summarizing the above, hypo- as well as hypersensitivity has been observed as a consequence of early life pain. The way in which sensory processing is altered by early life pain may be dependent on several factors. For example, the developmental time point at which injury is experienced can dictate the lasting effects of these injuries (see discussion of the "critical period" in the section "Early Life Exposure to Pain May Influence Fibromyalgia"). In addition, the type of noxious insult (surgical, burn, etc.) and therefore the relative proportion of nociceptors that are activated (i.e., C fibre or A delta nociceptors) may influence the exact nature of altered sensory processing. Finally, as detailed in the following section, it may be that discrete CNS systems are responsible for decreases in tactile and thermal sensitivities and the hyperalgesia seen in neonatally injured humans.

Animal models of neonatal pain show markedly similar effects as those seen in humans and are crucial to identify and understand cellular mechanisms that are impossible to study in humans. The development of nociceptive circuitry has been studied in-depth, and the neurobiology underlying long-lasting changes is becoming increasingly clear (see [3]).

14.2.2 THE NEUROBIOLOGY UNDERLYING LONG-TERM EFFECTS CAN BE STUDIED IN ANIMAL MODELS

The utility of animal models is illustrated by evidence showing that the generalised hyposensitivity to mechanical and thermal stimuli shown in humans with early life pain experience (e.g., [14]) is likely mediated by changes in the brainstem regions that modulate ascending afferent input, in particular the periaqueductal grey (PAG) and rostroventral medulla (RVM), which can either enhance or suppress nociceptive input from the spinal cord [22–25]. In the rat, these areas (PAG and RVM) mature over the first three weeks of age [26–28], which corresponds approximately to the time span from the third trimester of gestation to adolescence in human [29, 30]. La Prairie and Murphy [31] injected carrageenan (which causes short-term inflammation lasting around 24 hours) into the hindpaw of male and female rats on the day of birth and saw that, in adulthood, the animals showed decreased sensitivity to thermal stimulation in both the previously injured and uninjured paws. In addition, increased levels of endogenous opioid mRNA were seen in the PAG of the adult animal, and blockade of brain opioid receptors with naloxone abolished this decreased pain sensitivity, leading the authors to speculate that neonatal inflammation induces an upregulation in endogenous opioidergic tone that is maintained into adulthood, so that the adult displays a system that is constitutively "dampening" afferent spinal cord input, leading to decreased pain responses [32].

This suggests that early pain can alter endogenous pain inhibitory circuitry. However, FM patients show hyperalgesia rather than hyposensitivity [33–36]. Enhanced nociceptive responsiveness is consistently observed in neonatally injured animals upon adult reinjury and likely results from a number of changes in nociceptive circuitry induced by early pain exposure, all of which result in a nervous system "primed" to respond in an enhanced manner to a new insult. For example, neonatal inflammation causes thermal hyperalgesia in rat pups that lasts from several weeks up to adulthood [37–39]. This inflammation and concomitant release of inflammatory and trophic molecules results in enhanced spinal neuronal responses to paw pinch in the adult, as well as increased primary afferent nerve fibre innervation of the dorsal horn of the spinal cord [40]. Inflammation-induced alterations in the developmental connectivity of the spinal cord [41] and/ or sprouting of nerve fibres at the skin also result in an increased nociceptive response [42]. Neonatal skin wounds, like inflammation, cause drops in mechanical withdrawal thresholds at the site of injury and increase in dorsal horn receptive field size weeks after the wound had healed [43, 44], as well as causing release of nerve growth factors leading to hyperinnervation of the skin, and increased sensitivity to noxious stimuli in later life [45]. This is consistent with human studies showing hypersensitivity after injury in children, especially those with previous surgical history [19, 46, 47]. Therefore, the apparent contradiction between generalised decreased sensitivities and hyperalgesic responses to new noxious stimulation may in part be due to discrete CNS processing systems dictating behavioural responses, for example, enhanced endogenous pain inhibition at a brainstem level, but increased hypersensitivity/hyperinnervation at a spinal level.

14.2.3 EARLY LIFE EXPOSURE TO PAIN MAY INFLUENCE FIBROMYALGIA

As explained above, whilst La Prairie and Murphy [31] showed that animals subject to neonatal inflammation showed generalised hypoalgesia to thermal stimuli at baseline in adulthood, when animals were reinjured as adults, animals were more sensitive to noxious thermal stimulation. Importantly, all of these effects were greatest in female animals. This finding might be relevant to fibromyalgia syndrome, as the incidence of FM is greater in women [48]. Furthermore, disturbances in descending pain modulation have been reported in FM patients [49, 50] and patients show decreased blood serum levels of serotonin and lower CSF levels of serotonin and noradrenaline metabolites [51–54]. This may be relevant to

FM, as there is a discrete descending serotonergic system projecting from the RVM that modulates spinal cord excitability [55–59]. Furthermore, noradrenergic signalling, originating from the locus coeruleus, has a role centrally in feedback inhibition of pain (see [60]).

The developmental timing of any injury determines potential long-term effects, leading to the concept of a "critical period" of nociceptive development, within which pain experience permanently alters pain processing (see [5, 61]). To illustrate, giving a skin incision to the hindpaw of a neonatal rat at postnatal days (P) 3 or 6 produces an increased pain response to a repeat incision 2 weeks later. If, however, the initial incision is performed after the critical period (at P10, 21, or 40 followed by repeat incisions 2 weeks later), the enhanced hypersensitivity to the later incision is not seen [61]. Understanding the concept of a "critical period" of nociceptive development may be useful for determining some of the root causes of fibromyalgia—as human infants born prematurely display long-term alterations in pain processing, it is possible that early pain experience within this time window contributes to some of the adult pain in FM. At this time, there is very little literature that attempts to delve into the neo-natal and childhood life of current FM patients.

14.2.4 ADEQUATE PAIN MANAGEMENT OF NEONATES MAY DECREASE FM PREVALENCE IN ADULTS

In order to prevent these physiological disturbances, adequate pain management for human neonates is an important clinical issue. Pain management of neonates is a difficult area, as neonates cannot give verbal feedback on their pain experience, and are physiologically very different to the adult state that dictates dosage, metabolism and efficacy (i.e., [7]). Current treatments include morphine and benzodiazepines both for postsurgical pain and general sedation on the NICU [62], as well as nonpharmacological interventions such as the administration of sucrose for acute procedures including heel lance [63] (although the effectiveness of these interventions and the long-term effects of chronic exposure to sucrose and drugs such as morphine are not yet clear [64, 65]). The effects of these continue to be studied, and given the long-term effects of early pain as discussed above, adequate pain management for neonates might reduce various pain syndromes in later life, including fibromyalgia. The importance of adequate analgesia for neonatal procedures is illustrated in a seminal paper by Taddio et al. [46]. They performed a double-blind, randomized, controlled trial (RCT) on the effects of a topical anaesthetic (EMLA cream) used during male neonatal circumcision, which has traditionally been done without anaesthesia or analgesia. Looking at pain responses in the infants when they were later vaccinated at 4–6 months, boys who had been treated with the EMLA cream when circumcised showed lower pain responses than those who received no anaesthesia, and the circumcised groups both showed higher pain scores than uncircumcised controls.

14.3 RISK FACTOR: PREMATURE BIRTH AND RELATED STRESSORS

As discussed above, pain during the neonatal period can alter the nociceptive processing pathways of an organism for life, potentially impacting upon the development of fibromyalgia in later life. Many of the human studies mentioned in the previous section recruited infants born prematurely, needing the intensive care unit for survival. The NICU is a strange and abnormal environment in comparison to the womb, and premature infants are exposed to many stressful stimuli in addition to repeated nociceptive procedures, such as light, noise, tactile stimulation, surgery, medication, and maternal separation, all of which could feasibly affect development [66]. Indeed, even the act of nursing very premature infants (changing diapers etc.) causes increases in stress hormones [67]. In addition to the long-term effects of increased pain sensitivity in these children are effects upon stress regulatory systems, where a large body of evidence suggests that prematurity and the resulting experiences on the NICU can permanently alter in particular the hypothalamic-pituitary-adrenal (HPA) axis. As this is shown to be disturbed in FM patients [68–71], it is possible to speculate that premature birth in itself may influence the occurrence of FM in adults.

14.3.1 PREMATURE BIRTH IMPACTS UPON THE BODY'S RESPONSE TO STRESSORS

The HPA axis is the body's stress-response system. Cells of the hypothalamus produce corticotropin-releasing factor (CRF) in response to an environmental stressor, and a cascade of events ultimately causes adrenaline release and the production of the "stress hormone" cortisol. Under normal circumstances, adrenaline and cortisol release is terminated via a negative feedback circuit. In FM, however, HPA axis regulation appears to be abnormal. Whilst the precise dysfunctions in stress regulation via the HPA axis are not clear at this point (some studies find FM patients show hypocortisolism (see [72]), whilst others describe hypercortisolism and HPA hyperactivity [68, 69, 73]), what is clear is that HPA axis function is not normal in many patients. The discrepancy between findings of hyper- and hypocortisolism may be due to a number of factors. For example, disease-specific patient characteristics, such as symptom profiles and comorbidities, as well as disease-nonspecific characteristics, such as age, gender, personality traits, and socioeconomic background, can affect results. In addition, markers of HPA axis function differ between studies (e.g., basal levels or evoked cortisol responses) as do the time points of measurement (e.g., upon waking or following diurnal fluctuations), in addition to other technical details. In instances where the literature does not permit any conclusions on the direction, we refer to alterations in HPA axis function, rather than increases or decreases.

Nevertheless, the way in which premature birth alters cortisol levels and cortisol responses compared to term-born controls is relatively unambiguous. Grunau et al. [66] propose that early stressors such as those routinely experienced in the NICU can impact upon development of the HPA axis by causing consistent release of adrenaline and cortisol and increase the "allostatic load" of the neonate—the concept of "allostasis" explaining how an organism's physiological systems fluctuate over time in order to meet the demands of external stressors, in an attempt to regain homeostasis (bodily equilibrium). The impact of chronic stress and the accompanying neuroendocrine responses may also ultimately cause long-lasting damage to bodily organs and contribute to chronic disease development [74]. In preterm neonates, this may manifest as a life-long shift in HPA axis balance.

Basal cortisol levels are often low in neonates on the NICU in comparison to term infants, which is unexpected considering the length of time that infants spend there and the stressful procedures the still-developing neonate is subject to [75, 76]. Grunau and colleagues have published a series of studies giving convincing evidence that NICU experience causes "resetting" of the endocrine stress systems, by measuring cortisol levels after noxious experiences in ex-preterm infants, either whilst still on the NICU (short-term effects), or in the months to years following. When a clinically required heel lance was done whilst infants were still on the NICU, the earliest premature infants born at less than 28 weeks gestational age (i.e., approximately 3 months premature) showed a dampened cortisol response to heel lance. In addition, higher cumulative exposure to neonatal procedural pain over the length of stay in the NICU was related to lower cortisol release to standard nursing procedures [77]. When immunised at 2–4 months (corrected) age, low gestation age (LGA) boys (<32 weeks), but not girls, showed lower cortisol concentrations than full-term infants after injections, although facial and heart rate responses did not differ between groups [78].

When studied over a longer period of time, we see that this early dampened cortisol response in premature infants changes to elevations in cortisol levels and responses when the children are older. At 8 months old, infants born at extremely low gestational age (≤28 weeks) with previously low basal cortisol levels, showed elevated basal levels as well as greater increases in cortisol response to stressors, compared to term infants. In these children, greater increases in cortisol were associated with higher numbers of skin-breaking procedures experienced in the past on the NICU [79]. Grunau et al. [80] followed the time course of this "switch" from low to high stress hormone levels and found that at 3 months corrected age, basal cortisol levels were lower than term controls, but, at 8 and 18 months, the youngest ex-premature infants had significantly higher cortisol levels than term controls. The authors speculate that the HPA axis has been "reprogrammed" by NICU experience. Recent work has replicated the finding that premature children born onto the NICU later have higher basal cortisol levels compared to term-born controls at both 18 months [81] and upon waking in 8–14 year old ex-premature infants [82].

Animal studies support the existence of a developmental shift from low to high cortisol levels after perinatal corticosteroid exposure, and these higher levels of cortisol seen in older animals are associated with increased levels of corticotropin-releasing hormone (CRH) mRNA and glucocorticoid (GC) receptors in the amygdala (e.g., [83, 84]). In humans, this shift from low to high levels over development may be influenced by the fact that the mothers at risk of giving birth prematurely are routinely given corticosteroids to delay birth and enhance infant survival and lung function—an intervention which suppresses cortisol secretion in the infant when born [85] yet causes an increased cortisol response after heel lance at 24 hours after birth [86]. Further work is needed to address the impact of perinatal glucocorticoid exposure on the stress response axis in later life.

14.3.2 PREMATURITY MAY CONTRIBUTE TO ADULT FM SYMPTOMS VIA THE HPA AXIS

The above evidence highlights how premature birth and the stressors associated with it can influence the physiological response to stress and "reset" the balance of the HPA axis response. Seeing as FM patients routinely show imbalances in the stress response, it is reasonable to hypothesise that premature birth may be a risk factor for developing FM in later life. Indeed, Klingmann et al. [87] show that of 93 female FM patients, 62% reported a gestation length of <38 weeks, which was related to a lower cortisol response upon waking when compared to full-term FM patients. The authors speculate that enhanced glucocorticoid levels in the mother during pregnancy or in response to premature birth affect the development of the adrenal glands in the foetus/premature infant, rendering the HPA axis less capable of dampening stress responses to later stressors. This in turn may disinhibit responses to physical or psychological stress and affect brain function, resulting in enhanced responses to pain and increased fatigue levels. Support for this hypothesis comes from animal studies of prenatal glucocorticoid exposure, where dam rats are exposed to substances that increase HPA axis activity (such as glucocorticoid receptor agonists), in the third trimester of gestation. Results show that the adrenal glands and brain weight of the adult offspring are smaller, stress regulation is compromised, and cognitive dysfunction is seen in tests of memory as well as anxiety-like behaviour, with the effects exacerbated in female offspring [88–93].

14.3.3 COGNITIVE SYMPTOMS OF FM MAY ARISE FROM DIFFERENCES IN BRAIN DEVELOPMENT CAUSED BY PREMATURE BIRTH

One symptom of FM, colloquially called "fibro-fog" by sufferers, constitutes cognitive deficits, with patients complaining of difficulties in memory and attention that mimic the effects of an extra 20 years of ageing (e.g., [94, 95]). Additional evidence that prematurity may influence the development of FM comes from studies showing that the risks of cognitive and psychiatric impairment are much greater in ex-preterm infants. The EPICure cohorts (born at ≤25 weeks gestation) have recently had their cognitive abilities and psychiatric profiles investigated at 11 years of age. Results showed that the children born at the youngest preterm ages are at higher risk of ADHD, autism spectrum, and emotional disorders [96] and show increased incidences of learning impairments and poor academic attainment [97]. Other meta-analyses and epidemiological studies have confirmed the increased risk for psychiatric symptoms and poorer academic performance in older childhood after premature birth [98–100].

Brain imaging of ex-preterm infants compared to full-term controls has shown underlying changes in brain structure and function that may help explain some of these deficits. Cortical surface area is decreased at full-term in extremely preterm infants [101], and the incidence of white matter abnormalities persisting past 18 months (corrected) age is increased [102]. Thalamic volume is also reduced in preterm children at term-equivalent age [103] and at 2 years of age, and connectivity between the thalamus and cortex may be disrupted in ex-preterm children [104]. Seeing as the premature brain is still developing at a rapid pace and as myelination occurs during late preterm maturation [105], it is likely that prematurity influences white matter development, helping to explain why later cognitive deficits may arise. If premature birth becomes a proven risk factor for fibromyalgia, the neural bases of "fibro-fog" may become better understood in the adult.

14.4 RISK FACTOR: MATERNAL DEPRIVATION

As discussed above, early pain experience and prematurity may be risk factors for FM development in later life. Prematurity might be a risk factor in combination with the high exposure to additional stressors, including maternal deprivation. Relatively detailed information is available on the effects of maternal deprivation on the developing organism, and therefore maternal deprivation is discussed separately. Animal models of deprivation have proven extremely useful in illustrating the effects of deprivation from the primary caregiver (generally the mother) and the strong role for the fluidity of genetic expression during early life in the shaping of the adult phenotype. The study of epigenetics, or how the environment influences the activation and expression of different genes, has provided fascinating insights into this fluidity.

14.4.1 MATERNAL DEPRIVATION IN ANIMAL MODELS INFLUENCES LATER STRESS RESPONSES

Animal models of maternal deprivation often use rats and generally employ a paradigm whereby pups are separated from the mother for at least an hour per day, much longer than the 20–25 minutes of absence that dam (mother) rats are routinely away from the nest [106]. When neonatal rats are exposed to these prolonged periods of deprivation during the first weeks of life, a number of physiological and behavioural changes occur in the adult animal. For example, rats separated from the dam for 180 minutes per day from postnatal day (P) 2–14 show elevated levels of CRF mRNA as adults, which causes adrenaline and cortisol release via activation of the HPA axis [107, 108]. They also show more anxiety-like behaviours as adults and an increased propensity to consume alcohol [109, 110]. Accordingly, maternal deprivation has now been used to model various psychiatric states such as anxiety [111], addictive disorders

[112], and schizophrenia [113], and a recent study by Uhelski and Fuchs [114] showed that maternally deprived animals showed increased active avoidance of environments in which pain had been experienced, suggesting enhanced supraspinally mediated responses to pain. The importance of maternal presence is further illustrated by work with monkeys: infants reared in the absence of an adult caregiver but together with age-matched peers develop chronic anxiety-like behaviours and disordered cortisol levels to stressors (suggesting HPA axis imbalance), similar to FM symptoms [115]. Taken together, these data show that maternal deprivation has been associated with three important disturbances found among FM patients, namely, alterations in the HPA axis, increased anxiety, and increased pain responses [116].

Short-term separation of pups conversely causes opposite effects to longer maternal deprivation. If pups are subject to only 15 minutes of deprivation, the adult animals display more social contact and better stress-coping abilities compared to animals deprived for longer periods [117–121]. These effects seem to be mediated by the maternal style of the dam upon reunion with the pups—mothers of pups separated for short periods engage in more licking/grooming and arched-back nursing (LG-ABN) when reunited compared to dams of pups separated for more prolonged periods. Dams that engage in high levels of this nursing style produce offspring that show more efficient stress regulation, as measured by corticosterone (the animal equivalent of CRH) responses to stress and feedback sensitivity of the HPA axis [122, 123]. In fact, the maternal LG-ABN style (either high or low) causes individual differences in stress responsiveness and emotionality that remain stable in the adult offspring [124] and is in itself a trait that is passed on to female offspring. Cross-fostering studies, where pups from low or high LG-ABN dams are reared by dams showing the opposite LG-ABN behaviour illustrate that female offspring will show nursing styles akin to their "foster" dam rather than their biological dam [125]. Further evidence for the fluidity of these behavioural phenotypes comes from evidence that low LG-ABM dams rearing litters in a socially-enriched environment produce offspring showing enhanced exploration and licking/grooming behaviour of their own offspring [126].

The effects upon adult phenotype that depend on maternal style are regulated by changes in DNA methylation of the infant genome, leading to activation or silencing of certain genes, and alterations in levels of, for example, glucocorticoid receptors [122], neurotrophic factors and specific neurotransmitter receptors in the hippocampus [127]. Serotonin (5-HT) turnover (as seen by measures of 5-HT levels compared to levels of 5-HT metabolites) is also increased in maternally deprived animals [128], and expression and levels of serotonin receptors and transporter proteins altered [129, 130]. This is particularly relevant to FM, as the serotonergic system has been implicated in the affective components associated with FM—cerebrospinal fluid levels of 5-HT metabolites are decreased in patients [53], serotonin antagonists are effective drugs for some FM patients with no associated depressive comorbidities [131], and increased incidence of a specific genetic polymorphism in the 5-HT transporter gene has been identified in FM patients [132], although the exact disorders in serotonergic signalling are not yet clear [133]. However, some of the heterogeneity in FM patients may arise due to epigenetic alterations that occurred during early infancy. To date, no research has been conducted on this specific question.

14.4.2 QUALITY OF MATERNAL ATTACHMENT AFFECTS PAIN PROCESSING

The quality of the relationship between child and primary caregiver (generally the mother) can also dictate emotional reactivity throughout life and the type of attachment style that an individual will form with others throughout their life. Bowlby first developed the idea of "attachment theory", studying the bond between child and mother and suggested that a secure attachment style is the most beneficial for infant development [134, 135]. Since then, studies have shown that disordered attachment styles are linked to chronic pain and problems coping with pain (see [136]). For example, chronic pain patients with high levels of avoidant attachment self-scored pain intensity more highly, and patients with fearful attachment styles display increased levels of pain catastrophising, linked to anxiety levels [137, 138]. In acute pain tests, adults showing secure attachment styles rated pain as less intense and anxiogenic [139]. Importantly for FM, secure attachment formation is linked to the dopamine and opioidergic system in both animals and humans [140, 141], suggesting that a secure early attachment between infant and parent could be protective against developing FM in later life. Hallberg and Carlsson [142] indeed mention the overrepresentation of individuals with insecure attachment styles in the chronic pain patient population.

14.5 RISK FACTOR: CHILDHOOD PHYSICAL AND PSYCHOLOGICAL TRAUMA

Physical and sexual abuse during childhood are well-documented risk factors in the development of fibromyalgia [143–146], and two recent meta-analyses link childhood incidence of physical and sexual abuse with FM [147, 148]. Early life abuse carries with it the burden of a number of other behavioural and pathological problems, including increased incidence of depression, posttraumatic stress disorder, alcoholism, substance abuse, obesity, ill health, and suicide ([149; see 150]). A number of these are also comorbidities in FM. It is possible that the impact of early abuse and trauma contributes to FM via disruption of neurotransmitter systems such as the serotonergic and dopaminergic systems and impacts stress-management via the HPA axis [151–155].

14.5.1 A HIGH INCIDENCE OF CHILDHOOD ABUSE IS REPORTED IN FM PATIENTS

Self-report studies, where patients are questioned on their childhood history, consistently show increased early life adversity in FM patients. Goldberg et al. [156] found that three different groups of chronic pain patients (facial pain, myofascial pain, and FM), all had a history of abuse in nearly 50% of cases, rising to 65% in the fibromyalgia group. In particular, females with an alcoholic parent were likely to be members of the FM group. Hallberg and Carlsson [142] conducted in-depth interviews with 22 FM patients and describe "abundant examples of early loss (and...) high degree of responsibility early in life," and Anderberg et al. [157] found that, of 40 female FM patients, 51% had experienced very negative childhood or adolescence life events, compared to 28% in healthy age-matched women. Nicolson et al. [158] found an association of self-reported childhood abuse and neglect with FM patients' daily cortisol levels, finding the most disordered cortisol responses in the patients reporting the highest levels of sexual and emotional abuse. This suggests that early abuse further impacts upon the HPA axis, which shows aberrant functionality in FM in patients with no history of abuse [71]. Childhood rape has also been strongly associated with a lifetime diagnosis of FM [147].

The loss of a parent during early childhood is an emotionally traumatic event and is associated with altered daily cortisol levels in the adult, particularly in men [159, 160]. Poor quality family relationships during childhood also cause changes in cortisol release in response to a stressful event [161]. In addition, early-life stress and trauma has been shown to be a significant predictor of levels of CRF in cerebrospinal fluid (CSF) in non-FM subjects [162], and Danese et al. [163] found an association between early-life maltreatment and adult blood levels of C-reactive protein, a marker of inflammation. Specific to fibromyalgia, McLean et al. [164] showed that women with FM who reported sexual or physical abuse in their personal histories had differences in CSF levels of CRF compared to nonabused FM women, and Weissbecker and colleagues [165] found that a similar group of FM women abused in childhood had disordered diurnal cortisol levels.

However, studies based on the self-report of FM patients can be difficult to verify. Adult life experiences may bias memories from early life, and it is possible that FM patients, due to their current pain and comorbid states, become self-centred and preoccupied with the pain, and potentially overemphasise traumatic memories [166, 167]. Longitudinal studies following children and young people who have suffered from documented abuse, for example, children in social care, would be useful in order to avoid this confound. Epidemiological studies of socioeconomic position (SEP) during childhood have shown that lower SEP during childhood is a predictor of chronic widespread pain in later life [168]. Whilst socioeconomic status incorporates a large number of factors, trauma-related hospitalisations are more prevalent in children from lower socioeconomic statuses who have less access to appropriate healthcare [169–172].

14.6 RISK FACTOR: PERINATAL EXPOSURE TO SUBSTANCES OF ABUSE

Serotonin is not the only neurotransmitter system altered in fibromyalgia—dopamine and opioid neurotransmitter disturbances are also reported that may, in part, result from interference with these developing systems during pre- or early postnatal life. Dopamine responses of FM patients to painful stimuli are lower than those of healthy subjects [173], and low CSF levels of dopamine (as well as serotonin and norepinephrine) metabolites are seen in patients [52]. Furthermore, the opioid system is closely linked to dopaminergic signalling, constitutes an important endogenous antinociceptive system, and is altered in FM [174, 175]. Exposure to a number of substances during early life will impact upon the development of these neurotransmitter systems. For example, exposure of the foetus to alcohol induces dysfunction in dopaminergic and serotonergic systems and will persistently affect the development of the HPA axis [176–178], all of which are, as mentioned, disturbed in FM.

14.6.1 EARLY OPIOID EXPOSURE CAUSES LONG-LASTING CHANGES IN NOCICEPTIVE SYSTEMS

Early exposure to opioids, such as exposure of the foetus to heroin or methadone during pregnancy, or prolonged morphine administration after birth, for example, on the NICU, causes lasting changes in opioid signalling. If exposure occurs prenatally, it can result in the newly born infant undergoing withdrawal symptoms very soon after birth, the so-called neonatal abstinence syndrome (NAS) [179]—an effect mirrored in animal models of early drug dependence [180]. Morphine is used in the NICU to provide sedation in neonates requiring mechanical ventilation and to improve tolerance of ventilation and comfort of the infant. The lasting effects of this are not yet well-characterised [65, 181], although it is known that chronic morphine administration causes changes in mu-opioid receptor density and sensitivity that desensitises older animals to opiate analgesia [182, 183]. As the opioidergic and dopaminergic systems are crucial for endogenous modulation of pain, and this modulation seems to be disordered in fibromyalgia, longitudinal studies of early life exposure to opioids may help explain some of the disturbances in opioid function seen in FM. Positron emission tomography (PET) studies would be useful to investigate if the altered opioid receptor activity seen in FM patients is related to early life opioid exposure. Harris and colleagues [175, 184] have shown that FM patients show alterations in mu-opioid receptor availability, but a history of opioid use was one of the exclusion criterion for these studies, meaning that the impact of early life exposure to opioids is not yet known in terms of later FM prevalence.

14.6.2 DOPAMINE OVEREXPOSURE DURING EARLY LIFE AFFECTS HPA AXIS FUNCTION, AND MAY BE MEDIATED BY EPIGENETIC FACTORS

Dopaminergic drugs such as cocaine and amphetamine increase anxiety-like behaviours in animal models of abuse during pregnancy [185], and cocaine activates the HPA axis by potentiating adrenocorticotropin hormone (ACTH) release, which in turn stimulates the adrenal glands to produce adrenaline and cortisol [186]. Increasing evidence now suggests that prenatal exposure to drugs of abuse (including alcohol, cocaine, and amphetamine) may be toxic to developing dopamine-rich areas such as the basal ganglia (see [187]). After birth, studies on the development of the dopamine system suggest that functional connectivity in the young animal, particularly to frontal cortical areas, is not mature until adolescence or later [188, 189], and it is possible that the balance between tonic and phasic dopamine release will also be affected by exposure of the developing system to excessive dopamine activation, impacting upon later pain control mechanisms [190]. Therefore, early life exposure to increased levels of dopamine could help explain the aberrant dopaminergic functionality seen in FM patients [173, 191]. Interestingly, genetic polymorphisms linked to FM include the val158met polymorphism (in the gene coding for catechol-o-methyltranferase, an enzyme that metabolises dopamine) [192] and the dopamine D4 receptor [193]. Considering that epigenetic effects impact upon stress regulation in maternally deprived animals, it is possible to speculate that early life events cause epigenetic changes, which may interact with the above polymorphisms to produce an adult phenotype at increased risk of developing FM. Indeed, self-report studies of FM patients show higher levels of reported parental drug and alcohol abuse in comparison to other groups of other chronic pain patients [194, 195].

14.7 CONCLUSIONS

Fibromyalgia is classified as a disorder of pain processing and stress regulation, often along with comorbidities of anxiety and depression; the amount of evidence that links early life adversity to pain, stress, and emotional problems in later life strongly suggests that these adverse events and traumas could increase the risk of developing FM in adulthood. At this point, research has shown that FM patients self-report high levels of early adversity and indicates that early life circumstances affecting later pain processing and stress regulation may be more prevalent in FM. However, there is little evidence to conclusively link the two, for example, by proving that maternal deprivation or increased stress during pregnancy increases the incidence of FM.

This paper has focused on the evidence showing that painful procedures during early life cause long-lasting changes in pain processing and suggests that high exposure to painful experiences in early life may partially explain the increased pain sensitivity shown by FM patients. In addition, childhood adversities and maternal deprivation are also discussed in terms of the effects they have on stress-regulatory systems in the older organism, so again are potentially sources of the disorders in cortisol levels and stress response seen in FM patients. As previously mentioned, the precise dysfunctions in stress regulation via the HPA axis are not yet clear in FM patients, but what is clear is that the body's stress regulatory systems are compromised, and it is possible that the precise nature and combination of each individual's childhood experience may be contributing to the overall symptomatology of FM.

Exposure of the developing brain to perinatal stress and glucocorticoids during critical periods of development may affect the long-term function of areas involved in stress regulation such as the hippocampus and amygdala and help explain the "fibrofog" and anxiety disorders prevalent in FM. Furthermore, impairments in stress regulation caused by early exposure to stressors such as increased maternal cortisol levels, pain, or maternal deprivation may also partially explain the increased pain sensitivities seen in FM patients. Finally, as pain is itself a stressor, the pain experienced by FM patients may be acting in a positive feedback manner to further increase anxiety levels and impact upon stress regulation. Whilst FM is unlikely to be due to a single factor, it is possible that the factors outlined in this paper, concomitant with a number of other factors such as a genetic predisposition to enhanced pain sensitivity, stressful life experiences in adult life, and the influence of sex hormones, may combine or interact to create a phenotype at higher risk of developing this form of chronic pain. Teasing apart potential influences and their mechanisms may help treat sufferers or, in fact, decrease the risk of future suffering.

REFERENCES

1. F. Wolfe, K. Ross, J. Anderson, I. J. Russell, and L. Hebert, "The prevalence and characteristics of fibromyalgia in the general population," Arthritis and Rheumatism, vol. 38, no. 1, pp. 19–28, 1995.
2. D. J. Clauw and L. J. Crofford, "Chronic widespread pain and fibromyalgia: what we know, and what we need to know," Best Practice and Research, vol. 17, no. 4, pp. 685–701, 2003.
3. M. Fitzgerald, "The development of nociceptive circuits," Nature Reviews Neuroscience, vol. 6, no. 7, pp. 507–520, 2005.
4. M. Fitzgerald and S. M. Walker, "Infant pain management: a developmental neurobiological approach," Nature Clinical Practice Neurology, vol. 5, no. 1, pp. 35–50, 2009.
5. J. L. La Prairie and A. Z. Murphy, "Long-term impact of neonatal injury in male and female rats: sex differences, mechanisms and clinical implications," Frontiers in Neuroendocrinology, vol. 31, no. 2, pp. 193–202, 2010.
6. R. F. Howard, "Current status of pain management in children," Journal of the American Medical Association, vol. 290, no. 18, pp. 2464–2469, 2003.
7. S. M. Walker, "Pain in children: recent advances and ongoing challenges," British Journal of Anaesthesia, vol. 101, no. 1, pp. 101–110, 2008.
8. J. Langhoff-Roos, U. Kesmodel, B. Jacobsson, S. Rasmussen, and I. Vogel, "Spontaneous preterm delivery in primiparous women at low risk in Denmark: population based study," British Medical Journal, vol. 332, no. 7547, pp. 937–939, 2006.
9. S. H. P. Simons, M. Van Dijk, K. S. Anand, D. Roofthooft, R. A. Van Lingen, and D. Tibboel, "Do we still hurt newborn babies? A prospective study of procedural pain

and analgesia in neonates," Archives of Pediatrics and Adolescent Medicine, vol. 157, no. 11, pp. 1058–1064, 2003.

10. R. Carbajal, A. Rousset, C. Danan et al., "Epidemiology and treatment of painful procedures in neonates in intensive care units," Journal of the American Medical Association, vol. 300, no. 1, pp. 60–70, 2008.

11. S. M. Walker, L. S. Franck, M. Fitzgerald, J. Myles, J. Stocks, and N. Marlow, "Long-term impact of neonatal intensive care and surgery on somatosensory perception in children born extremely preterm," Pain, vol. 141, no. 1-2, pp. 79–87, 2009.

12. B. M. Schmelzle-Lubiecki, K. A. A. Campbell, R. H. Howard, L. Franck, and M. Fitzgerald, "Long-term consequences of early infant injury and trauma upon somatosensory processing," European Journal of Pain, vol. 11, no. 7, pp. 799–809, 2007.

13. T. F. Oberlander, R. E. Grunau, M. F. Whitfield, C. Fitzgerald, S. Pitfield, and J. P. Saul, "Biobehavioral pain responses in former extremely low birth weight infants at four months' corrected age," Pediatrics, vol. 105, no. 1, article e6, 2000.

14. R. V. E. Grunau, M. F. Whitfield, and J. H. Petrie, "Pain sensitivity and temperament in extremely low-birth-weight premature toddlers and preterm and full-term controls," Pain, vol. 58, no. 3, pp. 341–346, 1994.

15. C. Hermann, J. Hohmeister, S. Demirakça, K. Zohsel, and H. Flor, "Long-term alteration of pain sensitivity in school-aged children with early pain experiences," Pain, vol. 125, no. 3, pp. 278–285, 2006.

16. J. Hohmeister, S. Demirakça, K. Zohsel, H. Flor, and C. Hermann, "Responses to pain in school-aged children with experience in a neonatal intensive care unit: cognitive aspects and maternal influences," European Journal of Pain, vol. 13, no. 1, pp. 94–101, 2009.

17. A. Boehm, E. Eisenberg, and S. Lampel, "The contribution of social capital and coping strategies to functioning and quality of life of patients with fibromyalgia," The Clinical Journal of Pain, vol. 27, pp. 233–239, 2011.

18. H. M. Abdulkader, Y. Freer, E. M. Garry, S. M. Fleetwood-Walker, and N. McIntosh, "Prematurity and neonatal noxious events exert lasting effects on infant pain behaviour," Early Human Development, vol. 84, no. 6, pp. 351–355, 2008.

19. J. W. B. Peters, R. Schouw, K. J. S. Anand, M. Van Dijk, H. J. Duivenvoorden, and D. Tibboel, "Does neonatal surgery lead to increased pain sensitivity in later childhood?" Pain, vol. 114, no. 3, pp. 444–454, 2005.

20. I. Wollgarten-Hadamek, J. Hohmeister, S. Demirakça, K. Zohsel, H. Flor, and C. Hermann, "Do burn injuries during infancy affect pain and sensory sensitivity in later childhood?" Pain, vol. 141, no. 1-2, pp. 165–172, 2009.

21. D. Buskila, L. Neumann, E. Zmora, M. Feldman, A. Bolotin, and J. Press, "Pain sensitivity in prematurely born adolescents," Archives of Pediatrics and Adolescent Medicine, vol. 157, no. 11, pp. 1079–1082, 2003.

22. M. M. Heinricher and K. Drasner, "Lumbar intrathecal morphine alters activity of putative nociceptive modulatory neurons in rostral ventromedial medulla," Brain Research, vol. 549, no. 2, pp. 338–341, 1991.

23. H. L. Fields and A. I. Basbaum, "Brainstem control of spinal pain-transmission neurons," Annual Review of Physiology, vol. 40, pp. 217–248, 1978.

24. A. I. Basbaum and H. L. Fields, "Endogenous pain control mechanisms: review and hypothesis," Annals of Neurology, vol. 4, no. 5, pp. 451–462, 1978.

25. M. M. Heinricher, N. M. Barbaro, and H. L. Fields, "Putative nociceptive modulating neurons in the rostral ventromedial medulla of the rat: firing of on- and off-cells is related to nociceptive responsiveness," Somatosensory and Motor Research, vol. 6, no. 4, pp. 427–439, 1989.

26. T. Boucher, E. Jennings, and M. Fitzgerald, "The onset of diffuse noxious inhibitory controls in postnatal rat pups: a c-fos study," Neuroscience Letters, vol. 257, no. 1, pp. 9–12, 1998.

27. H. Van Praag and H. Frenk, "The development of stimulation-produced analgesia (SPA) in the rat," Developmental Brain Research, vol. 64, no. 1-2, pp. 71–76, 1991.

28. G. J. Hathway, S. Koch, L. Low, and M. Fitzgerald, "The changing balance of brainstem-spinal cord modulation of pain processing over the first weeks of rat postnatal life," Journal of Physiology, vol. 587, no. 12, pp. 2927–2935, 2009.

29. B. Clancy, B. L. Finlay, R. B. Darlington, and K. J. S. Anand, "Extrapolating brain development from experimental species to humans," NeuroToxicology, vol. 28, no. 5, pp. 931–937, 2007.

30. B. Clancy, B. Kersh, J. Hyde, R. B. Darlington, K. J. S. Anand, and B. L. Finlay, "Web-based method for translating neurodevelopment from laboratory species to humans," Neuroinformatics, vol. 5, no. 1, pp. 79–94, 2007.

31. J. L. La Prairie and A. Z. Murphy, "Female rats are more vulnerable to the long-term consequences of neonatal inflammatory injury," Pain, vol. 132, supplement 1, pp. S124–S133, 2007.

32. J. L. La Prairie and A. Z. Murphy, "Neonatal injury alters adult pain sensitivity by increasing opioid tone in the periaqueductal gray," Frontiers in Behavioral Neuroscience, vol. 3, article 31, 2009.

33. A. J. McDermid, G. B. Rollman, and G. A. McCain, "Generalized hypervigilance in fibromyalgia: evidence of perceptual amplification," Pain, vol. 66, no. 2-3, pp. 133–144, 1996.

34. R. Staud, C. J. Vierck, R. L. Cannon, A. P. Mauderli, and D. D. Price, "Abnormal sensitization and temporal summation of second pain (wind-up) in patients with fibromyalgia syndrome," Pain, vol. 91, no. 1-2, pp. 165–175, 2001.

35. R. H. Gracely, F. Petzke, J. M. Wolf, and D. J. Clauw, "Functional magnetic resonance imaging evidence of augmented pain processing in fibromyalgia," Arthritis and Rheumatism, vol. 46, no. 5, pp. 1333–1343, 2002.

36. F. Petzke, D. J. Clauw, K. Ambrose, A. Khine, and R. H. Gracely, "Increased pain sensitivity in fibromyalgia: effects of stimulus type and mode of presentation," Pain, vol. 105, no. 3, pp. 403–413, 2003.

37. K. J. S. Anand, V. Coskun, K. V. Thrivikraman, C. B. Nemeroff, and P. M. Plotsky, "Long-term behavioral effects of repetitive pain in neonatal rat pups," Physiology and Behavior, vol. 66, no. 4, pp. 627–637, 1999.

38. S. M. Walker, J. Meredith-Middleton, C. Cooke-Yarborough, and M. Fitzgerald, "Neonatal inflammation and primary afferent terminal plasticity in the rat dorsal horn," Pain, vol. 105, no. 1-2, pp. 185–195, 2003.

39. M. S. Lidow, Z. M. Song, and K. Ren, "Long-term effects of short-lasting early local inflammatory insult," NeuroReport, vol. 12, no. 2, pp. 399–403, 2001.

40. M. A. Ruda, Q. D. Ling, A. G. Hohmann, Y. B. Peng, and T. Tachibana, "Altered nociceptive neuronal circuits after neonatal peripheral inflammation," Science, vol. 289, no. 5479, pp. 628–630, 2000.

41. S. M. Walker, J. Meredith-Middleton, C. Cooke-Yarborough, and M. Fitzgerald, "Neonatal inflammation and primary afferent terminal plasticity in the rat dorsal horn," Pain, vol. 105, no. 1-2, pp. 185–195, 2003.

42. K. Ren, V. Anseloni, S. P. Zou et al., "Characterization of basal and re-inflammation-associated long-term alteration in pain responsivity following short-lasting neonatal local inflamatory insult," Pain, vol. 110, no. 3, pp. 588–596, 2004.

43. M. L. Reynolds and M. Fitzgerald, "Long-term sensory hyperinnervation following neonatal skin wounds," Journal of Comparative Neurology, vol. 358, no. 4, pp. 487–498, 1995.

44. C. Torsney and M. Fitsgerald, "Spinal dorsal horn cell receptive field size is increased in adult rats following neonatal hindpaw skin injury," Journal of Physiology, vol. 550, no. 1, pp. 255–261, 2003.

45. J. De Lima, D. Alvares, D. J. Hatch, and M. Fitzgerald, "Sensory hyperinnervation after neonatal skin wounding: effect of bupivacaine sciatic nerve block," British Journal of Anaesthesia, vol. 83, no. 4, pp. 662–664, 1999.

46. A. Taddio, J. Katz, A. L. Ilersich, and G. Koren, "Effect of neonatal circumcision on pain response during subsequent routine vaccination," The Lancet, vol. 349, no. 9052, pp. 599–603, 1997.

47. K. Andrews and M. Fitzgerald, "Wound sensitivity as a measure of analgesic effects following surgery in human neonates and infants," Pain, vol. 99, no. 1-2, pp. 185–195, 2002.

48. F. Wolfe, K. Ross, J. Anderson, and I. J. Russell, "Aspects of fibromyalgia in the general population: sex, pain threshold, and fibromyalgia symptoms," Journal of Rheumatology, vol. 22, no. 1, pp. 151–156, 1995.

49. N. Julien, P. Goffaux, P. Arsenault, and S. Marchand, "Widespread pain in fibromyalgia is related to a deficit of endogenous pain inhibition," Pain, vol. 114, no. 1-2, pp. 295–302, 2005.

50. E. Normand, S. Potvin, I. Gaumond, G. Cloutier, J.-F. Corbin, and S. Marchand, "Pain inhibition is deficient in chronic widespread pain but normal in major depressive disorder," Journal of Clinical Psychiatry, vol. 72, no. 2, pp. 219–224, 2011. View at Publisher · View at Google Scholar

51. I. J. Russell, J. E. Michalek, G. A. Vipraio, E. M. Fletcher, M. A. Javors, and C. A. Bowden, "Platelet 3H-imipramine uptake receptor density and serum serotonin levels in patients with fibromyalgia/fibrositis syndrome," Journal of Rheumatology, vol. 19, no. 1, pp. 104–109, 1992.

52. I. J. Russell, H. Vaeroy, M. Javors, and F. Nyberg, "Cerebrospinal fluid biogenic amine metabolites in fibromyalgia/fibrositis syndrome and rheumatoid arthritis," Arthritis and Rheumatism, vol. 35, no. 5, pp. 550–556, 1992.

53. E. Legangneux, J. J. Mora, O. Spreux-Varoquaux et al., "Cerebrospinal fluid biogenic amine metabolites, plasma-rich platelet serotonin and [3H]imipramine reuptake in the primary fibromyalgia syndrome," Rheumatology, vol. 40, no. 3, pp. 290–296, 2001.

54. K. C. Light, E. E. Bragdon, K. M. Grewen, K. A. Brownley, S. S. Girdler, and W. Maixner, "Adrenergic dysregulation and pain with and without acute beta-blockade in women with fibromyalgia and temporomandibular disorder," Journal of Pain, vol. 10, no. 5, pp. 542–552, 2009.

55. A. H. Dickenson, J. P. Rivot, and A. Chaouch, "Diffuse noxious inhibitory controls (DNIC) in the rat with or without pCPA pretreatment," Brain Research, vol. 216, no. 2, pp. 313–321, 1981.

56. N. El-Yassir and S. M. Fleetwood-Walker, "A 5-HT1-type receptor mediates the antinociceptive effect of nucleus raphe magnus stimulation in the rat," Brain Research, vol. 523, no. 1, pp. 92–99, 1990.

57. T. J. Grudt, J. T. Williams, and R. A. Travagli, "Inhibition by 5-hydroxytryptamine and noradrenaline in substantia gelatinosa of guinea-pig spinal trigeminal nucleus," Journal of Physiology, vol. 485, no. 1, pp. 113–120, 1995.

58. R. Suzuki, S. Morcuende, M. Webber, S. P. Hunt, and A. H. Dickenson, "Superficial NK1-expressing neurons control spinal excitability through activation of descending pathways," Nature Neuroscience, vol. 5, no. 12, pp. 1319–1326, 2002.

59. W. Rahman, R. Suzuki, M. Webber, S. P. Hunt, and A. H. Dickenson, "Depletion of endogenous spinal 5-HT attenuates the behavioural hypersensitivity to mechanical and cooling stimuli induced by spinal nerve ligation," Pain, vol. 123, no. 3, pp. 264–274, 2006.

60. A. Pertovaara, "Noradrenergic pain modulation," Progress in Neurobiology, vol. 80, no. 2, pp. 53–83, 2006.

61. S. M. Walker, K. K. Tochiki, and M. Fitzgerald, "Hindpaw incision in early life increases the hyperalgesic response to repeat surgical injury: critical period and dependence on initial afferent activity," Pain, vol. 147, no. 1–3, pp. 99–106, 2009.

62. E. Walter-Nicolet, D. Annequin, V. Biran, D. Mitanchez, and B. Tourniaire, "Pain management in newborns: from prevention to treatment," Pediatric Drugs, vol. 12, no. 6, pp. 353–365, 2010.

63. S. Gibbins, B. Stevens, E. Hodnett, J. Pinelli, A. Ohlsson, and G. Darlington, "Efficacy and safety of sucrose for procedural pain relief in preterm and term neonates," Nursing Research, vol. 51, no. 6, pp. 375–382, 2002.

64. R. Slater, L. Cornelissen, L. Fabrizi et al., "Oral sucrose as an analgesic drug for procedural pain in newborn infants: a randomised controlled trial," The Lancet, vol. 376, no. 9748, pp. 1225–1232, 2010.

65. R. Bellù, K. A. de Waal, and R. Zanini, "Opioids for neonates receiving mechanical ventilation," Cochrane Database of Systematic Reviews, no. 1, Article ID CD004212, 2008.

66. R. E. Grunau, L. Holsti, and J. W. B. Peters, "Long-term consequences of pain in human neonates," Seminars in Fetal and Neonatal Medicine, vol. 11, no. 4, pp. 268–275, 2006.

67. L. Holsti, J. Weinberg, M. F. Whitfield, and R. E. Grunau, "Relationships between adrenocorticotropic hormone and cortisol are altered during clustered nursing care in preterm infants born at extremely low gestational age," Early Human Development, vol. 83, no. 5, pp. 341–348, 2007.

68. G. A. McCain and K. S. Tilbe, "Diurnal hormone variation in fibromyalgia syndrome: a comparison with rheumatoid arthritis," Journal of Rheumatology, vol. 16, supplement 19, pp. 154–157, 1989.

69. E. N. Griep, J. W. Boersma, and E. R. De Kloet, "Altered reactivity of the hypothalamic-pituitary-adrenal axis in the primary fibromyalgia syndrome," Journal of Rheumatology, vol. 20, no. 3, pp. 469–474, 1993.

70. L. J. Crofford, S. R. Pillemer, K. T. Kalogeras et al., "Hypothalamic-pituitary-adrenal axis perturbations in patients with fibromyalgia," Arthritis and Rheumatism, vol. 37, no. 11, pp. 1583–1592, 1994.

71. D. Catley, A. T. Kaell, C. Kirschbaum, and A. A. Stone, "A naturalistic evaluation of cortisol secretion in persons with fibromyalgia and rheumatoid arthritis," Arthritis Care and Research, vol. 13, no. 1, pp. 51–61, 2000.

72. E. Fries, J. Hesse, J. Hellhammer, and D. H. Hellhammer, "A new view on hypocortisolism," Psychoneuroendocrinology, vol. 30, no. 10, pp. 1010–1016, 2005.

73. L. J. Crofford, E. A. Young, N. C. Engleberg et al., "Basal circadian and pulsatile ACTH and cortisol secretion in patients with fibromyalgia and/or chronic fatigue syndrome," Brain, Behavior, and Immunity, vol. 18, no. 4, pp. 314–325, 2004.

74. B. S. McEwen and E. Stellar, "Stress and the individual: mechanisms leading to disease," Archives of Internal Medicine, vol. 153, no. 18, pp. 2093–2101, 1993.

75. E. F. Fernandez, R. Montman, and K. L. Watterberg, "ACTH and cortisol response to critical illness in term and late preterm newborns," Journal of Perinatology, vol. 28, no. 12, pp. 797–802, 2008.

76. C. E. Hanna, L. D. Keith, M. A. Colasurdo et al., "Hypothalamic pituitary adrenal function in the extremely low birth weight infant," Journal of Clinical Endocrinology and Metabolism, vol. 76, no. 2, pp. 384–387, 1993.

77. R. E. Grunau, L. Holsti, D. W. Haley et al., "Neonatal procedural pain exposure predicts lower cortisol and behavioral reactivity in preterm infants in the NICU," Pain, vol. 113, no. 3, pp. 293–300, 2005.

78. R. E. Grunau, M. T. Tu, M. F. Whitfield et al., "Cortisol, behavior, and heart rate reactivity to immunization pain at 4 months corrected age in infants born very preterm," Clinical Journal of Pain, vol. 26, no. 8, pp. 698–704, 2010.

79. R. E. Grunau, J. Weinberg, and M. F. Whitfield, "Neonatal procedural pain and preterm infant cortisol response to novelty at 8 months," Pediatrics, vol. 114, no. 1, pp. e77–e84, 2004.

80. R. E. Grunau, D. W. Haley, M. F. Whitfield, J. Weinberg, W. Yu, and P. Thiessen, "Altered basal cortisol levels at 3, 6, 8 and 18 months in infants born at extremely low gestational age," Journal of Pediatrics, vol. 150, no. 2, pp. 151–156, 2007.

81. S. Brummelte, R. E. Grunau, A. Zaidman-Zait, J. Weinberg, D. Nordstokke, and I. L. Cepeda, "Cortisol levels in relation to maternal interaction and child internalizing behavior in preterm and full-term children at 18 months corrected age," Developmental Psychobiology, vol. 53, no. 2, pp. 184–195, 2011. View at Publisher · View at Google Scholar

82. A. Buske-Kirschbaum, S. Krieger, C. Wilkes, W. Rauh, S. Weiss, and D. H. Hellhammer, "Hypothalamic-pituitary-adrenal axis function and the cellular immune response in former preterm children," Journal of Clinical Endocrinology and Metabolism, vol. 92, no. 9, pp. 3429–3435, 2007.

83. L. A. M. Welberg and J. R. Seckl, "Prenatal stress, glucocorticoids and the programming of the brain," Journal of Neuroendocrinology, vol. 13, no. 2, pp. 113–128, 2001.

84. I. Zohar and M. Weinstock, "Differential effect of prenatal stress on the expression of cortiocotrophin-releasing hormone and its receptors in the hypothalamus and amygdala in male and female rats," Journal of Neuroendocrinology, vol. 23, no. 4, pp. 320–328, 2011. View at Publisher · View at Google Scholar

85. R. Karlsson, J. Kallio, J. Toppari, M. Scheinin, and P. Kero, "Antenatal and early postnatal dexamethasone treatment decreases cortisol secretion in preterm infants," Hormone Research, vol. 53, no. 4, pp. 170–176, 2000.

86. E. P. Davis, F. Waffarn, and C. A. Sandman, "Prenatal treatment with glucocorticoids sensitizes the hpa axis response to stress among full-term infants," Developmental Psychobiology, vol. 53, no. 2, pp. 175–183, 2011. View at Publisher · View at Google Scholar

87. P. O. Klingmann, I. Kugler, T. S. Steffke, S. Bellingrath, B. M. Kudielka, and D. H. Hellhammer, "Sex-specific prenatal programming: a risk for fibromyalgia?" Annals of the New York Academy of Sciences, vol. 1148, pp. 446–455, 2008.

88. S. T. DeKosky, A. J. Nonneman, and S. W. Scheff, "Morphologic and behavioral effects of perinatal glucocorticoid administration," Physiology and Behavior, vol. 29, no. 5, pp. 895–900, 1982.

89. J. P. Vicedomini, A. J. Nonneman, S. T. DeKosky, and S. W. Scheff, "Perinatal glucocorticoids disrupt learning: a sexually dimorphic response," Physiology and Behavior, vol. 36, no. 1, pp. 145–149, 1986.

90. M. Fameli, E. Kitraki, and F. Stylianopoulou, "Effects of hyperactivity of the maternal hypothalamic-pituitary-adrenal (HPA) axis during pregnancy on the development of the HPA axis and brain monoamines of the offspring," International Journal of Developmental Neuroscience, vol. 12, no. 7, pp. 651–659, 1994.

91. K. Muneoka, M. Mikuni, T. Ogawa et al., "Prenatal dexamethasone exposure alters brain monoamine metabolism and adrenocortical response in rat offspring," American Journal of Physiology, vol. 273, no. 5, pp. R1669–R1675, 1997.

92. S. A. Ferguson and R. R. Holson, "Neonatal dexamethasone on day 7 causes mild hyperactivity and cerebellar stunting," Neurotoxicology and Teratology, vol. 21, no. 1, pp. 71–76, 1999.

93. S. A. Ferguson, M. G. Paule, and R. R. Holson, "Neonatal dexamethasone on Day 7 in rats causes behavioral alterations reflective of hippocampal, but not cerebellar, deficits," Neurotoxicology and Teratology, vol. 23, no. 1, pp. 57–69, 2001.

94. J. M. Glass, "Fibromyalgia and cognition," Journal of Clinical Psychiatry, vol. 69, supplement 2, pp. 20–24, 2008.

95. A. Kuchinad, P. Schweinhardt, D. A. Seminowicz, P. B. Wood, B. A. Chizh, and M. C. Bushnell, "Accelerated brain gray matter loss in fibromyalgia patients: premature aging of the brain?" Journal of Neuroscience, vol. 27, no. 15, pp. 4004–4007, 2007.

96. S. Johnson, C. Hollis, P. Kochhar, E. Hennessy, D. Wolke, and N. Marlow, "Psychiatric disorders in extremely preterm children: longitudinal finding at age 11 years in the EPICure study," Journal of the American Academy of Child and Adolescent Psychiatry, vol. 49, no. 5, pp. 453–463, 2010.

97. S. Johnson, E. Hennessy, R. Smith, R. Trikic, D. Wolke, and N. Marlow, "Academic attainment and special educational needs in extremely preterm children at 11 years of age: the EPICure study," Archives of Disease in Childhood, vol. 94, no. 4, pp. F283–F289, 2009.

98. A. T. Bhutta, M. A. Cleves, P. H. Casey, M. M. Cradock, and K. J. S. Anand, "Cognitive and behavioral outcomes of school-aged children who were born preterm: a meta-analysis," Journal of the American Medical Association, vol. 288, no. 6, pp. 728–737, 2002.

99. M. S. Indredavik, T. Vik, S. Heyerdahl, S. Kulseng, P. Fayers, and A. M. Brubakk, "Psychiatric symptoms and disorders in adolescents with low birth weight," Archives of Disease in Childhood, vol. 89, no. 5, pp. F445–F450, 2004.

100. S. V. Desai, T. J. Law, and D. M. Needham, "Long-term complications of critical care," Critical Care Medicine, vol. 39, no. 2, pp. 371–379, 2011. View at Publisher · View at Google Scholar

101. M. Ajayi-Obe, N. Saeed, F. M. Cowan, M. A. Rutherford, and A. D. Edwards, "Reduced development of cerebral cortex in extremely proterm infants," The Lancet, vol. 356, no. 9236, pp. 1162–1163, 2000.

102. L. E. Dyet, N. Kennea, S. J. Counsell et al., "Natural history of brain lesions in extremely preterm infants studied with serial magnetic resonance imaging from birth and neurodevelopmental assessment," Pediatrics, vol. 118, no. 2, pp. 536–548, 2006.

103. J. P. Boardman, S. J. Counsell, D. Rueckert et al., "Abnormal deep grey matter development following preterm birth detected using deformation-based morphometry," NeuroImage, vol. 32, no. 1, pp. 70–78, 2006.

104. S. J. Counsell, L. E. Dyet, D. J. Larkman et al., "Thalamo-cortical connectivity in children born preterm mapped using probabilistic magnetic resonance tractography," NeuroImage, vol. 34, no. 3, pp. 896–904, 2007.

105. J. I. Berman, P. Mukherjee, S. C. Partridge et al., "Quantitative diffusion tensor MRI fiber tractography of sensorimotor white matter development in premature infants," NeuroImage, vol. 27, no. 4, pp. 862–871, 2005.

106. P. G. Croskerry, G. K. Smith, and M. Leon, "Thermoregulation and the maternal behaviour of the rat," Nature, vol. 273, no. 5660, pp. 299–300, 1978.

107. P. M. Plotsky and M. J. Meaney, "Early, postnatal experience alters hypothalamic corticotropin-releasing factor (CRF) mRNA, median eminence CRF content and stress-induced release in adult rats," Molecular Brain Research, vol. 18, no. 3, pp. 195–200, 1993.

108. P. M. Plotsky, K. V. Thrivikraman, C. B. Nemeroff, C. Caldji, S. Sharma, and M. J. Meaney, "Long-term consequences of neonatal rearing on central corticotropin- releasing factor systems in adult male rat offspring," Neuropsychopharmacology, vol. 30, no. 12, pp. 2192–2204, 2005.

109. C. Caldji, D. Francis, S. Sharma, P. M. Plotsky, and M. J. Meaney, "The effects of early rearing environment on the development of GABA(A) and central benzodiazepine receptor levels and novelty-induced fearfulness in the rat," Neuropsychopharmacology, vol. 22, no. 3, pp. 219–229, 2000.

110. R. L. Huot, K. V. Thrivikraman, M. J. Meaney, and P. M. Plotsky, "Development of adult ethanol preference and anxiety as a consequence of neonatal maternal separa-

tion in Long Evans rats and reversal with antidepressant treatment," Psychopharmacology, vol. 158, no. 4, pp. 366–373, 2001.

111. W. M. U. Daniels, C. Y. Pietersen, M. E. Carstens, and D. J. Stein, "Maternal separation in rats leads to anxiety-like behavior and a blunted ACTH response and altered neurotransmitter levels in response to a subsequent stressor," Metabolic Brain Disease, vol. 19, no. 1-2, pp. 3–14, 2004.

112. K. Matthews and T. W. Robbins, "Early experience as a determinant of adult behavioural responses to reward: the effects of repeated maternal separation in the rat," Neuroscience and Biobehavioral Reviews, vol. 27, no. 1-2, pp. 45–55, 2003.

113. B. A. Ellenbroek and M. A. Riva, "Early maternal deprivation as an animal model for schizophrenia," Clinical Neuroscience Research, vol. 3, no. 4-5, pp. 297–302, 2003.

114. M. L. Uhelski and P. N. Fuchs, "Maternal separation stress leads to enhanced emotional responses to noxious stimuli in adult rats," Behavioural Brain Research, vol. 212, no. 2, pp. 208–212, 2010.

115. C. Fahlke, J. G. Lorenz, J. Long, M. Champoux, S. J. Suomi, and J. D. Higley, "Rearing experiences and stress-induced plasma cortisol as early risk factors for excessive alcohol consumption in nonhuman primates," Alcoholism, vol. 24, no. 5, pp. 644–650, 2000.

116. L. A. Bradley, "Pathophysiology of fibromyalgia," American Journal of Medicine, vol. 122, no. 12, pp. S22–S30, 2009.

117. M. J. Meaney, D. H. Aitken, and S. R. Bodnoff, "The effects of postnatal handling on the development of the glucocorticoid receptor systems and stress recovery in the rat," Progress in Neuro-Psychopharmacology and Biological Psychiatry, vol. 9, no. 5-6, pp. 731–734, 1985.

118. M. J. Meaney, D. H. Aitken, C. Van Berkel, S. Bhatnagar, and R. M. Sapolsky, "Effect of neonatal handling on age-related impairments associated with the hippocampus," Science, vol. 239, no. 4841, pp. 766–768, 1988.

119. M. J. Meaney, D. H. Aitken, S. Bhatnagar, and R. M. Sapolsky, "Postnatal handling attenuates certain neuroendocrine, anatomical, and cognitive dysfunctions associated with aging in female rats," Neurobiology of Aging, vol. 12, no. 1, pp. 31–38, 1991.

120. C. R. Pryce, D. Bettschen, N. I. Bahr, and J. Feldon, "Comparison of the effects of infant handling, isolation, and nonhandling on acoustic startle, prepulse inhibition, locomotion, and HPA activity in the adult rat," Behavioral Neuroscience, vol. 115, no. 1, pp. 71–83, 2001.

121. C. O. Ladd, R. L. Huot, K. V. Thrivikraman, C. B. Nemeroff, M. J. Meaney, and P. M. Plotsky, "Long-term behavioral and neuroendocrine adaptations to adverse early experience," Progress in Brain Research, vol. 122, pp. 81–103, 2000.

122. D. Liu, J. Diorio, B. Tannenbaum et al., "Maternal care, hippocampal glucocorticoid receptors, and hypothalamic- pituitary-adrenal responses to stress," Science, vol. 277, no. 5332, pp. 1659–1662, 1997.

123. C. R. Pryce, D. Bettschen, and J. Feldon, "Comparison of the effects of early handling and early deprivation on maternal care in the rat," Developmental Psychobiology, vol. 38, no. 4, pp. 239–251, 2001.

124. C. Caldji, B. Tannenbaum, S. Sharma, D. Francis, P. M. Plotsky, and M. J. Meaney, "Maternal care during infancy regulates the development of neural systems mediating the expression of fearfulness in the rat," Proceedings of the National Academy of Sciences of the United States of America, vol. 95, no. 9, pp. 5335–5340, 1998.

125. D. Francis, J. Diorio, D. Liu, and M. J. Meaney, "Nongenomic transmission across generations of maternal behavior and stress responses in the rat," Science, vol. 286, no. 5442, pp. 1155–1158, 1999.

126. F. A. Champagne and M. J. Meaney, "Transgenerational effects of social environment on variations in maternal care and behavioral response to novelty," Behavioral Neuroscience, vol. 121, no. 6, pp. 1353–1363, 2007.

127. D. Liu, J. Diorio, J. C. Day, D. D. Francis, and M. J. Meaney, "Maternal care, hippocampal synaptogenesis and cognitive development in rats," Nature Neuroscience, vol. 3, no. 8, pp. 799–806, 2000.

128. G. Rentesi, K. Antoniou, M. Marselos, A. Fotopoulos, J. Alboycharali, and M. Konstandi, "Long-term consequences of early maternal deprivation in serotonergic activity and HPA function in adult rat," Neuroscience Letters, vol. 480, no. 1, pp. 7–11, 2010.

129. S. Oreland, C. Pickering, C. Gökturk, L. Oreland, L. Arborelius, and I. Nylander, "Two repeated maternal separation procedures differentially affect brain 5-hydroxytryptamine transporter and receptors in young and adult male and female rats," Brain Research, vol. 1305, pp. S37–S49, 2009.

130. A. Vicentic, D. Francis, M. Moffett et al., "Maternal separation alters serotonergic transporter densities and serotonergic 1A receptors in rat brain," Neuroscience, vol. 140, no. 1, pp. 355–365, 2006.

131. M. F. Seidel and W. Müller, "Differential pharmacotherapy for subgroups of fibromyalgia patients with specific consideration of 5-HT3 receptor antagonists," Expert Opinion on Pharmacotherapy, vol. 12, no. 9, pp. 1381–1391, 2011. View at Publisher · View at Google Scholar

132. M. Offenbaecher, B. Bondy, S. De Jonge et al., "Possible association of fibromyalgia with a polymorphism in the serotonin transporter gene regulatory region," Arthritis and Rheumatism, vol. 42, no. 11, pp. 2482–2488, 1999.

133. E. Paul-Savoie, S. Potvin, K. Daigle et al., "A deficit in peripheral serotonin levels in major depressive disorder but not in chronic widespread pain," Clinical Journal of Pain, vol. 27, no. 6, pp. 529–534, 2011. View at Publisher · View at Google Scholar

134. J. Bowlby, "The nature of the child's tie to his mother," The International Journal of Psycho-Analysis, vol. 39, no. 5, pp. 350–373, 1958.

135. J. Bowlby, "Attachment theory and its therapeutic implications," Adolescent Psychiatry, vol. 6, pp. 5–33, 1978.

136. L. S. Porter, D. Davis, and F. J. Keefe, "Attachment and pain: recent findings and future directions," Pain, vol. 128, no. 3, pp. 195–198, 2007.

137. P. Ciechanowski, M. Sullivan, M. Jensen, J. Romano, and H. Summers, "The relationship of attachment style to depression, catastrophizing and health care utilization in patients with chronic pain," Pain, vol. 104, no. 3, pp. 627–637, 2003.

138. P. Meredith, J. Strong, and J. A. Feeney, "Adult attachment, anxiety, and pain self-efficacy as predictors of pain intensity and disability," Pain, vol. 123, no. 1-2, pp. 146–154, 2006.

139. P. J. Meredith, J. Strong, and J. A. Feeney, "The relationship of adult attachment to emotion, catastrophizing, control, threshold and tolerance, in experimentally-induced pain," Pain, vol. 120, no. 1-2, pp. 44–52, 2006.

140. T. R. Insel, "Is social attachment an addictive disorder?" Physiology and Behavior, vol. 79, no. 3, pp. 351–357, 2003.

141. A. Moles, B. L. Kieffer, and F. R. D'Amato, "Deficit in attachment behavior in mice lacking the μ-opioid receptor gene," Science, vol. 304, no. 5679, pp. 1983–1986, 2004.

142. L. R. M. Hallberg and S. G. Carlsson, "Psychosocial vulnerability and maintaining forces related to fibromyalgia: in-depth interviews with twenty-two female patients," Scandinavian Journal of Caring Sciences, vol. 12, no. 2, pp. 95–103, 1998.

143. E. A. Walker, D. Keegan, G. Gardner, M. Sullivan, D. Bernstein, and W. J. Katon, "Psychosocial factors in fibromyalgia compared with rheumatoid arthritis: II. Sexual, physical, and emotional abuse and neglect," Psychosomatic Medicine, vol. 59, no. 6, pp. 572–577, 1997.

144. R. W. Alexander, L. A. Bradley, G. S. Alarcón et al., "Sexual and physical abuse in women with fibromyalgia: association with outpatient health care utilization and pain medication usage," Arthritis Care and Research, vol. 11, no. 2, pp. 102–115, 1998.

145. J. McBeth, G. J. Macfarlane, S. Benjamin, S. Morris, and A. J. Silman, "The association between tender points, psychological distress, and adverse childhood experiences: a community based study," Arthritis and Rheumatism, vol. 42, no. 7, pp. 1397–1404, 1999.

146. B. Van Houdenhove, E. Neerinckx, R. Lysens et al., "Victimization in chronic fatigue syndrome and fibromyalgia in tertiary care: a controlled study on prevalence and characteristics," Psychosomatics, vol. 42, no. 1, pp. 21–28, 2001.

147. M. L. Paras, M. H. Murad, L. P. Chen et al., "Sexual abuse and lifetime diagnosis of somatic disorders: a systematic review and meta-analysis," Journal of the American Medical Association, vol. 302, no. 5, pp. 550–561, 2009.

148. W. Häuser, M. Kosseva, N. Üceyler, P. Klose, and C. Sommer, "Emotional, physical, and sexual abuse in fibromyalgia syndrome: a systematic review with meta-analysis," Arthritis Care and Research, vol. 63, no. 6, pp. 808–820, 2011. View at Publisher · View at Google Scholar

149. V. J. Felitti, R. F. Anda, D. Nordenberg et al., "Relationship of childhood abuse and household dysfunction to many of the leading causes of death in adults: the adverse childhood experiences (ACE) study," American Journal of Preventive Medicine, vol. 14, no. 4, pp. 245–258, 1998.

150. B. S. McEwen, "Understanding the potency of stressful early life experiences on brain and body function," Metabolism, vol. 57, no. 2, pp. S11–S15, 2008.

151. H. Steiger, L. Gauvin, M. Israël et al., "Association of serotonin and cortisol indices with childhood abuse in bulimia nervosa," Archives of General Psychiatry, vol. 58, no. 9, pp. 837–843, 2001.

152. J. Kim-Cohen, A. Caspi, A. Taylor et al., "MAOA, maltreatment, and gene-environment interaction predicting children's mental health: new evidence and a meta-analysis," Molecular Psychiatry, vol. 11, no. 10, pp. 903–913, 2006.

153. K. Ritchie, I. Jaussent, R. Stewart et al., "Association of adverse childhood environment and 5-HTTLPR genotype with late-life depression," Journal of Clinical Psychiatry, vol. 70, no. 9, pp. 1281–1288, 2009.

154. J. M. Miller, E. L. Kinnally, R. T. Ogden, M. A. Oquendo, J. J. Mann, and R. V. Parsey, "Reported childhood abuse is associated with low serotonin transporter binding in vivo in major depressive disorder," Synapse, vol. 63, no. 7, pp. 565–573, 2009.

155. R. Tikkanen, F. Ducci, D. Goldman et al., "MAOA alters the effects of heavy drinking and childhood physical abuse on risk for severe impulsive acts of violence among alcoholic violent offenders," Alcoholism, vol. 34, no. 5, pp. 853–860, 2010.

156. R. T. Goldberg, W. N. Pachas, and D. Keith, "Relationship between traumatic events in childhood and chronic pain," Disability and Rehabilitation, vol. 21, no. 1, pp. 23–30, 1999. View at Publisher · View at Google Scholar

157. U. M. Anderberg, I. Marteinsdottir, T. Theorell, and L. Von Knorring, "The impact of life events in female patients with fibromyalgia and in female healthy controls," European Psychiatry, vol. 15, no. 5, pp. 295–301, 2000.

158. N. A. Nicolson, M. C. Davis, D. Kruszewski, and A. J. Zautra, "Childhood maltreatment and diurnal cortisol patterns in women with chronic pain," Psychosomatic Medicine, vol. 72, no. 5, pp. 471–480, 2010.

159. N. A. Nicolson, "Childhood parental loss and cortisol levels in adult men," Psychoneuroendocrinology, vol. 29, no. 8, pp. 1012–1018, 2004.

160. A. R. Tyrka, L. Wier, L. H. Price et al., "Childhood parental loss and adult hypothalamic-pituitary-adrenal function," Biological Psychiatry, vol. 63, no. 12, pp. 1147–1154, 2008.

161. L. J. Luecken, "Childhood attachment and loss experiences affect adult cardiovascular and cortisol function," Psychosomatic Medicine, vol. 60, no. 6, pp. 765–772, 1998.

162. L. L. Carpenter, A. R. Tyrka, C. J. McDougle et al., "Cerebrospinal fluid corticotropin-releasing factor and perceived early-life stress in depressed patients and healthy control subjects," Neuropsychopharmacology, vol. 29, no. 4, pp. 777–784, 2004.

163. A. Danese, C. M. Pariante, A. Caspi, A. Taylor, and R. Poulton, "Childhood maltreatment predicts adult inflammation in a life-course study," Proceedings of the National Academy of Sciences of the United States of America, vol. 104, no. 4, pp. 1319–1324, 2007.

164. S. A. McLean, D. A. Williams, P. K. Stein et al., "Cerebrospinal fluid corticotropin-releasing factor concentration is associated with pain but not fatigue symptoms in patients with fibromyalgia," Neuropsychopharmacology, vol. 31, no. 12, pp. 2776–2782, 2006.

165. I. Weissbecker, A. Floyd, E. Dedert, P. Salmon, and S. Sephton, "Childhood trauma and diurnal cortisol disruption in fibromyalgia syndrome," Psychoneuroendocrinology, vol. 31, no. 3, pp. 312–324, 2006.

166. L. R. M. Hallberg and S. G. Carlsson, "Coping with fibromyalgia—a qualitative study," Scandinavian Journal of Caring Sciences, vol. 14, no. 1, pp. 29–36, 2000.

167. J. McBeth, S. Morris, S. Benjamin, A. J. Silman, and G. J. Macfarlane, "Associations between adverse events in childhood and chronic widespread pain in adult-

hood: are they explained by differential recall?" Journal of Rheumatology, vol. 28, no. 10, pp. 2305–2309, 2001.

168. C. Power, K. Atherton, D. P. Strachan et al., "Life-course influences on health in British adults: effects of socio-economic position in childhood and adulthood," International Journal of Epidemiology, vol. 36, no. 3, pp. 532–539, 2007.

169. M. O'Donnell, N. Nassar, H. Leonard, R. Mathews, Y. Patterson, and F. Stanley, "Monitoring child abuse and neglect at a population level: patterns of hospital admissions for maltreatment and assault," Child Abuse and Neglect, vol. 34, no. 11, pp. 823–832, 2010.

170. A. Jud, U. Lips, and M. A. Landolt, "Characteristics associated with maltreatment types in children referred to a hospital protection team," European Journal of Pediatrics, vol. 169, no. 2, pp. 173–180, 2010.

171. S. Howell, N. J. Talley, S. Quine, and R. Poulton, "The irritable bowel syndrome has origins in the childhood socioeconomic environment," American Journal of Gastroenterology, vol. 99, no. 8, pp. 1572–1578, 2004.

172. E. L. Rangel, R. S. Burd, and R. A. Falcone Jr., "Socioeconomic disparities in infant mortality after nonaccidental trauma: a multicenter study," Journal of Trauma, vol. 69, no. 1, pp. 20–25, 2010.

173. P. B. Wood, P. Schweinhardt, E. Jaeger et al., "Fibromyalgia patients show an abnormal dopamine response to pain," European Journal of Neuroscience, vol. 25, no. 12, pp. 3576–3582, 2007.

174. J. N. Baraniuk, G. Whalen, J. Cunningham, and D. J. Clauw, "Cerebrospinal fluid levels of opioid peptides in fibromyalgia and chronic low back pain," BMC Musculoskeletal Disorders, vol. 5, article 48, 2004.

175. R. E. Harris, D. J. Clauw, D. J. Scott, S. A. McLean, R. H. Gracely, and J. K. Zubieta, "Decreased central μ-opioid receptor availability in fibromyalgia," Journal of Neuroscience, vol. 27, no. 37, pp. 10000–10006, 2007.

176. M. J. Druse, N. Tajuddin, A. Kuo, and M. Connerty, "Effects of in utero ethanol exposure on the developing dopaminergic system in rats," Journal of Neuroscience Research, vol. 27, no. 2, pp. 233–240, 1990.

177. M. J. Druse, A. Kuo, and N. Tajuddin, "Effects of in utero ethanol exposure on the developing serotonergic system," Alcoholism, vol. 15, no. 4, pp. 678–684, 1991.

178. X. Zhang, J. H. Sliwowska, and J. Weinberg, "Prenatal alcohol exposure and fetal programming: effects on neuroendocrine and immune function," Experimental Biology and Medicine, vol. 230, no. 6, pp. 376–388, 2005.

179. L. P. Finnegan, J. F. Connaughton, R. E. Kron, and J. P. Emich, "Neonatal abstinence syndrome: assessment and management," Addictive Diseases, vol. 2, no. 1-2, pp. 141–158, 1975.

180. K. L. Jones and G. A. Barr, "Ontogeny of morphine withdrawal in the rat," Behavioral Neuroscience, vol. 109, no. 6, pp. 1189–1198, 1995.

181. J. C. Rozé, S. Denizot, R. Carbajal et al., "Prolonged sedation and/or analgesia and 5-year neurodevelopment outcome in very preterm infants: results from the EPIPAGE cohort," Archives of Pediatrics and Adolescent Medicine, vol. 162, no. 8, pp. 728–733, 2008.

182. A. Tempel, J. Habas, W. Paredes, and G. A. Barr, "Morphine-induced downregulation of mu-opioid receptors in neonatal rat brain," Brain Research, vol. 469, no. 1-2, pp. 129–133, 1988.

183. S. R. Thornton and F. L. Smith, "Long-term alterations in opiate antinociception resulting from infant fentanyl tolerance and dependence," European Journal of Pharmacology, vol. 363, no. 2-3, pp. 113–119, 1998.

184. R. E. Harris, J. K. Zubieta, D. J. Scott, V. Napadow, R. H. Gracely, and D. J. Clauw, "Traditional Chinese acupuncture and placebo (sham) acupuncture are differentiated by their effects on μ-opioid receptors (MORs)," NeuroImage, vol. 47, no. 3, pp. 1077–1085, 2009.

185. T. Sithisarn, H. S. Bada, H. Dai, D. C. Randall, and S. J. Legan, "Effects of perinatal cocaine exposure on open field behavior and the response to corticotropin releasing hormone (CRH) in rat offspring," Brain Research, vol. 1370, pp. 136–144, 2011. View at Publisher · View at Google Scholar

186. G. Battaglia and T. M. Cabrera, "Potentiation of 5-HT(1A) receptor-mediated neuroendocrine responses in male but not female rat progeny after prenatal cocaine: evidence for gender differences," Journal of Pharmacology and Experimental Therapeutics, vol. 271, no. 3, pp. 1453–1461, 1994.

187. F. Roussotte, L. Soderberg, and E. Sowell, "Structural, metabolic, and functional brain abnormalities as a result of prenatal exposure to drugs of abuse: evidence from neuroimaging," Neuropsychology Review, vol. 20, no. 4, pp. 376–397, 2010.

188. F. I. Tarazi and R. J. Baldessarini, "Comparative postnatal development of dopamine D1, D2 and D4 receptors in rat forebrain," International Journal of Developmental Neuroscience, vol. 18, no. 1, pp. 29–37, 2000.

189. P. B. Wood, "Mesolimbic dopaminergic mechanisms and pain control," Pain, vol. 120, no. 3, pp. 230–234, 2006. View at Publisher · View at Google Scholar

190. P. B. Wood, "Mesolimbic dopaminergic mechanisms and pain control," Pain, vol. 120, no. 3, pp. 230–234, 2006.

191. P. B. Wood, M. F. Glabus, R. Simpson, and J. C. Patterson II, "Changes in gray matter density in fibromyalgia: correlation with dopamine metabolism," Journal of Pain, vol. 10, no. 6, pp. 609–618, 2009.

192. S. Gürsoy, E. Erdal, H. Herken, E. Madenci, B. Alaşehirli, and N. Erdal, "Significance of catechol-O-methyltransferase gene polymorphism in fibromyalgia syndrome," Rheumatology International, vol. 23, no. 3, pp. 104–107, 2003.

193. D. Buskila, H. Cohen, L. Neuman, and R. P. Ebstein, "An association between fibromyalgia and the dopamine D4 receptor exon III repeat polymorphism and relationship to novelty seeking personality traits," Molecular Psychiatry, vol. 9, no. 8, pp. 730–731, 2004.

194. M. H. Boisset-Pioro, J. M. Esdaile, and M. A. Fitzcharles, "Sexual and physical abuse in women with fibromyalgia syndrome," Arthritis and Rheumatism, vol. 38, no. 2, pp. 235–241, 1995.

195. K. Imbierowicz and U. T. Egle, "Childhood adversities in patients with fibromyalgia and somatoform pain disorder," European Journal of Pain, vol. 7, no. 2, pp. 113–119, 2003.

PART IV

INTERVENTIONS

CHAPTER 15

LISTEN PROTECT CONNECT FOR TRAUMATIZED SCHOOLCHILDREN: A PILOT STUDY OF PSYCHOLOGICAL FIRST AID

MARIZEN RAMIREZ, KARISA HARLAND, MAISHA FREDERICK, RHODA SHEPHERD, MARLEEN WONG, AND JOSEPH E. CAVANAUGH

15.1 BACKGROUND

Trauma is defined as incidents experienced, witnessed or learned about that 1) involve "actual or threatened death or serious injury, or other threat to one or another's physical integrity" and 2) elicit intense "fear, helplessness or horror" (American Psychiatric Association 2000)". Trauma is common in youth, impacting as many as 80% of children worldwide (Sharma-Patel et al. 2011). In a US-based longitudinal study, 68.8% of children were exposed to one or more traumatic events by their 16th birthday (Copeland et al. 2007). Children experience a variety of traumas, including learning

Listen Protect Connect for Traumatized Schoolchildren: A Pilot Study of Psychological First Aid. © *Ramirez M, Harland K, Frederick M, Shepherd R, Wong M, and Cavanaugh JE; licensee BioMed Central Ltd.* BMC Psychology *1,26 (2013), doi:10.1186/2050-7283-1-26. Licensed under the Creative Commons Attribution 2.0 Generic License, http://creativecommons.org/licenses/by/2.0/.*

about traumatic experiences of relatives or friends (62%), sudden death of friends or relatives (60%), assaults (38%), motor vehicle crash (28%), and natural disasters (i.e., tornados, fire, flood or earthquake) (17%) (Breslau et al. 1998). Although extremely rare, school shootings, such as the recent shooting at Sandy Hook Elementary School in Newtown, Connecticut, also represent the types of trauma that may directly impact children.

Exposure to trauma may trigger adverse psychological responses, of which post-traumatic stress disorder (PTSD) and depression are most prominent (Kenardy et al. 2006). Studies show great variability in the rates of PTSD, depending on the type of, severity of, and time elapsed since a traumatic event. Between 23-70% of children exposed to natural disasters and 10-80% of children witnessing violence display symptoms of PTSD (McDermott et al. 2005; Neuner et al. 2006; Vernberg et al. 1996; Pynoos 1993; Goenjian et al. 2005; Nader et al. 1990; Hoven 2005; Ahmad et al. 2000). PTSD also tends to co-occur with other types of psychiatric disorders, particularly depression. Thirty-seven to 47% of PTSD diagnosis in children is accompanied by a diagnosis of depression. In an urban population of 1,007 youths exposed to violence, 23.6% developed PTSD and among those, 36.6% had major depression (Breslau et al. 1991). Among child witnesses of violent crime with PTSD, 47% were also found to be diagnosed with depression in comparison to 16% without PTSD (Muesar and Taub 2008).

Schools are a place where children often exhibit signs of trauma-related distress, and can therefore serve as a successful point of contact and treatment (President's New Freedom Commission on Mental Health 2003). Currently, the state of school mental health practice focuses on referring students who are at high risk for developing mental health disorders to a school psychologist for individual care (Dowdy et al. 2010; Cash & Nealis 2004). Of therapeutic methods used in schools, Cognitive Behavior Therapy (CBT) and trauma/grief-informed psychotherapy have been found to effectively reduce symptoms of depression and PTSD among trauma-exposed youth (Goenjian et al. 2005; Stein et al. 2003; Layne 2001). CBT and psychotherapy are both time-intensive modalities supported by professional mental health clinicians and intended for use among individuals with full-blown PTSD (symptoms after 30 days).

An important gap of service exists in the areas of triage and early intervention, which are the critical first steps that can direct trauma-exposed students to advanced care. The most common early intervention treatment in school mental health practice is Psychological Debriefing, a community-based early psychological intervention delivered to trauma exposed individuals. It was initially concluded as effective in reducing an array of psychopathology symptoms (Flannery and Everly 2004). However, recent randomized controlled trials conducted among adults, children and adolescents demonstrated that Psychological Debriefing failed to improve outcomes when compared with a control group (Stallard et al. 2006; Hobbs et al. 1996). As a result of these contradicting findings, the Task Force on Community Preventive Services recommended against the use of this therapy among trauma-exposed children and adolescents (Wethington et al. 2008). To date, there are no evidence-based triage and early interventions delivered by non-mental health professionals for trauma-exposed students.

To address this service gap, Listen, Protect & Connect (LPC) was developed as an intervention program of Psychological First Aid. Psychological First Aid, which is analogous to physical First Aid, involves post-trauma contact and engagement, safety and comfort, stabilization, information gathering, practical assistance, connection with social supports, information on coping support and linking to services (Ruzak et al. 2007). Informed by research on posttraumatic resilience (Kataoka et al. 2012; Wong 2008), LPC was initially designed for delivery by a non-professional to provide information, education, comfort and support to traumatized youth after a community disaster or emergency. However, the elements of LPC could also be used to support children impacted by personal traumas. The effectiveness of LPC in improving children's recovery from trauma has not been scientifically evaluated. Hence, we began a small-scale study of LPC delivered by school nurses in a school district in Iowa (US). Our implementation and outcome evaluation of LPC was conducted to (1) describe the acceptability and barriers of this program, and (2) measure the extent to which LPC reduces symptoms of psychological distress and improves social support and school connectedness.

15.2 METHODS

15.2.1 PARTICIPANTS

A pilot quasi-experiment was conducted with 20 middle and high school students enrolled in four middle and two high schools from a single urban school district in the Midwest from May 2009 through 2011. These subjects were recruited from two consecutive school years.

Our year one eligible population was comprised of students directly impacted by the 2008 Great Flood of Iowa, identified from the school district's list of relocated students. Due to IRB delays, time required for training and district approvals, eligible students were recruited approximately 10 months after the flood.

To increase our sample size in year two, additional students potentially traumatized by other types of traumas (such as violence or death of a loved one) were recruited from the same schools involved in year one. A number of indicators were used to identify these students, based on prior research on factors associated with trauma (Caffo and Belaise 2003). To be eligible, students either 1) had to be seen at the nurse's office for nonspecific physical symptoms (e.g., vague headaches, stomachaches) or behavioral problems at least 1×/week for three consecutive weeks or 2×/week for two weeks, 2) had to have reported a personal trauma or expressed distress to the nurse or school staff, or 3) had to have 3–5 consecutive days of unexcused absences.

Eligible students and their parent(s) were mailed an introductory letter about the study and asked to return a self-addressed postcard if interested in participating. Interested families were mailed an information sheet, informed assent/consent documents and an enrollment and contact form. Our passive recruitment efforts yielded a sample of 8 flood-affected students and 12 students with a history of individual trauma. This study was approved by the University of Iowa Institutional Review Board. We obtained parental consent and child assent for participation.

15.2.2 PROCEDURE

Nurses from the six middle and high schools received three hours of training in LPC by the developer (M. Wong) and Principal Investigator (M. Ramirez) in year one. Participants were provided basic information on trauma and its psychological impacts on children. The three required steps of Listen, Protect, and Connect were each described in detail. Manuals, worksheets, and pocket cards summarizing these key steps were provided; nurses participated in role playing to increase familiarization with the steps. In year two, a two hour refresher course was provided to review LPC steps and materials with a focus on individual traumas such as interpersonal violence and injury.

After obtaining assent and consent, baseline questionnaires were distributed to the students by the University of Iowa research team online or in-person. To confirm exposure to trauma, students were asked to report the types of traumas experienced, witnessed, or learned about through the Life Events Checklist (LEC), a scale with adequate reliability and validity. Respondents reported their traumatic experiences on a 5-point scale (1 = happened to me, 2 = witnessed it, 3 = learned about it, 4 = not sure, and 5 = does not apply) (Gray et al. 2004). All students had personally experienced, witnessed or learned about a traumatic event therefore meeting PTSD Criterion A.

Within one week of baseline survey completion, an LPC session was scheduled with the school nurse. After completion of the LPC session, both the nurse and student completed LPC session evaluation forms. Follow-up questionnaires were completed by the student at 2-, 4- and 8-weeks following the initial LPC session.

15.2.3 INTERVENTION

Listen Protect Connect (LPC) is composed of three basic steps designed to specifically target PTSD symptomatic reactions (Kataoka et al. 2012; Wong 2008).

15.2.3.1 LISTEN STEP 1

Interventionists use reflective listening skills and non-invasive questions to elicit responses about a student's specific traumatic experiences. For example, the interventionist asks the students "How, What, or Tell me more…" questions to begin an open dialogue of the student's concerns.

15.2.3.2 PROTECT STEP 2

The interventionist conducts a brief screener of non-specific distress using the six-item K6 screener (Furukawa et al. 2003). The interventionist is taught to identify cognitive, physiological, and psychological reactions to trauma, and engages in open discussion with the student about their fears and worries. Through assessment and honest discussion, the interventionist "protects" students by identifying potentially high risk children who score high on the K6 screener or reveal maladaptive reactions to trauma. The interventionist is therefore equipped with critical information indicating need for additional services. During this step, as concerns and worries surface, the interventionist engages in open discussion about the crisis and actions taken by schools, families and schools to keep the traumatized child safe. This includes discussions about school safety protocols, support provided by parents and families or by the local community or school, or assistance provided by professionals such as counselors and nurses on campus.

15.2.3.3 CONNECT STEP 3

The PFA interventionist uses information from Steps 1 and 2 to identify students who may be at risk for potential distress. The interventionist then facilitates access to resources and advanced mental health care. Furthermore, the interventionist encourages the student to re-connect with friends, family and to re-engage in previously enjoyed activities. LPC is a flexible program that could be implemented repeatedly and as short or as long as the interventionist and student desire.

15.2.4 IMPLEMENTATION EVALUATION WITH NURSES AND STUDENTS

Using post-session evaluation forms, school nurses reported the number and length of each LPC session, as well as the ease or difficulty in completing each LPC step using a 5-point scale where 1 is very easy to 5 is very difficult. School nurses also reported use of the program materials (worksheets, manual, screener, pocket card) and how helpful these materials were during the delivery of the session (1 = not at all to 5 = very helpful). Using a similar 5-point scale (1 = not at all to 5 = very comfortable), students were asked to provide comfort level while communicating with the school nurse, and comfort with the length of the LPC session. Both the nurse and student were asked to report their perceived helpfulness of the LPC session (1 = not at all helpful to 5 = very helpful). School nurses described perceived barriers and suggestions to improve the program.

15.2.5 OUTCOME EVALUATION

15.2.5.1 INSTRUMENTS

We collected the following measures from students in baseline and follow-up questionnaires:

The modified Child PTSD Symptom Scale is the 17-item child version of the adult posttraumatic diagnostic scale with scores ranging from 0 to 51 (Chronbach $\alpha = 0.89$) (Foa et al. 2001). A cut point of 14 was used to classify children as symptomatic for PTSD (Stein et al. 2003).

To measure depressive symptoms, we utilized the Center for Epidemiologic Studies Depression Scale (CES-D), a 20-item self-rating scale assessing frequency of symptoms. A child with a score of 16 or higher was categorized as displaying depressive symptoms (Radloff 1977).

The Multidimensional Scale of Perceived Support (MSPSS) is a 12-item scale measuring perceived social support from family, friends and a significant other (Chronbach $\alpha = 0.93$) (Zimet et al. 1990; Canty-Mitchell and Zimet 2000; Bruwer et al. 2008).

To assess the extent to which students feel connected to their school, we used selected items from the Healthy Kids Resilience Measure of School Connectedness that measure students' perceived connectedness with adults at their school and the strength of these relationships (Constantine et al. 1999). All items in this scale showed strong internal consistency (Cronbach $\alpha = 0.87$).

Age, gender and ethnicity, potential confounders identified from prior research, were also collected from students at baseline (Davis and Siegel 2000).

15.2.5.2 DATA ANALYSIS

For the implementation evaluation, we performed descriptive analysis to describe the ease/difficulty and perceived helpfulness of LPC steps and materials. For open-ended questions about barriers, we used content analysis to identify and create categories of themes.

For our outcome evaluation, we analyzed changes in psychological symptoms, social support and school connectedness over time. To control for the correlation among longitudinal responses collected on the same student and among responses collected within the same school, we first fit hierarchical mixed effects linear regression models that included random effects to induce clustering at both the student level and the school level. Compared with standard repeated measures ANOVA, the hierarchical mixed effects model is a more flexible approach to account for irregular time measurement points, missing observations and time-dependency (Gueorguieva and Krystal 2004). We fit our initial models with an autoregressive correlation structure at the student level to allow for the magnitude of the correlation between two measurements to depend on the time period between the measurements. (For example, observations taken at

2- and 4- weeks follow-up are assumed to be more highly correlated than measurements taken at 2- and 8-weeks follow-up.) At the school level, we employed an exchangeable correlation structure. However, based on the variance component estimates for these models, there was no evidence of school-level clustering. Therefore, we used a simpler mixed effects model that accounted only for the correlation among longitudinal responses collected on the same student. Age, gender, ethnicity and types of trauma were included in the model as covariates and potential confounders. Statistical significance was set at a p-value <0.05 but, given the small sample size, p-values of <0.10 are also documented as suggestive of change.

All study activities were approved by the University of Iowa Institutional Review Board. All data analysis was conducted using SAS® software, Version 9.2 of the SAS System for Microsoft, SAS Institute Inc., Cary, NC, USA.

15.3 RESULTS

15.3.1 DESCRIPTION OF SUBJECTS AT BASELINE

A total of 20 students completed LPC sessions over two phases. Approximately, 80% of the students were male between the ages 12–17 years and represented grades 6th through 11th (Table 1). Twenty students completed baseline and 2-week follow-up questionnaires, 18 completed the 4-week follow-up, and 15 completed the final 8-week follow-up. There were a total of 71 repeated measures among the students.

At baseline, 60% (n=12) and 55% (n=11) of students were symptomatic for depression and PTSD, respectively. The average depressive (M, SD=23.5, 13.4) and PTSD (M, SD =16.4, 12.8) symptoms scores exceeded the cut point for demonstrating clinical symptomology of these conditions (Table 1).

Prior to the intervention, students were neutral or somewhat agreed (mean=3.6, range 1–5) to having individuals in their lives to support or help them. A moderate level of school connectedness was reported by the students (mean=58.4, median=52.5).

TABLE 1: Demographics and baseline symptomology of students experiencing trauma and receiving Psychological First Aid, Iowa, 2009–2010 (N=20)

Demographics	All (N=20)	
	n (Col %)	Mean (SD)
Age		
12	7 (35.0)	
13	4 (20.0)	
14	3 (15.0)	
15	4 (20.0)	
≥16	2 (10.0)	
Gender		
Male	16 (80.0)	
Female	4 (20.0)	
Race		
White	7 (35.0)	
Hispanic	1 (5.0)	
African-American	1 (5.0)	
Asian/Pacific Islander	8 (40.0)	
Other race	3 (15.0)	
Grade		
6th	7 (35.0)	
7th	4 (20.0)	
8th	2 (10.0)	
9th	2 (10.0)	
10th	4 (20.0)	
11th	1 (5.0)	
Symptoms		
Depressive symptoms[1]	60% (n=12)	23.5 (13.4)
PTSD symptoms[2]	55% (n=11)	16.4 (12.8)
# of Traumas experienced[3]		
Happened to student		4.0 (2.0)
Student witnessed		2.4 (2.8)
Student learned about		2.1 (3.4)
Social support[4]		
Overall		3.6 (1.0)
Family		3.5 (1.2)

TABLE 1: *Cont.*

Demographics	All (N=20)	
	n (Col %)	Mean (SD)
Friends		3.7 (1.0)
Significant other		3.8 (0.9)
School connectedness[5]		58.4 (10.7)

[1]*Center for Epidemiologic Studies Depression Scale (CES-D, 20 item), range (0–60), score ≥16 considered exhibiting mild/moderate/major depressive symptoms.* [2]*Post-Traumatic Stress Disorder Checklist (civilian), score ≥14 demonstrates showing PTSD symptoms.* [3]*Life Events Checklist.* [4]*Multidimensional Scale of Perceived Social Support, Overall = average of all 12 items, Subscales (family, friends, significant other) = average of 4 items within each subscale.* [5]*Healthy Kids Resilience Measure.*

15.3.2 IMPLEMENTATION EVALUATION

15.3.2.1 SCHOOL NURSE FEEDBACK TO IMPLEMENTING LPC

The school nurse only provided one session of LPC to each student in the school nurse office. Sessions lasted on average 25 minutes long (range 10 to 40 minutes). Programmatic steps of LPC were somewhat to very easy to implement by school nurses during almost all sessions (Table 2). No nurses reported any steps being very difficult. Nurses indicated that the Protect tools were effective in assisting the students to "convey their issues" and determine solutions. The Connect step was reportedly a "great closing conversation" to discuss coping strategies and resources. One nurse reported that this step "helped the student to organize his feelings to know what he truly needed". In only one session did the nurse report struggling through the Protect and Connect steps.

During most of the LPC sessions, nurses reported using intervention materials (e.g., worksheets, manual and pocket card) and found them to be somewhat to very helpful.

TABLE 2: School Nurses' feedback on the effectiveness and adaptability of Psychological First Aid steps during the sessions

Did you use:	How easy or difficult were these steps? n (%)			
	Very easy	Somewhat easy	Neither easy nor difficult	Somewhat difficult
The Listen Step?	13 (65%)	5 (25%)	2 (10%)	0
The Protect Step?	11 (55%)	6 (30%)	2 (10%)	1 (5%)
The Connect Step?	9 (45%)	6 (30%)	4 (20%)	1 (5%)

(N = 20)

15.3.2.2 STUDENT FEEDBACK

Ninety-five percent of the students thought the sessions lasted about the right amount of time. Sixty percent felt very comfortable speaking with the school nurse; only one student reported feeling very uncomfortable. The sessions were rated as somewhat helpful in dealing with the recent events in their lives by 40% of the students. Students reported enjoying the opportunity to "share my feelings with someone else" and having someone to "listen to my problems". One student reported that the nurse was "awesome and helped me by talking and understanding me".

15.3.3 BARRIERS

Although a largely successful intervention, two nurses reported the length of the sessions and the scheduling of sessions as barriers. One nurse reported difficulty in scheduling a time most convenient for the student. Furthermore, students experienced a range of traumatic experiences (e.g. bullying, difficulty socializing with peers, a shooting, flood), and this heterogeneity was somewhat challenging for nurses. Nurses were first trained in LPC in response to the flood, and two found it somewhat difficult to relate some intervention materials to the students' current situation.

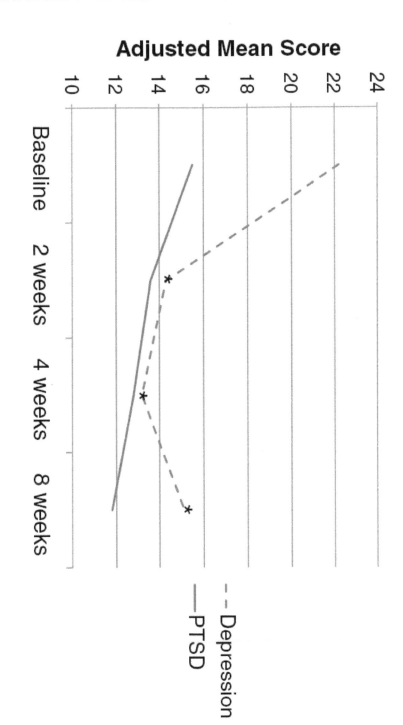

FIGURE 1: Mean depressive and PTSD symptoms, Iowa PFA pilot, N=71 measurements1.

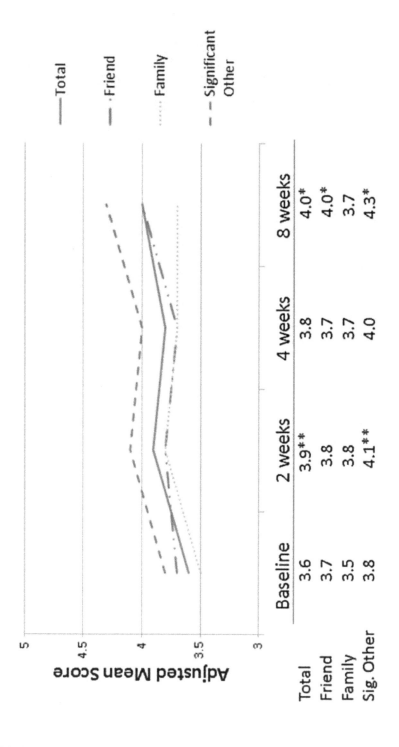

	Baseline	2 weeks	4 weeks	8 weeks
Total	3.6	3.9**	3.8	4.0*
Friend	3.7	3.8	3.7	4.0*
Family	3.5	3.8	3.7	3.7
Sig. Other	3.8	4.1**	4.0	4.3*

FIGURE 2: Means for total social support and by source of social support, Iowa PFA pilot, N = 71 measurements[1]

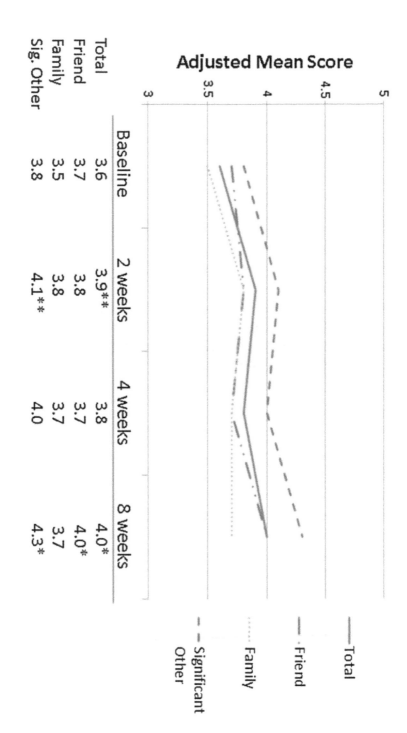

Adjusted Mean Score

	Baseline	2 weeks	4 weeks	8 weeks
Total	3.6	3.9**	3.8	4.0*
Friend	3.7	3.8	3.7	4.0*
Family	3.5	3.8	3.7	3.7
Sig. Other	3.8	4.1**	4.0	4.3*

——Total

—— Friend

······ Family

— — · Significant
Other

FIGURE 3: Mean school connectedness, Iowa PFA pilot, N=71 measuerment1.

15.3.4 OUTCOME EVALUATION

After controlling for students' race, gender, sex and type of trauma (flood or individual trauma), a significant decline in depressive symptoms was seen from baseline to each follow-up period (Figure 1). The adjusted baseline mean depressive score was 22.2; this dropped to 14.3 (2-weeks, $p<0.01$), 13.2 (4-weeks, $p<0.01$) and increased just slightly to 15.2 (8-weeks, $p<0.01$), all levels below the clinical cut point for depression. PTSD symptoms decreased 3.7 points from baseline to the 8-week follow-up, although this change was not statistically significant (range 15.5-11.8; $p=0.09$). Total social support (Figure 2) increased from baseline to the 2-week follow-up ($p=0.08$), and increased significantly from baseline to the 8-week follow-up ($p<0.01$). The increase in average social support from significant others bordered on significance at 2-weeks ($p=0.09$), but a strong, significant increase was seen by 8-weeks ($p<0.01$). Students felt more connected to their school at 2- (mean$=63.8$, $p=0.06$) and 4-weeks (mean$=68.9$, $p<0.01$) than at baseline (mean$=58.6$), but this relationship diminished by 8-weeks (Figure 3).

15.4 DISCUSSION

After recent tragedies, like the Boston Marathon Bombings, the Sandy Hook School Shooting and the many individual traumas experienced by students, school must quickly implement crisis intervention programs to support traumatized students. Psychological First Aid, one type of crisis intervention recommended by the National Child Traumatic Stress Network, is a promising approach but still lacks an evidence base.

Our pilot study is the first evaluation of a school-based specific form of Psychological First Aid, Listen Protect Connect (LPC), and our research showed that LPC was well-received by school interventionists and youth. Moreover, LPC has some promising evidence of effectiveness even in this pilot study, as treated youth had improved symptoms of psychological distress and increased school connectedness and social support.

Although marginally significant, the decrease in PTSD symptoms over time suggests an appropriate level of impact for an early intervention program delivered in a one-encounter session. As a first level of defense, LPC provides initial relief with a trusted adult at school. These findings are noteworthy, given that more complex school-based interventions report similar reductions in PTSD among children traumatized by violence, war or an earthquake (Goenjian et al. 2005; Stein et al. 2003; Layne 2001).

The significant decrease in depressive symptoms after treatment through LPC was particularly notable. On one hand, findings may have resulted from the natural regression to the mean, which cannot be ruled out given our lack of a control group. On the other hand, it is possible that LPC did indeed improve depressive symptoms. If LPC had no effect, depressive symptoms would have persisted at similar levels as baseline or increase, as demonstrated in prior studies of traumatized children who received no treatment (Goenjian et al. 2005; Stein et al. 2003). At the same time, a slight but notable increase in depressive symptoms was observed from 4 to 8 weeks. This also coincided with a decrease in school connectedness by 8 weeks, suggesting need for re-delivery of LPC with traumatized youth between 4–8 weeks after initial dosage. Despite these encouraging findings, clearly, a large trial is needed to gather more evidence of LPC's effectiveness and its adequate dosage for maximal impact.

The LPC program teaches children to "re-establish social connectedness with family, teachers and peers", and through improved social support, adopt positive coping and increase resilience (Kataoka et al. 2012; Wong 2008). Through these proposed mechanisms, it is feasible that the risk for developing PTSD and depression may be reduced. The increased social support and school connectedness reported among treated students were, therefore, consistent with our hypothesized effect of LPC.

Feedback from the school nurses who delivered LPC was overwhelmingly positive. Overall, LPC steps and related materials were perceived as helpful and easy to use during delivery. The few nurses who reported challenges in the delivery of LPC suggested minimal improvements in our training. Our revised manual will include specific response strategies to a variety of traumatic events and ideas for scheduling sessions so they are least disruptive (e.g., lunch time, after school). Students who received LPC also reported high acceptance of the intervention, with 40% of them likely to adopt coping strategies provided during the LPC session. These findings underscore a great potential for successful dissemination and implementation of LPC by schools.

LPC, a promising school-based post-traumatic intervention, is an alternative to practices currently in place and without an evidence base. For decades, school personnel, including nurses, have utilized Psychological Debriefing, a modality which has mixed evidence of effectiveness. If delivered as prescribed, Psychological Debriefing involves extensive probing within 48–72 hours of trauma exposure. In two randomized trials, debriefing did not improve post-trauma stress (Stallard et al. 2006; Hobbs et al. 1996), and it has been hypothesized that the deep probing involved in debriefing may lead to re-traumatization (Wethington et al. 2008). In contrast, other studies found that Psychological Debriefing significantly reduced PTSD symptoms among firefighters and alcohol use among soldiers (Mitchell et al. 1999; Deahl et al. 2000). As an alternative to debriefing, the LPC program uses reflective listening skills without deep probing. Our results show that LPC is not harmful to children; rather, the intervention facilitates identification of trauma and related distress, provision of initial relief, and referral to advanced care when necessary.

15.4.1 LIMITATIONS

LPC is a program that is meant to be delivered soon after a traumatic experience. For our research study, it was not feasible to implement LPC after the flood due to IRB constraints and the time needed to adequately train school nurses in the LPC protocol. In the case of non-flood affected students, we were able to deliver LPC after a traumatized child was identified by the school nurse and informed consent was obtained. The time lag between exposure to trauma and treatment, which was not captured and controlled, may have impacted the effectiveness of LPC intervention and partially accounted for our inability to find a significant change in PTSD over time. Of note, even with this limitation, our analysis still suggested a decrease in symptomology. In fact, our analysis included students with both normative and elevated levels of depression and PTSD at baseline. We examined a sub-sample of students with elevated levels of depression and PTSD at baseline, and results were consistent. Therefore, based on these findings, LPC could potentially serve as a coping tool for all traumatized children, including those who might "fall through the cracks" with moderate distress.

For this pilot quasi-experiment, we also lacked a control group. To address this limitation, we adjusted for potential confounders such as age and gender. Ultimately, large-scale randomized trials are needed to add to LPC's evidence base. Our small study sample had limited generalizability because recruitment was restricted to either self-referred children or students identified by the school nurse. Despite the small sample, we were able to conduct statistical tests through our repeated measures modeling framework. Our study is also prone to reporting bias due to the nature of self-administered questionnaires.

Finally, we examined LPC's effect on different types of trauma—a situation also dictated by our small sample. Although we controlled for the type of traumatic experience (community disaster versus individual trauma), there may be unaccounted differing effects of the intervention on various individual traumas experienced by subjects (e.g., violent trauma, death of a loved one, severe injury). However, in practice, schools often encounter children with different types of traumatic experiences. An efficient approach to care is to provide school personnel with a repertoire of skills that can be modified to respond to different traumas. Lastly, we were unable to determine how many students sought formal mental health services following LPC—an important outcome of LPC. Our pilot study is the first evaluation of LPC, but future studies clearly are needed to address possible heterogeneity of effect, measure time lags from trauma to treatment, and post-LPC treatments.

15.5 CONCLUSIONS

Trauma is commonly experienced by youth, and schools are a setting where trauma symptoms may surface and persist if untreated. Based on this pilot study, LPC, a form of Psychological First Aid delivered by school personnel, was found to be a promising response strategy. With reduced resources available for school-based mental health services, LPC is an efficient first-level of defense that identifies children in distress, provides initial support from a trusted adult, and links those most in need of advanced care.

REFERENCES

1. Ahmad A, Sofi MA, Sundelin-Wahlsten V, von Knorring AL: Posttraumatic stress disorder in children after the military operation "Anfal" in Iraq Kurdistan. European Child and Adolescent Psychiatry 2000, 9:235-243.
2. American Psychiatric Association: Diagnostic and Statistical Manual of Mental Disorders. 4th ed, text revision. Washington, DC: American Psychiatric Association; 2000.
3. Breslau N, Davis G, Andreski P, Peterson E: Traumatic events and posttraumatic stress disorder in an urban population of young adults. Archives of General Psychiatry 1991, 48:216-222.
4. Breslau N, Kessler RC, Chilcoat HD, Schultz LR, Davis GC, Andreski P: Trauma and posttraumatic stress disorder in the community: the 1996 Detroit Area Survey of Trauma. Archives of General Psychiatry 1998, 55(7):626-632.
5. Bruwer B, Emsley R, Kidd M, Lochner C, Seedat S: Psychometric properties of the Multidimensional Scale of Perceived Social Support in youth. Comprehensive Psychiatry 2008, 49(2):195-201.
6. Caffo E, Belaise C: Psychological aspects of traumatic injury in children and adolescents. Child and Adolescent Psychiatric Clinics of North America 2003, 12(3):493-535.
7. Canty-Mitchell J, Zimet G: Psychometric properties of the Multidimensional Scale of Perceived Social Support in urban adolescents. American Journal of Community Psychology 2000, 28(3):391-400.
8. Cash RE, Nealis LK: Mental health in the schools: It's a matter of public policy. Washington, DC: Paper presented at the National Association of School Psychologists Public Policy Institute; 2004.
9. Constantine NA, Benard B, Diaz M: Measuring Protective Factors and Resilience Traits in Youth: The Healthy Kids Resilience Assessment. New Orleans, LA: Paper presented at the Seventh Annual Meeting of the Society for Prevention Research; 1999.
10. Copeland WE, Keeler G, Angold A, Costello EJ: Traumatic events and posttraumatic stress in childhood. Archives of General Psychiatry 2007, 64:577-584.
11. Davis L, Siegel L: Posttraumatic stress disorder in children and adolescents: a review and analysis. Clinical Child and Family Psychology Review 2000, 3(3):135-154.
12. Deahl M, Srinivasan M, Jones N, Thomas J, Neblett C, Jolly A: Preventing psychological trauma in soldiers: The role of operational stress training and psychological debriefing. British Journal of Medical Psychology 2000, 73(1):77-85.
13. Dowdy E, Ritchey K, Kamphaus RW: School-based screening: a population-based approach to inform and monitor children's mental health needs. School Mental Health 2010, 2(4):166-176.

14. Flannery RB, Everly GS: Critical Incident Stress Management (CISM): Updated review of findings, 1998–2002. Aggression and Violent Behavior 2004, 9:319-329.

15. Foa E, Johnson K, Feeny N, Treadwell K: The Child PTSD Symptom Scale: A Preliminary Examination of its Psychometric Properties. Journal of Clinical Child Psychology 2001, 30(3):376-384.

16. Furukawa TA, Kessler RC, Slade T, Andrews G: The performance of the K6 and K10 screening scales for psychological distress in the Australian national survey of mental health and well-being. Psychological Medicine 2003, 33(2):357-362.

17. Goenjian A, Walling D, Steinberg A, Karayan I, Najarian L, Pynoos R: A prospective study of posttraumatic stress and depressive reactions among treated and untreated adolescents 5 years after a catastrophic disaster. The American Journal of Psychiatry 2005, 162:2302-2308.

18. Gray MJ, Litz BT, Hsu JL, Lombardo TW: Psychometric properties of life events checklist. Assessment Dec 2004, 11(4):330-341.

19. Gueorguieva R, Krystal J: Move over ANOVA: Progress in analyzing repeated-measures data and its reflections in papers. Archives of General Psychiatry 2004, 61:310-317.

20. Hobbs M, Mayou R, Harrison B, Warlock P: A randomised controlled trial of psychological debriefing for victims of road traffic accidents. BMJ 1996, 313:1438.

21. Hoven CW: Psychopathology among New York City public school children 6 months after september 11. Archives of General Psychiatry 2005, 62(5):545.

22. Kataoka S, Langley AK, Wong M, Baweja S, Stein BD: Responding to students with PTSD in schools. Child and Adolescent Psychiatric Clinics of North America 2012, 21(1):119-133.

23. Kenardy J, Spence S, Macleod A: Screening for posttraumatic stress disorder in children after accidental injury. Pediatrics 2006, 118(3):1002-1009.

24. Layne CM: Trauma/grief-focused group psychotherapy: School-based postwar intervention with traumatized bosnian adolescents. Group Dynamics 2001, 5(4):277.

25. McDermott B, Child C, Lee E, Judd M, Gibbon P: Posttraumatic stress disorder and general psychopathology in children and adolescents following a wildfire disaster. Canadian Journal of Psychiatry 2005, 50(3):137-143.

26. Mitchell JT, Schiller G, Eyler VA, Everly GS: Community crisis intervention: the Coldenham tragedy revisited. International Journal of Emergency Mental Health 1999, 1(4):227.

27. Muesar K, Taub J: Trauma and PTSD among adolescents with severe emotional disorders involved in multiple service systems. Psychiatric Services 2008, 59(6):627-634.

28. Nader K, Pynoos R, Fairbanks L, Frederick C: Children's PTSD reactions one year after a sniper attack at their school. The American Journal of Psychiatry 1990, 147(11):1526-1530.

29. Neuner F, Schauer E, Catani C, Ruf M, Elber T: Post-tsunami stress: A study of posttraumatic stress disorder in children living in three severely affected regions in Sri Lanka. Journal of Traumatic Stress 2006, 19(3):339-347.

30. President's New Freedom Commission on Mental Health: Achieving the Promise: Transforming Mental Health Care in America. 2003. Accessed November 19, 2013,

at http://govinfo.library.unt.edu/mentalhealthcommission/reports/FinalReport/toc.
html

31. Pynoos RS: Post-traumatic stress reactions in children after the 1988 Armenian
     earthquake. The British Journal of Psychiatry 1993, 163(2):239.

32. Radloff LS: The CES-D Scale: A self-report depression scale for research in the gen-
     eral population. Applied Psychological Measurement 1977, 1(3):385-401.

33. Ruzak JI, Brymer MJ, Jacobs AK, Layne CM, Vernberg EM, Watson PJ: Psycho-
     logical First Aid. Journal of Mental Health Counseling 2007, 29(1):17-49.

34. Sharma-Patel K, Filton B, Brown E, Zlotnik D, Campbell C, Yedlin J: Pediatric post-
     traumatic stress disorder. In Handbook of Child and Adolescent Anxiety Disorders,
     2011. Edited by McKay D, Storch EA. New York: Springer; 2011:303-322.

35. Stallard P, Velleman R, Salter E, Howse I, Yule W, Taylor G: A randomised con-
     trolled trial to determine the effectiveness of an early psychological intervention
     with children involved in road traffic accidents. Journal of Child Psychology and
     Psychiatry 2006, 47(2):127-134.

36. Stein BD, Jaycox LH, Kataoka SH, Wong M, Tu W, Elliott MN, et al.: A mental
     health intervention for schoolchildren exposed to violence: a randomized controlled
     trial. JAMA 2003, 290(5):603-611.

37. Vernberg E, La Greca A, Silverman W, Prinstein M: Prediction of posttraumatic
     stress symptoms in children after Hurricane Andrew. Journal of Abnormal Psychol-
     ogy 1996, 105(2):237-248.

38. Wethington H, Hahn R, Fuqua Whitley D, et al.: The effectiveness of interventions to
     reduce psychological harm from traumatic events among children and adolescents: A
     systematic review. American Journal of Preventive Medicine 2008, 35(3):287-313.

39. Wong M: Interventions to reduce psychological harm from traumatic events among
     children and adolescents: A commentary on the application of findings to the real
     world of school. American Journal of Preventive Medicine 2008, 35(4):398-400.

40. Zimet G, Powell S, Farley G, Werkman S, Berkoff K: Psychometric characteristics
     of the Multidimensional Scale of Perceived Social Support. Journal of Personality
     Assessment 1990, 55(3–4):610-617.

AUTHOR NOTES

CHAPTER 1

Competing Interests
The authors declare that they have no competing interests.

Author Contributions
MB participated in drafting and revising the manuscript, data analysis, interpretation of data, made a substantial contribution to the acquisition of data, and ensured that questions related to the accuracy of the work are appropriately resolved. DR participated in the design of the study, performed statistical analysis, critically revising the manuscript, and made a substantial contribution on the acquisition of data (all clinical data). SB participated in the design of the study, data acquisition, and critically revising the manuscript. SS conceptualized the study and made substantial contributions to the conception and design, analysis of data, revised the manuscript, and provided the final approval of the version to be published. All authors read and approved the final manuscript.

Acknowledgments
This work is based on the research supported by the South African Research Chairs Initiative of the Department of Science and Technology and the National Research Foundation of South Africa.

CHAPTER 2

Competing Interests
The authors declare they have no competing interests.

Author Contributions
MMcD, MD, AP and MF designed the study, MMcD, AP and MD collected the data, MD, AP and CC analysed the data and drafted the paper. All authors contributed to writing, and read and approved the final manuscript.

Acknowledgments

We wish to acknowledge the valuable contribution to this study of our former colleague Dr Patrick McCrystal who sadly died before this paper could be completed for publication. We also acknowledge the contributions of Mr Joe Martin and Mr Jack Walls of the Western Education & Library Board and Mr David Bolton of Sperrin Lakeland Health & Social Care Trust for facilitating the approval and procedures for the data collection within school settings.

CHAPTER 3

Author Contributions

Conceived and designed the experiments: IK FL MH DC CF DWM MC. Performed the experiments: IK PF FL MH DC. Analyzed the data: HR IK MCC MC. Contributed reagents/materials/analysis tools: DWM. Wrote the manuscript: HR. Critical revisions: IK PF MCC FL MH DC CF DWM MC.

CHAPTER 5

Conflict of Interests

The authors declare that there is no conflict of interests regarding the publication of this paper.

Acknowledgment

This study was partly supported by an investigator-initiated grant from Astra Zeneca awarded to Dr. Rajamannar Ramasubbu and the Cuthbertson and Fischer Chair (Dr. Frank P. MacMaster).

CHAPTER 6

Funding

This work was supported by a VIDI grant (grant number 016•085•353) awarded by the Netherlands Wetenschaps Organisatie (NWO) to the principal investigator BME. The funders had no role in study design, data collection and analysis, decision to publish, or preparation of the manuscript.

Competing Interests

The authors have declared that no competing interests exist.

Acknowledgments

We would like to thank Carolien Giessen and Charlotte van Schie for their help with data acquisition, and Helena de Klerk for her help with data processing.

Author Contributions

Conceived and designed the experiments: ALvH BGM BME PS KH AEB. Performed the experiments: ALvH KH BGM. Analyzed the data: ALvH BGM EAC. Wrote the paper: ALvH KH AEB EAC PS BME.

CHAPTER 7

Funding

The authors sincerely thank the support of funds from the Major Project of Chinese National Programs for Fundamental Research and Development (973 program, 2009CB918303 to LJL), the National Natural Science Foundation of China (30830046, 81171286 & 91232714 to LJL), the Innovation Project of Hunan Graduate Education (CX2012B100 to SJL), and National Hi-Tech Research and Development Program of China (863 program, 2008AA02Z413 to ZJZ). The funders had no role in study design, data collection and analysis, decision to publish, or preparation of the manuscript.

Competing Interests

The authors have declared that no competing interests exist.

Acknowledgments

The authors would like to thank all participants who took part in this study, and the experts at the Magnetic Resonance Center of the Second Xiangya Hospital for providing scan time and technical assistant.

Author Contributions

Conceived and designed the experiments: SJL LJL ZJZ. Performed the experiments: SJL ZGW WWW. Analyzed the data: WJG ML. Contributed reagents/materials/analysis tools: WJG YQD. Wrote the paper: SJL LJL.

CHAPTER 8

Funding

This study was supported by the National Natural Science Foundation of China (30830046 & 81171286 & 91232714 to Lingjiang Li, 81171291 to Linyan Su, 81101004 to Yan Zhang), the National 973 Program of China (2009CB918303 to Lingjiang Li) and Program of Chinese Ministry of Education (20090162110011 to Lingjiang Li). The funders had no role in study design, data collection and analysis, decision to publish, or preparation of the manuscript.

Competing Interests

The authors have declared that no competing interests exist.

Acknowledgments

We would like to thank Professor Baoci Shan for his advice on data analyzing and Professor Yuqiang Ding for his help on language editing.

Author Contributions

Conceived and designed the experiments: LL LS. Performed the experiments: ML FY YZ ZH SL WW. Analyzed the data: ML ZL. Contributed reagents/materials/analysis tools: MS TJ. Wrote the paper: ML FY.

CHAPTER 9

Funding

This study was supported by RO1 awards from the United States of America National Institute of Mental Health (MH-066222, MH-091391) http://www.nimh.nih.gov/index.shtml and National Institute of Drug Abuse (DA-016934, DA-017846) http://www.drugabuse.gov/to MHT, and Grant-in-Aid for Scientific Research to AT from Japan-United States

Brain Research Cooperation Programhttp://www.nips.ac.jp/jusnou/eng/. The funders had no role in study design, data collection and analysis, decision to publish, or preparation of the manuscript.

Competing Interests
The authors have declared that no competing interests exist.

Acknowledgments
We thank Ms. Cynthia E. McGreenery, Keren Rabi, M.A., Hanako Suzuki, M.A., Carryl P. Navalta, Ph.D. and Daniel Webster R.N., M.S., C.S., for recruitment and interviewing of subjects.

Author Contributions
Conceived and designed the experiments: MHT. Performed the experiments: MHT AP. Analyzed the data: AT CMA MHT. Wrote the paper: AT MHT.

CHAPTER 10

Funding
Dr. Thacker's efforts on this research were supported in part by National Institutes of Health (NIH) Grant # UL1TR00058 from the National Centers for Advancing Translational Sciences. The funders had no role in study design, data collection, and analysis, decision to publish, or preparation of the manuscript. No additional external funding was received for this study.

Competing Interests
The authors have declared that no competing interests exist.

Acknowledgments
Conceived and designed the experiments: MJB LRT SAC. Performed the experiments: MJB LRT SAC. Analyzed the data: MJB LRT SAC. Contributed reagents/materials/analysis tools: MJB. Wrote the paper: MJB. Revised the manuscript: MJB LRT SAC.

Author Contributions
Conceived and designed the experiments: MJB LRT SAC. Performed the experiments: MJB LRT SAC. Analyzed the data: MJB LRT SAC. Con-

tributed reagents/materials/analysis tools: MJB. Wrote the paper: MJB. Revised the manuscript: MJB LRT SAC.

CHAPTER 11

Competing Interests
The authors declare that they have no competing interests.

Author Contributions
MKI was involved in the conception and design of the study, analysing and interpreting the data, drafted the manuscript and made modifications. PG, CD, and TL were involved in the conception and design of the study, analysing and interpreting the data, and revising the manuscript. BL and DD analysed and interpreted the analyses, and revised the manuscript. RL, NC, MB, and DB interpreted the analyses and revised the manuscript. The first author had full access to all of the data in the study and takes full responsibility for the integrity of the data and the accuracy of the data analysis. All authors approved the current version.

Acknowledgments
Thank you to the Centre for Longitudinal Studies (CLS), Institute of Education for the use of the NCDS data and to the Economic and Social Data Service (ESDS) for making them available. However, neither CLS nor ESDS bear any responsibility for the analysis or interpretation of these data.

Funding
This work was initially supported by the French Institut national du Cancer (INCa). Project title: Inégalités sociales et cancer: approche par chaînes de causalité. Michelle Kelly-Irving was funded to carry out this work by the charitable organisation "Ligue nationale contre le cancer".

CHAPTER 12

Competing Interests
The authors declare that they have no competing interests.
Authors' contributions

CP performed the statistical analysis, participated in the acquisition of data, and drafted the manuscript. DDO and JC participated in design and coordination of the present study and contributed to revising the manuscript. TJW was responsible for the design and coordination of the present study, conceptualization of the analysis, oversight of the statistical analysis, and assisted in drafting the manuscript. All authors read and approved the final manuscript.

Acknowledgments

This study was funded by the Heart and Stroke Foundation of Ontario (SDA6237) CP was supported by graduate fellowships from both the Canadian Institutes of Health Research and Ontario Graduate Scholarship program. DDO was supported by the Canadian Foundation for Innovation (10118), TJW was supported by the Canada Research Chairs program, JC is supported by an endowed professorship through the Department of Family Medicine, McMaster University. The authors thank Frank Putnam for comments on a previous version of the manuscript.

CHAPTER 13

Competing Interests

The authors declare that they have no competing interests.

Author Contributions

WR and TT conceived the study aims and design. TT, CH and WR contributed to the systematic review, data extraction, quality assessment and interpretation of the results. TT wrote the manuscript, supervised by CH and WR. All authors approved the final version of the manuscript.

Acknowledgments

The work is part of the Diabetes Competence Network including the DI-AB-CORE (Collaborative Research of Epidemiologic Studies) Consortium which is funded by the German Ministry of Education and Research. Further support was obtained from The German Federal Ministry of Health and the Ministry of Innovation, Science, Research and Technology of the state North Rhine-Westphalia (Düsseldorf, Germany).

CHAPTER 15

Competing Interests
The authors declare that they have no competing interests.

Author Contributions
MR conceived, designed and carried out the study, interpreted the analysis, and drafted and revised the manuscript. KH, MF and JC conducted and interpreted analysis, and drafted and revised the manuscript. MW and RS helped conceive and design the study, acquired data, and drafted and revised the manuscript. All authors read and approved the final manuscript.

Acknowledgments
We would like to acknowledge Cedar Rapids Community School District nurses Sally Immerfall, Sharon Neilly, Monica Piersall, Susan Rummelhart, Jan Schneider, and Connie Trautman, and the parents and students from Cedar Rapids who participated in this study. We also thank Danielle Pettit-Majewski, MPH, our research assistant, and Corinne Peek-Asa, MPH, PhD, the Director of the University of Iowa Injury Prevention Research Center, who provided feedback to M.Ramirez in the design of this study. This research was funded by the University of Iowa Injury Prevention Research Center.

INDEX